Handbook of Botany

Volume II

Handbook of Botany
Volume II

Edited by **Austin Balfour**

New York

Published by Callisto Reference,
106 Park Avenue, Suite 200,
New York, NY 10016, USA
www.callistoreference.com

Handbook of Botany: Volume II
Edited by Austin Balfour

International Standard Book Number: 978-1-63239-376-0 (Hardback)

Printed in the United States of America.

Contents

Preface   IX

Chapter 1   **Growth, Root Formation, and Nutrient Value of Triticale Plants Fertilized with Biosolids**   **1**
Wendy Mercedes Rauw, Michael Bela Teglas, Sudeep Chandra and Matthew Lewis Forister

Chapter 2   **Occurrence of Morphological and Anatomical Adaptive Traits in Young and Adult Plants of the Rare Mediterranean Cliff Species *Primula palinuri* Petagna**   **8**
Veronica De Micco and Giovanna Aronne

Chapter 3   **Antioxidant Potential and Oil Composition of *Callistemon viminalis* Leaves**   **18**
Muhammad Zubair, Sadia Hassan, Komal Rizwan, Nasir Rasool, Muhammad Riaz, M. Zia-Ul-Haq and Vincenzo De Feo

Chapter 4   **Influence of Abscisic Acid and Sucrose on Somatic Embryogenesis in Cactus *Copiapoa tenuissima* Ritt. forma *mostruosa***   **26**
J. Lema-Rumińska, K. Goncerzewicz and M. Gabriel

Chapter 5   **Induction of MAP Kinase Homologues during Growth and Morphogenetic Development of Karnal Bunt (*Tilletia indica*) under the Influence of Host Factor(s) from Wheat Spikes**   **33**
Atul K. Gupta, J. M. Seneviratne, G. K. Joshi and Anil Kumar

Chapter 6   **Effects of CO_2 Enrichment on Growth and Development of *Impatiens hawkeri***   **45**
Fan-Fan Zhang, Yan-Li Wang, Zhi-Zhe Huang, Xiao-Chen Zhu, Feng-Jiao Zhang, Fa-Di Chen, Wei-Min Fang and Nian-Jun Teng

Chapter 7   **Localisation of Abundant and Organ-Specific Genes Expressed in *Rosa hybrida* Leaves and Flower Buds by Direct *In Situ* RT-PCR**   **54**
Agata Jedrzejuk, Heiko Mibus and Margrethe Serek

Chapter 8   **mRNA Expression of EgCHI1, EgCHI2, and EgCHI3 in Oil Palm Leaves (*Elaeis guineesis* Jacq.) after Treatment with *Ganoderma boninense* Pat. And *Trichoderma harzianum* Rifai**   **63**
Laila Naher, Soon Guan Tan, Chai Ling Ho, Umi Kalsom Yusuf, Siti Hazar Ahmad and Faridah Abdullah

Chapter 9   **The Effect of High Concentrations of Glufosinate Ammonium on the**
Yield Components of Transgenic Spring Wheat (*Triticum aestivum* L.)
Constitutively Expressing the *bar* Gene   69
Zoltán Áy, Róbert Mihály, Mátyás Cserháti, Éva Kótai and János Pauk

Chapter 10   **Pollen Viability, Pistil Receptivity, and Embryo Development in**
Hybridization of *Nelumbo nucifera* Gaertn   78
Yan-Li Wang, Zhi-Yong Guan, Fa-Di Chen, Wei-Min Fang and Nian-Jun Teng

Chapter 11   **Erratic Male Meiosis Resulting in 2*n* Pollen Grain Formation in a**
4x Cytotype (2*n* = 28) of *Ranunculus laetus* Wall. ex Royle   86
Puneet Kumar and Vijay Kumar Singhal

Chapter 12   **NADP-Dependent Isocitrate Dehydrogenase from *Arabidopsis* Roots**
Contributes in the Mechanism of Defence against the Nitro-Oxidative
Stress Induced by Salinity   95
Marina Leterrier, Juan B. Barroso, Raquel Valderrama, José M. Palma
and Francisco J. Corpas

Chapter 13   **Biological Effects of Weak Electromagnetic Field on Healthy and Infected**
Lime (*Citrus aurantifolia*) Trees with Phytoplasma   104
Fatemeh Abdollahi, Vahid Niknam, Faezeh Ghanati, Faribors Masroor and
Seyyed Nasr Noorbakhsh

Chapter 14   **Identification of Xylem Occlusions Occurring in Cut *Clematis* (*Clematis***
***L., fam. Ranunculaceae* Juss.) Stems during Their Vase Life**   110
Agata Jedrzejuk, Julia Rochala, Jacek Zakrzewski and Julita Rabiza-Świder

Chapter 15   **Carrizo citrange Plants Do Not Require the Presence of Roots to Modulate**
the Response to Osmotic Stress   122
Rosa M. Pérez-Clemente, Almudena Montoliu, Sara I. Zandalinas,
Carlos de Ollas, and Aurelio Gómez-Cadenas

Chapter 16   **Botanicals to Control Soft Rot Bacteria of Potato**   135
M. M. Rahman, A. A. Khan, M. E. Ali, I. H. Mian, A. M. Akanda and
S. B. Abd Hamid

Chapter 17   **Growth and Anatomical Parameters of Adventitious Roots Formed on**
Mung Bean Hypocotyls Are Correlated with Galactoglucomannan
Oligosaccharides Structure   141
K. Kollárová, I. Zelko, M. Henselová, P. Capek and D. Lišková

Chapter 18   **Antimicrobial Activity and Phytochemical Screening of *Buchenavia***
***tetraphylla* (Aubl.) R. A. Howard (Combretaceae: Combretoideae)**   148
Ygor Lucena Cabral de Oliveira, Luís Cláudio Nascimento da Silva,
Alexandre Gomes da Silva, Alexandre José Macedo, Janete Magali de Araújo,
Maria Tereza dos Santos Correia and Márcia Vanusa da Silva

Chapter 19   **Nemesia Root Hair Response to Paper Pulp Substrate for Micropropagation**   154
Pascal Labrousse, David Delmail, Raphaël Decou, Michel Carlué,
Sabine Lhernould and Pierre Krausz

Chapter 20   **Identification and Mechanism of *Echinochloa crus-galli* Resistance to Fenoxaprop-p-ethyl with respect to Physiological and Anatomical Differences**
Amany Hamza, Aly Derbalah and Mohamed El-Nady   **161**

Chapter 21   **Physiological and Growth Responses of Six Turfgrass Species Relative to Salinity Tolerance**
Md. Kamal Uddin, Abdul Shukor Juraimi, Mohd. Razi Ismail,
Md. Alamgir Hossain, Radziah Othman and Anuar Abdul Rahim   **169**

Chapter 22   **Pollen, Tapetum, and Orbicule Development in *Colletia paradoxa* and *Discaria Americana* (Rhamnaceae)**
M. Gotelli, B. Galati and D. Medan   **179**

Chapter 23   ***Ephedra alte* (Joint Pine): An Invasive, Problematic Weedy Species in Forestry and Fruit Tree Orchards in Jordan**
Jamal R. Qasem   **187**

Permissions

List of Contributors

Preface

Botany, as we all are well aware, is the science of plants. Botany is essentially a sub-category of Biology. Though fungi and algae have some non-green species, their study is also included in botany. It is believed that approximately 4 lac species are studied under this field of biology. Botany happens to be one of the most ancient fields of science, as it has been studied since the evolution of human beings. Humans started identifying harmful and useful plants at a very early age. It was only in early 15th century that they started cultivating edible and medicinal plants. Padua botanical garden is one of the most ancient gardens which studied plants with written records and botanical study parameters.

In the present age of advancement in technology and technical assistance used to study plants, some of the most advanced techniques available today include electron and optical microscopy, cell imaging, enzymes analysis, genetic engineering, etc. The advancements in the field of botany are such that today, it is possible to create hybrids of plants these days. This is also because botanists have invented techniques to explore molecular genetics and DNA sequencing to classify species more specifically.

This book has been produced with the motive of exploring and spreading knowledge related to botanical sciences, and to help students as well as botanists in their studies and researches. I especially wish to acknowledge the contributing authors, without whom a work of this magnitude would clearly not have been realizable. I thank all our authors for allocating much of their scarce time to this project. Not only do I appreciate their participation, but also their adherence as a group to the time parameters set for this publication.

Editor

Growth, Root Formation, and Nutrient Value of Triticale Plants Fertilized with Biosolids

Wendy Mercedes Rauw,[1] Michael Bela Teglas,[2] Sudeep Chandra,[3] and Matthew Lewis Forister[4]

[1] Departamento de Mejora Genética Animal, Instituto Nacional de Investigación y Tecnología Agraria y Alimentaria, 28040 Madrid, Spain
[2] Department of Agriculture, Nutrition and Veterinary Sciences, University of Nevada, Mail Stop 202, Reno, NV 89557, USA
[3] Department of Natural Resources and Environmental Science, University of Nevada, Mail Stop 186, Reno, NV 89512, USA
[4] Department of Biology, University of Nevada, Reno, Mail Stop 314, Reno, NV 89557, USA

Correspondence should be addressed to Wendy Mercedes Rauw, rauw.wendy@inia.es

Academic Editors: R. Clemente and H. A. Torbert

Biosolids are utilized as nutrient rich fertilizer. Little material is available on benefits to forage crops resulting from fertilization with biosolids. This paper aimed to compare the effects of fertilization with biosolids versus commercial nitrogen fertilizer on growth, root formation, and nutrient value of triticale plants in a greenhouse experiment. Per treatment, five pots were seeded with five triticale seeds each. Treatments included a nonfertilized control, fertilization with 100, 200, 300, 400, and 500 ml biosolids per pot, and fertilization with a commercial nitrogen fertilizer at the recommended application rate and at double that rate. Biomass production, root length, root diameter, nitrogen, phosphorus, and potassium concentration were analyzed at harvest. Fertilization with biosolids increased triticale production ($P < 0.001$); production was similar for the 100 to 400 mL treatments. Root length, nitrogen, and phosphorus concentration increased, and potassium concentration decreased linearly with application rate. At the recommended rate, biomass production was similar between fertilization with biosolids and commercial fertilizer. However, plants fertilized with commercial fertilizer had considerably longer roots ($P < 0.001$), higher nitrogen concentration ($P < 0.05$), and lower potassium concentration ($P < 0.01$) than those fertilized with biosolids. Our results indicate that at the recommended application rate, biomass production was similar between fertilization with biosolids and with commercial nitrogen fertilizer, indicating the value of biosolids fertilization as a potential alternative.

1. Introduction

Biosolids are derived from the treatment of domestic sewage sludge at publically owned treatment works. The term biosolids generally refers to sewage sludge treated to meet the land-application standards outlined in the Code of Federal Regulations, Title 40 (Part 503) under section 405 (d) of the United States Clean Water Act [1, 2]. Because of the increasing costs of sewage sludge disposal (e.g., landfilling) and the increasing desire to reuse waste residuals wherever possible, land application of biosolids is increasingly chosen as a disposal practice [3]. In addition, organic compounds, plant nutrients, and trace elements in biosolids make it a valuable resource for land application [4]. In North America, over half of the biosolids produced (approximately 3 to 4 million Mg) are applied to land as nutrient rich fertilizer [5, 6]. Much of this is used in agriculture, including animal production systems. The use of biosolids in animal production systems is widespread in North America as well as in other countries, such as the United Kingdom, Australia, New Zealand, and Pakistan. Biosolids can be used to produce forage and feed crops or can be used on pastures and range for grazing animals [7, 8]. Biosolids can impact domestic animals through feeding on vegetation grown on biosolid-amended soil or by direct consumption of the soil attached to vegetation [9].

Despite the potential advantages of biosolid application in agriculture, land application of biosolids may potentially pose a risk to public health from heavy metals or toxic organics that might enter the food chain and from pathogens that might be present in the biosolids [10]. There is a considerable body of literature available on risk assessment of biosolids application (e.g., [11, 12]). However, less material is available on benefits to forage crops resulting from fertilization with biosolids. In the present study, Triticale (X *Triticosecale* Wittmack) is chosen as the subject of study due to its popularity as a forage crop in livestock production systems. It is a product of the cross between wheat (*Triticum, spp.*) and rye (*Secale cereale* L.), resulting in a crop that is environmentally more flexible than most other cereal crops and has been shown to have superior yields and tolerance to many diseases and pests relative to its parental species or distant relatives [13]. It can fulfill the needs of grazing, ensilage, hay, and grain for feed [14]. Breeding programs of triticale mainly focus on the improvement of economic traits such as yield, biomass, nutritional factors, plant height, early maturity, and grain volume weight [15]. Nutritional values of 15 high-yielding cultivars and lines of triticale were evaluated by Heger and Eggum [16]. The biological value of triticale protein was superior to that of wheat protein (65.3 versus 61.6); the utilizable protein yields for most triticale cultivars were higher than those for wheat [16]. The agronomic advantages of triticale grains over wheat make it an attractive option for increasing global food production, in particular for marginal and stress-prone growing conditions [15].

The aim of the present paper was to compare the influence of fertilization with biosolids versus fertilization with a commercial nitrogen fertilizer on growth, root formation, and nutrient value of triticale plants in a controlled greenhouse experiment.

2. Materials and Methods

2.1. Experimental Procedures. Class B biosolids from the Truckee Meadow Water Reclamation Facility, Reno, NV, were used in a greenhouse experiment between July and September 2009. Pots of 3.8 liters were seeded with 5 triticale (*Triticosecale)* seeds each. Data were collected on 5 pots for each of the following treatments: 0 (control), 100, 200, 300, 400, and 500 mL biosolids per pot, mixed with untreated soil originating from the Main Station Field Laboratory fields (MSFL; University of Nevada, Reno), and two treatments of Main Station Field Laboratory soil mixed with Best Ammonium Sulfate 21-0-0 commercial fertilizer at a recommended application rate of 10 mL (F1) per pot and at double that rate, 20 mL/pot (F2). The soil at the MSFL fields primarily consists of Truckee silt loam (fluvaquentic haploxeroll; [17]). Fertilizer application rates were calculated based on the estimated amount of plant-available nitrogen. According to the analytical report of the Truckee Meadow Water Reclamation Facility, biosolids contained 61792 mg/kg total Kjeldahl N, 10665 mg/kg Ammonium N, 313 mg/kg Nitrate N, and 51127 mg/kg Organic N. The amount of fertilizer needed to produce a yield goal of 7 tons/0.405 ha on MSFL soil was equivalent to 40.9 kilo liters of biosolids or 113 kg N fertilizer per 0.405 ha. This is translated to a recommended application rate of 185 mL of biosolids and 10 mL commercial fertilizer per pot. Therefore, the 200 mL biosolids treatment is comparable to recommended field application rates.

Plants were grown within a single greenhouse bay at the University of Nevada, Reno, under the same temperature and light intervals and all pots were watered daily until the plants were harvested for analysis. Plant height (HEIGHT) and number of leaves (LEAVES) were measured weekly for each individual plant until 4 weeks of age at which time they were harvested. Plant height was measured by stretching the tallest leaf on each plant to its full length. Growth rate (HRATE) and rate of leaf emergence (LRATE) were estimated by fitting a linear regression equation to data, averaged by pot, on HEIGHT and LEAVES as a function of age:

$$\text{TRAIT}_{\text{Age}} = a + (b \times \text{Age}), \qquad (1)$$

where $\text{TRAIT}_{\text{Age}}$ is the HEIGHT and LEAVES at a specific age (wk), a is the intercept, and b is the regression coefficient representing HRATE and LRATE per week, respectively.

After harvest, plants were dried and above-ground plant biomass production was recorded for each pot (WEIGHT). Plant roots were separated from the soil, washed carefully, and individually analyzed using the WinRhizo 2007 root scanning program (Regent Instruments Inc, Montreal, Quebec), resulting in measurements of average root length (ROOTL) and diameter (ROOTD).

A single pot in the control group was excluded from the analysis due to abnormal plant growth. Of the remaining 195 plants, 9 plants were identified as exhibiting unusually slow growth (two plants from the 100 mL treatment, two from the 200 mL treatment, two from the 400 mL treatment, one from the 500 mL treatment, and two from the F2 treatment) both in terms of height and in terms of leaf emergence and were therefore not considered in the analyses of HEIGHT, LEAVES, HRATE, and LRATE. However, these plants could not be identified in the root data; therefore, root data was analyzed for the full dataset of 195 plants.

After harvest, plant samples were sent to A & L Western Agricultural Labs Inc (Modesto, CA, http://www.al-labs-west.com/) and analyzed for total nitrogen (by automated combustion at 900°C), and phosphorus and potassium concentration (by inductively coupled plasma emission spectrometry (ICP)) on dry matter (DM) basis. A & L Western Agricultural Labs Inc follows the North American Proficiency Testing (NAPT) Program (http://www.naptprogram.org/).

2.2. Data Handling and Statistical Analysis. The SAS program [18] was used for statistical analysis of all traits. The model used to describe the data on weekly, individual, measurements of HEIGHT, LEAVES, ROOTL, and ROOTD was as follows:

$$Y_{ijk} = \mu + \text{Treatment}_i + \text{Pot (Treatment)}_j + e_{ijk}, \qquad (2)$$

where μ is the population intercept, Treatment_i is the effect of treatment i (control, 100, 200, 300, 400, 500, F1, F2),

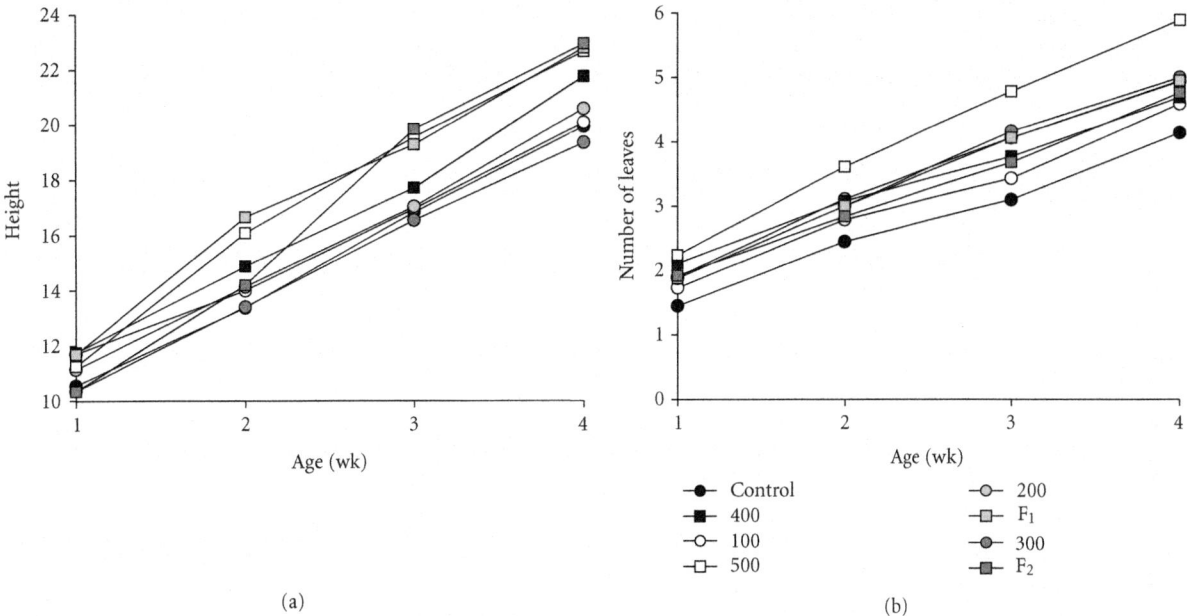

FIGURE 1: Average plant height (a) and number of leaves (b) from 1 to 4 weeks of age, by treatment (Control, 0, 100, 200, 300, 400, and 500 mL biosolids, and F1 and F2).

Pot (Treatment)$_j$ is the effect of pot (1 to 5) nested within treatment j, and e_{ijk} is the residual error term of plant k, $e_{ijk} \sim \text{NID}(0,\sigma_e^2)$. The effect of treatment was considered fixed; the residual error term and the effect of pot nested within treatment were considered random. The traits HEIGHT, LEAVES, ROOTL, and ROOTD were denoted by Y_{ijk}, as measured on plant k of treatment i, in pot j.

The model used to describe the data, averaged by pot, on HRATE, LRATE, and WEIGHT was

$$Y_{ij} = \mu + \text{Treatment}_i + e_{ij}, \qquad (3)$$

where μ, Treatment$_i$, and e_{ij} are as in model (1). The traits HRATE, LRATE, and WEIGHT were denoted by Y_{ij}, as measured on plant j of treatment i.

Phenotypic correlations were calculated from measurements averaged by pot: HEIGHT and LEAVES at 4 weeks of age, HRATE, LRATE, WEIGHT, ROOTL, and ROOTD, and nitrogen, phosphorus, and potassium concentration, after adjusting the values for the effect of treatment with model (3).

3. Results

3.1. Plant Growth. Figure 1 presents least squares means of plant height and leaf emergence between one and four weeks of age, adjusted for the effect of pot, for each treatment. The effect of treatment on HEIGHT and LEAVES was significant between two and four weeks of age ($P < 0.001$). At two weeks of age, plants from the 500 mL treatment were taller than those from the 0, 100, 200, 300 mL, and F2 treatments ($P < 0.05$), and plants from the F1 treatment were taller than those from the 0, 100, 200, 300, 400 mL, and F2 treatments ($P < 0.05$). At three weeks of age, plants from the 500 mL,

F1 and F2 treatments were taller than those from the 0, 100, 200, 300, and 400 mL treatments ($P < 0.05$). At four weeks of age, plants from the 400 mL treatment were taller than those from the 300 mL treatment ($P < 0.01$) and plants from the 500 mL, F1, and F2 treatments were taller than those from the 0, 100, 200, and 300 mL treatments ($P < 0.05$).

At one week of age, plants from the control treatment had fewer leaves than those from the 400 and 500 mL treatments ($P < 0.05$) and plants from the 500 mL treatment had more leaves than those from the 100 mL treatment ($P < 0.05$). At two weeks of age, plants from the control line had fewer leaves than those from the 200, 400, and 500 mL treatments and plants from the 500 mL treatment had more leaves than those from the 100, 300 mL, F1, and F2 treatments ($P < 0.05$). At three weeks of age, plants from the control treatment had fewer leaves than those from the 200, 300, 500 mL, and F1 treatments ($P < 0.01$), plants from the 100 mL treatment had fewer leaves than those from the 300 and 500 mL treatments ($P < 0.05$), and plants from the 200, 400 mL, F1, and F2 treatments had fewer leaves than those from the 500 mL treatment ($P < 0.05$). At four weeks of age, plants from the control treatment had fewer leaves than those from the 200, 300, 500 mL, and F1 treatments ($P < 0.05$), and plants from the 500 mL treatment had more leaves than those from the 100, 200, 300, 400 mL, F1, and F2 treatments ($P < 0.05$).

Average HRATE, LRATE, WEIGHT, ROOTL, and ROOTD are presented in Table 1 for each treatment. Coefficients of determination (R^2) of curves fitting (1) per pot were 93 to 100% for HRATE and 79 to 100% for LRATE. Plants from the F2 group grew faster than those from the 100, 200, and 300 mL treatments. Biomass production was lower for the control treatment than that for all other treatments

TABLE 1: Least squares means and standard errors of the least squares means of growth rate (HRATE), rate of leaf emergence (LRATE), above-ground biomass production (WEIGHT), average root length (ROOTL), root diameter (ROOTD), and nitrogen, phosphorus, and potassium percentage on DM basis of *Triticosecale* plants, by treatment[1].

	Control	100 mL	200 mL	300 mL	400 mL	500 mL	F1	F2	s.e.[2]	s.e.[3]
HRATE (cm/wk)	3.16^{ab}	2.81^a	3.12^a	3.01^a	3.28^{ab}	3.77^{ab}	3.59^{ab}	4.34^b	0.46	0.41
LRATE (number/wk)	0.87^a	0.92^a	1.01^a	1.05^a	0.85^a	1.21^a	1.03^a	0.94^a	0.15	0.14
WEIGHT (gr)	1.30^a	4.07^b	5.08^{bc}	4.89^{bc}	5.24^{bc}	6.15^c	4.46^b	6.11^c	0.59	0.53
ROOTL (cm)	21.8^a	34.6^{ab}	39.3^b	44.6^b	85.3^c	70.8^d	116.6^e	117.9^e	5.16	4.62
ROOTD (mm)	0.26^{ab}	0.25^a	0.23^a	0.29^{bd}	0.34^{ce}	0.30^{de}	0.33^{ce}	0.31^e	0.013	0.012
Nitrogen (%)	4.93^{ab}	4.77^a	5.05^{abc}	5.10^{bc}	5.25^{cd}	5.43^{de}	5.62^e	6.14^f	0.11	0.10
Phosphorus (%)	0.66^{ad}	0.56^b	0.68^{ad}	0.74^{ac}	0.78^{ce}	0.82^e	0.60^{db}	0.54^b	0.03	0.03
Potassium (%)	5.23^{ab}	5.70^a	5.68^a	4.93^{bc}	4.53^{cd}	4.51^{cd}	4.68^{cd}	4.37^d	0.20	0.18

[1] 0 (control), 100, 200, 300, 400, and 500 mL biosolids mixed with untreated farm soil, and 5 mL (F1) and 10 mL (F2) commercial fertilizer mixed with untreated soil. [2] Standard errors of the Control group. [3] Standard errors of the 100, 200, 300, 400, and 500 mL biosolids groups and the F1 and F2 groups.
a,b,c,d,e,f Values with different superscripts are significantly different ($P < 0.05$).

($P < 0.001$), and higher for the 500 mL and F2 treatments than that for the 100 mL and F1 treatments ($P < 0.05$).

3.2. Root Development.

Average root length increased with the level of biosolid application between 0 and 500 mL. A regression performed on root length as a function of the level of biosolid application gave a response of 12 cm per 100 mL biosolids with an R^2 of 81%.

The average root length in the control treatment was shorter than in all the other treatments ($P < 0.05$); average root length was shorter in the 100, 200, and 300 mL treatments than that in the 400, 500 mL, F1, and F2 treatments ($P < 0.0001$), shorter in the 400 mL treatment than that in the 500 mL, F1 and F2 treatments ($P < 0.05$), and shorter in the 500 mL treatment than that in the F1 and F2 treatments ($P < 0.0001$).

On average, roots were thinner in the control treatment than those in the 400, 500 mL, F1, and F2 treatments ($P < 0.05$), thinner in the 100 and 200 mL treatments than those in the 300, 400, 500 mL, F1, and F2 treatments ($P < 0.05$), thinner in the 300 mL treatment than those in the F1 treatment ($P < 0.01$), and thinner in the 500 mL treatment than those in the F1, treatment ($P < 0.05$).

3.3. Forage Nutrient Value.

Average nitrogen and phosphorus concentration increased and average potassium concentration decreased with the level of biosolid application between 100 and 500 mL. A regression performed on nitrogen, phosphorus, and potassium concentration as a function of the level of biosolid application gave a response of 0.15, 0.06, and −0.35% change per 100 mL biosolids with an R^2 of 96, 95, and 89%, respectively.

Concentrations of nitrogen were lower in the control treatment than those in the 400, 500 mL, F1, and F2 treatments ($P < 0.05$), lower in the 100 mL treatment than those in the 300, 400, 500 mL, F1, and F2 treatments ($P < 0.05$), lower in the 200 and 300 mL treatments than those in the 500 mL, F1, and F2 treatments ($P < 0.05$), lower in the 400 mL treatment than those in the F1 and F2 treatments ($P < 0.05$), lower in the 500 mL treatment than those in the F2 treatment ($P < 0.0001$), and lower in the F1 treatment than those in the F2 treatment ($P < 0.001$).

Concentrations of phosphorus in the control treatment were higher than those in the 100 mL treatment ($P < 0.05$) and lower than those in the 400, 500 mL, and F2 treatments ($P < 0.01$). Phosphorus concentrations were lower in the 100 mL treatment than those in the 200, 300, 400, and 500 mL treatments, lower in the 200 mL treatment than those in the 400, 500 mL, and F2 treatments ($P < 0.01$), lower in the 300 mL treatment than those in the 500 mL, F1, and F2 treatments ($P < 0.05$), and lower in the 400 and 500 mL treatments than those in the F1 and F2 treatments ($P < 0.0001$).

Concentrations of potassium were higher in the control treatment than those in the 400, 500 mL, F1, and F2 treatments ($P < 0.05$), higher in the 100 and 200 mL treatments than those in the 300, 400, 500 mL, F1, and F2 treatments ($P < 0.01$), and higher in the 300 mL treatment than those in the F2 treatment ($P < 0.05$).

3.4. Phenotypic Correlations.

Table 2 presents phenotypic correlations, adjusted for the effect of treatment, between HEIGHT and LEAVES at 4 weeks of age, HRATE, LRATE, WEIGHT, ROOTL, ROOTD, and nitrogen, phosphorus, and potassium concentration. Plants that grew taller grew at a faster rate, with a larger number of leaves that emerged at a faster speed, and produced more above-ground biomass. Plants with those properties had a lower nitrogen and phosphorus concentration. Plants that grew taller and at a faster rate had longer roots. Plants that had a lower potassium concentration were those that had a slower leaf emergence and tended to be those that had fewer leaves. Plants that had longer roots also had wider root diameters and a lower nitrogen concentration. Plants containing more nitrogen contained more phosphorus and plants containing more phosphorus contained more potassium.

4. Discussion

4.1. Biosolid Application Rate.

In the present experiment, plants receiving the 500 mL biosolids treatment grew tallest

TABLE 2: Phenotypic correlations between plant height (HEIGHT) and number of leaves (LEAVES) at four weeks of age, growth rate (HRATE), rate of leaf emergence (LRATE), above-ground biomass production (WEIGHT), root length (ROOTL), root diameter (ROOTD), and nitrogen, phosphorus, and potassium percentage on DM basis in triticale plants.

	LEAVES	HRATE	LRATE	WEIGHT	ROOTL	ROOTD	Nitrogen	Phosphorus	Potassium
HEIGHT	0.59***	0.65***	0.47***	0.48***	0.17*	−0.09	−0.58***	−0.30***	−0.11
LEAVES		0.37***	0.64***	0.34***	0.06	0.09	−0.35***	−0.35***	−0.12†
HRATE			0.32***	0.35***	0.20**	−0.07	−0.51***	−0.21**	−0.06
LRATE				0.36***	0.07	−0.03	−0.50***	−0.44***	−0.16*
WEIGHT					0.09	−0.06	−0.33***	−0.25***	−0.03
ROOTL						0.26***	−0.19**	−0.06	−0.01
ROOTD							−0.05	−0.08	−0.02
Nitrogen								0.42***	0.11
Phosphorus									0.63***

*$P < 0.05$; **$P < 0.01$; ***$P < 0.001$.

with significantly more leaves, at the highest rate of leaf emergence. These traits resulted in a biomass production that was highest for the 500 mL treatment and indicates that in the present greenhouse experiment, fertilization with biosolids resulted in increased triticale production.

In contrast with growth traits, root length increased linearly with biosolid application rate, each 100 ml of biosolids adding 12 cm to their length. Root diameter increased with application rate, but this effect was not as pronounced. Plants with an increased root length grew faster and ended up being taller at four weeks of age. Tschaplinski and Blake [19] observed a positive relationship between early root production (number, length, and dry weight) and accumulation of aboveground biomass in hybrid poplar. In the present study, however, although taller plants produced more biomass, plants with longer roots had a higher biomass production but this was not significant. The relationship between application rate and root length did not translate to a linear effect in plant growth rate and length.

Plant nutrient concentration is determined by the stage of development, the species, variety or hybrid, the plant organ, and various parts or tissues of the organs. Among abiotic factors, application of fertilizers has the greatest effect on the nutrient concentration, in addition to genetic factors such as the capacity and dynamics of nutrient uptake and the utilization and distribution of assimilates [20]. Phosphorus concentration in the present experiment was higher than that observed by Brown et al. [21], who observed a triticale total P concentration of 0.18 to 0.53% with a mean of 0.33%. In the present experiment, increased levels of biosolid fertilization added 0.15% to the nitrogen concentrations and 0.06% to the phosphorus concentrations and subtracted 0.35% to the potassium concentrations for each 100 mL of biosolids added. Bennett et al. [22] observed that the application of nitrogen significantly increased the percent of nitrogen in corn leaves in all of eight experiments and that phosphorus percentage in the leaf was significantly increased in certain experiments. Also in the study of Lasztity [20], the P concentration increased simultaneously with a rising fertilizer dose (N, P, and K); the K concentration increased as the result of K application.

The configuration and growth rate of the root system influence nutrient uptake by plants [23]. This could explain the observation that both root length and the N and P concentration increased linearly with increasing biosolid application rates. However, root length was not significantly correlated with phosphorus or potassium concentration and was *negatively* correlated with nitrogen concentration, after adjustment for the effect of treatment. The relationship between root length and nitrogen concentration is depicted in Figure 2(a): although nitrogen concentration increases with increasing biosolid application rates and accompanying root lengths, it decreases with increasing root lengths within treatment, resulting in an overall negative correlation after adjustment for the effect of treatment (Figure 2(b)). In addition, a negative correlation was observed between growth traits and nitrogen, phosphorus, and potassium concentrations. This observation might be explained by the fact that concentrations of nutrients in triticale, such as N, P, K, Ca, and Mg, decrease during the vegetation period [24]. Therefore, although a higher rate of fertilization resulted in longer roots and an increased plant nutrient concentration, within application rate, faster growing plants with longer roots may be physiologically more mature resulting in a lower nutrient concentration when the results are adjusted for the effect of treatment.

4.2. Biosolids versus Commercial Fertilizer. In order to compare the effects of biosolid application with the application of commercial fertilizer, we compare the two recommended application rates, that is, the 200 mL biosolids treatment and the F1 commercial fertilizer treatment, with the control treatment. Our results indicate that growth rate and rate of leaf emergence were very similar for all three treatment groups, but that commercially fertilized plants were taller at four weeks of age than plants fertilized with biosolids and nonfertilized plants; nonfertilized plants had fewer leaves at four weeks of age than fertilized plants. Overall biomass production was significantly increased with fertilization but was similar for biosolids and commercial fertilizer treatments. This supports the findings by Priestly [25] who compared fertilization of wheat on agricultural land including two rates

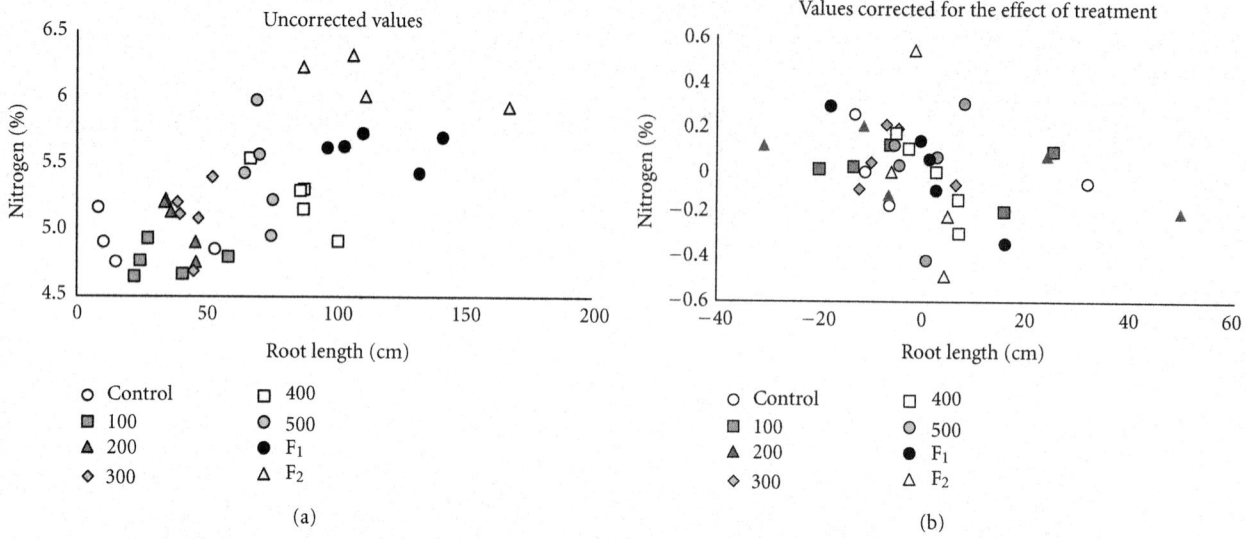

FIGURE 2: Relationship between root length and nitrogen concentration, by treatment (Control, 0, 100, 200, 300, 400, and 500 mL biosolids, and F1 and F2) (a), and relationship between root length and nitrogen concentration after correction for the effect of treatment, by treatment (Control, 0, 100, 200, 300, 400, and 500 mL biosolids, and F1 and F2) (b).

of biosolids and application of commercial fertilizers, Agstar and Urea. Results indicated no significant differences in grain yields between the treatments and it was concluded that biosolid application in wheat is likely to result in the same production levels as commercial fertilizers.

However, in the present study, commercial fertilizer appeared to have a large positive effect on root length and root diameter. Those traits were considerably larger than in the control and 200 mL biosolid treatments. Nitrogen concentration was significantly higher and potassium concentration significantly lower in the F1 treatment compared with the control and 200 mL biosolid treatments. The two fertilizers had similar effects on the phosphorus concentration of the forage which was lower in both treatments than in the control treatment.

4.3. Double Dose Applications. In order to compare the influence of double-dose applications, we compare the two recommended applications rates of 200 mL biosolids and F1 commercial fertilizer with double the amount at 400 mL biosolids and F2 commercial fertilizer. Doubling the application rate did not significantly affect growth rate, rate of leaf emergence, or plant height and number of leaves at four weeks of age. However, overall biomass production was significantly higher in the F2 treatment than in the F1 treatment. Root length and root diameter were similar for the single application rates as for the double-application rates for both the biosolids treatment and the commercial fertilizer treatment. Doubling the application rate did not significantly influence nitrogen and potassium concentration of the triticale in the biosolids treatment or phosphorus and potassium concentration in the commercial fertilizer treatment, but it did result in higher phosphorus concentration in the biosolids treatment and higher nitrogen concentration in the commercial fertilizer treatment.

5. Conclusions

In conclusion, the results of the present greenhouse study indicate that fertilization with biosolids resulted in increased triticale production compared with nonfertilized plants; biomass production was very similar for the different application rates. Root length, nitrogen concentration, and phosphorus concentration increased, and potassium concentrations decreased linearly with application rate. At the recommended application rate, biomass production was similar between fertilization with biosolids and with commercial nitrogen fertilizer, indicating the value of biosolids fertilization as a potential alternative. However, plants fertilized with commercial fertilizer had considerably longer roots, higher nitrogen concentration, and lower potassium concentration than those fertilized with biosolids. In addition, preliminary results on a field study on triticale plants fertilized at the recommended application rate versus plants grown on non-fertilized fields indicate that fertilization with biosolids resulted in a lower dry mass and ash concentration, but higher N, crude protein, NDF, and ADF concentrations (results not presented). Further research is needed to verify all results in the field.

Acknowledgments

This work was funded by a grant from the United States Department of Agriculture, project no. 2008-35101-19119, "Detection of nutrient flow in a managed agroecosystem following treatment with biosolids". The authors would like to thank Charlotte Konken for her excellent work, the Leger Lab (Department of Natural Resources and Environmental Science, UNR) for the use of root scanning equipment, Stuart Taylor at Main Station Field Laboratory, for his assistance in the application of this project, and the anonymous reviewers for their useful comments.

References

[1] USEPA, *A Guide to the Biosolids Risk Assessment for the EPA Part 503 Rule*, USEPA, Washington, DC, USA, 1993.

[2] NRC, *Biosolids Applied to Land-Advancing Standards and Practices*, The National Academies Press, Washington, DC, USA, 2002.

[3] Eastern Research Group, *Land Application of Biosolids-Process Design Manual*, CRC Press, Boca Raton, Fla, USA, 1997.

[4] E. Epstein, *Land Application of Sewage Sludge and Biosolids*, Lewis Publishers, Boca Raton, Fla, USA, 2003.

[5] G. A. O'Connor, H. A. Elliott, N. T. Basta et al., "Sustainable land application: an overview," *Journal of Environmental Quality*, vol. 34, no. 1, pp. 7–17, 2005.

[6] USEPA, "Frequently asked questions," 2007, http://water.epa .gov/polwaste/wastewater/treatment/biosolids/genqa.cfm.

[7] M. E. Tiffany, L. R. McDowell, G. A. O'Connor et al., "Effects of residual and reapplied biosolids on forage and soil concentrations over a grazing season in North Florida. I. Macrominerals, crude protein, and in vitro digestibility," *Communications in Soil Science and Plant Analysis*, vol. 32, no. 13-14, pp. 2189–2209, 2001.

[8] B. M. Wallace, M. Krzic, T. A. Forge, K. Broersma, and R. F. Newman, "Biosolids increase soil aggregation and protection of soil carbon five years after application on a crested wheatgrass pasture," *Journal of Environmental Quality*, vol. 38, no. 1, pp. 291–298, 2009.

[9] R. L. Chaney, J. A. Ryan, and G. A. O'Connor, "Organic contaminants in municipal biosolids: risk assessment, quantitative pathways analysis, and current research priorities," *Science of the Total Environment*, vol. 185, no. 1–3, pp. 187–216, 1996.

[10] P. R. Ponugoti, M. F. Dahab, and R. Surampalli, "Effects of different biosolids treatment systems on pathogen and pathogen indicator reduction," *Water Environment Research*, vol. 69, no. 7, pp. 1195–1206, 1997.

[11] C. P. Gerba, I. L. Pepper, and L. F. Whitehead III, "A risk assessment of emerging pathogens of concern in the land application of biosolids," *Water Science and Technology*, vol. 46, no. 10, pp. 225–230, 2002.

[12] R. A. Schoof and D. Houkal, "The evolving science of chemical risk assessment for land-applied biosolids," *Journal of Environmental Quality*, vol. 34, no. 1, pp. 114–121, 2005.

[13] N. L. Darvey, H. Naeem, and J. P. Gustafson, "Triticale: production and utilization," in *Handbook of Cereal Science and Technology*, K. Kulp and J. G. Ponte, Eds., pp. 257–274, Marcel Dekker, New York, NY, USA, 2000.

[14] G. R. Fohner and A. H. Sierra, "Triticale marketing: strategies for matching crop capabilities to user needs," in *Triticale Improvement and Production*, M. Mergoum and H. Gomez-Macpherson, Eds., Food and Agriculture Organization of the United Nations, Rome, Italy, 2004.

[15] M. Mergoum, P. K. Singh, R. J. Peña et al., "Triticale: a "new" crop with old challanges," in *Handbook of Plant Breeding*, vol. 3, pp. 1–21, 2009.

[16] J. Heger and B. O. Eggum, "The nutritional values of some high-yielding cultivars of triticale," *Journal of Cereal Science*, vol. 14, pp. 63–71, 1991.

[17] NCSS, "Truckee series," 2012, https://soilseries.sc.egov.usda .gov/OSD_Docs/T/TRUCKEE.html.

[18] Statistical Analysis Systems Institute, *SAS User's Guide: Statistics*, Statistical Analysis Systems Institute, Cary, NC, USA, 5th edition, 1985.

[19] T. J. Tschaplinski and T. J. Blake, "Correlation between early root production, carbohydrate metabolism, and subsequent biomass production in hybrid poplar," *Canadian Journal of Botany*, vol. 67, no. 7, pp. 2168–2174, 1989.

[20] B. Lasztity, "Fertilizers and nutrient relations in some genotypes of cereal," in *Genetic Aspects of Plant Nutrition*, M. R. Saric and B. C. Loughman, Eds., pp. 359–364, Martinus Nijhoff, The Hague, The Netherlands, 1983.

[21] B. Brown, J. Dalton, M. Chahine, B. Hazen, S. Jensen, and S. Etter, "Phosphorus removal with triticale in manured fields," *Nutrient Digest*, vol. 1, pp. 1–6, 2009.

[22] W. F. Bennett, G. Stanford, and L. Dumenil, "Nitrogen, phosphorus and potassium content of the corn leaf and grain as related to nitrogen fertilization and yield," *Soil Science Society of America Journal*, vol. 17, pp. 252–258, 1953.

[23] S. Itoh and S. A. Barber, "A numerical solution of whole plant nutrient uptake for soil-root systems with root hairs," *Plant and Soil*, vol. 70, no. 3, pp. 403–413, 1983.

[24] B. Lasztity, "The variation of element contents in triticale during vegetative growth," *Nutrient Cycling in Agroecosystems*, vol. 13, no. 2, pp. 155–159, 1987.

[25] M. H. Priestly, *The effect of biosolids application on wheat establishment, growth and yield*, BSc thesis, Muresk Institute of Agriculture, Curtin University of Technology, 1998.

Occurrence of Morphological and Anatomical Adaptive Traits in Young and Adult Plants of the Rare Mediterranean Cliff Species *Primula palinuri* Petagna

Veronica De Micco and Giovanna Aronne

Laboratorio di Botanica ed Ecologia Riproduttiva, Dipartimento di Arboricoltura, Botanica e Patologia Vegetale, Università degli Studi di Napoli Federico II, Viale Università 100, 80055 Portici, Italy

Correspondence should be addressed to Giovanna Aronne, aronne@unina.it

Academic Editor: Aurelio Gómez-Cadenas

Cliffs worldwide are known to be reservoirs of relict biodiversity. Despite the presence of harsh abiotic conditions, large endemic floras live in such environments. *Primula palinuri* Petagna is a rare endemic plant species, surviving on cliff sites along a few kilometres of the Tyrrhenian coast in southern Italy. This species is declared at risk of extinction due to human impact on the coastal areas in question. Population surveys have shown that most of the plants are old individuals, while seedlings and plants at early stages of development are rare. We followed the growth of *P. palinuri* plants from seed germination to the adult phase and analysed the morphoanatomical traits of plants at all stages of development. Our results showed that the pressure of cliff environmental factors has been selected for seasonal habitus and structural adaptive traits in this species. The main morphoanatomical modifications are suberized cell layers and accumulation of phenolic compounds in cell structures. These features are strictly related to regulation of water uptake and storage as well as defence from predation. However, we found them well established only in adult plants and not in juvenile individuals. These findings contribute to explain the rare recruitment of the present relict populations, identifying some of the biological traits which result in species vulnerability.

1. Introduction

Few ecological studies have been carried out on plants living on vertical cliffs. However, it is repeatedly reported that these habitats act as reservoirs of relict biodiversity because they are refuges from competition, predation, fire, human activities, or climatic change in the surrounding landscape [1].

Throughout the world, cliffs have large endemic floras because species are subjected to intense selective pressure by the extreme abiotic conditions [2]. The concentration of rare species in cliff habitats reflects their intolerance of competition and disturbance and their adaptation to a harsh physical environment. These species show very slow growth and sporadic seedling recruitment. Their survival strategies are based on high genetic variability of the populations and extreme longevity of single individuals [1, 3]. These features

allow cliff species to deal with the different sets of environmental factors which are distinctive of each microsite in a cliff. Therefore, conservation of these rare species is strictly dependent on the long-term maintenance of the fragile as well as stable equilibrium of surrounding abiotic conditions [1]. Consequently, environmental changes constrain the survival of single plants and start the process of consecutive decline of individuals within the population.

The conservation of rare plants cannot be guaranteed solely by their preservation through national parks and protected areas: any strategies for plant conservation need to be based on the real knowledge of the life cycle of single species and of possible interactions with other organisms, both invasive plant species and animals. The "endangered status" is worsened by continuous reproductive failure, given that sexual reproduction is the sole natural system able to improve

Occurrence of Morphological and Anatomical Adaptive Traits in Young and Adult Plants of the Rare Mediterranean
Cliff Species Primula palinuri Petagna

9

genetic variability which is basic for the adaptation process to changing environmental conditions [4].

Completion of the reproductive cycle and high reproductive success (i.e., a large number of viable seeds produced per plant) is no guarantee of species survival in the long term. Even allowing for the accomplishment of seed dispersal and germination, seedling establishment and survival during the early stages of plant development are considered delicate phases in the regeneration process by sexual reproduction [5]. These early developmental phases can be constrained by many ecological factors of the seed microsite which can trigger phenomena of adaptation [6, 7]. Adaptation at seedling level seems to involve various strategic tradeoffs that restrict a species to optimal performance at the narrow range of the successional gradient, thus affecting community structure and dynamics [8]. Seedling survival depends on their ability to cope with numerous environmental factors such as water availability, temperature, radiation, pathogens, herbivory, and competitive interactions [9]. Space limitation due to restricted availability of microsites suitable for growth is a further constraint for cliff recruitment [1]. In a number of species of various environments, adaptive strategies for seedling survival have been shown to be based on morphoanatomical modifications aiming to regulate water relations, to protect from high levels of radiation and defend from biotic factors [7, 10–12].

Primula palinuri Petagna is a rare endemic species, found only in a few restricted populations on cliff sites along the Tyrrhenian coast of southern Italy. It has been inserted as Near Threatened species in the IUCN Red List; human impact, fire, competition with invasive alien species and landslides are the main threats to the species [13].

Although it is the symbol of the National Park of Cilento and Diano Valley, little is known of its biology and there is no management plan for the areas where it grows. Our preliminary studies evidenced that the species is able to form viable seeds; moreover, seedling establishment under controlled conditions is enhanced by the development of hypocotyl hairs which have a role in the control of water uptake and mechanical support [14]. However, our preliminary field surveys evidenced that the remnant populations have a structure based on old individuals, while seedlings and small plants at early stages of development are rare. Based on these considerations, the aim of this study was to monitor the Morphoanatomical development of *P. palinuri* plants from seed germination up to their adult phase in order to look for any trait that would explain the fragility of plants at early stages of development as opposed to the high survival capability of adult plants under the environmental constraints of Mediterranean cliffs.

2. Materials and Methods

2.1. Study Sites and Plant Material. Observations of plants of *Primula palinuri* Petagna were performed on populations at the Cilento National Park, along the Tyrrhenian coast of southern Italy. The species is mainly found on limestone coastal cliffs, in soil pockets inside rock fractures. The area has a Mediterranean climate [15, 16] with an annual rainfall

of about 700 mm. Precipitation is concentrated in autumn and winter, while a dry period occurs from May to September.

Phenology was monitored in the field on adult plants. The presence of seedlings in the surroundings of the adult individuals was also checked every other week throughout the years 2010-2011. Plants of different ages were sampled in the field, for anatomical analyses, during vegetative growth and the summer rest. Rhizome yearly growth was measured in the field on 30 plants from three different populations. Despite the large quantity of seeds produced by the plants in all populations, no seedlings and only a limited number of young plants (1–5 years old) were found in the field. To increase the number of specimens to be used for Morphoanatomical analyses, we sampled seeds in the field and grew new plants under controlled conditions. To measure germination percent, 100 seeds from two provenances were mixed, placed in petri dishes (layered with three sheets of filter paper moistened with distilled water) and kept in the dark at 15°C. The test was replicated three times. Germinated seeds were transplanted into 0.5 dm^3 pots with a mix of peat and field soil. Plants were grown from November 2010 to June 2011 under the natural conditions of light and temperature, while they were watered daily to completely replace the water lost by transpiration.

2.2. Microscopy. In order to analyse the morphofunctional traits of the different vegetative organs, we dissected samples of leaves, roots, and rhizome from adult plants as well as the epicotyl, hypocotyl, and roots from plants at early stages of development. Samples were fixed in FAA (40% formaldehyde : glacial aceticacid : 50% ethanol—5 : 5 : 90 by volume) for subsequent microscopy analyses.

After fixation in FAA for several days, samples were dehydrated in an ethanol series, embedded in the acrylic resin JB4 (Polysciences, Warrington, PA, USA) and sectioned through a rotative microtome to obtain both cross and longitudinal semithin sections (5 mm). The sections were stained with 0.5% toluidine blue [17] and mounted with Canadian balsam; unstained sections were mounted with mineral oil for fluorescence microscopy. Stained sections were observed under a transmitted light microscope (BX60, Olympus, Germany), while unstained sections were analyzed under the same microscope equipped with a mercury lamp, band pass filter BP 330–385, dichromatic mirror >400 nm, and barrier filter >420 nm. Epifluorescence mode allowed us to investigate the presence of suberized and/or lignified structures as well as simple phenolics that are autofluorescent at these wavelengths [18, 19]. Photomicrographs at different magnifications were obtained with a digital camera (CAMEDIA C4040, Olympus).

3. Results

P. palinuri develops highly viable seeds: germination percent was 86.17 (SD ± 6.14). After germination, seedling establishment is facilitated by the development of a ring of hypocotyl hairs in the region between the hypocotyl and radicle (Figures 1(a) and 2(a)). Plant growth continues

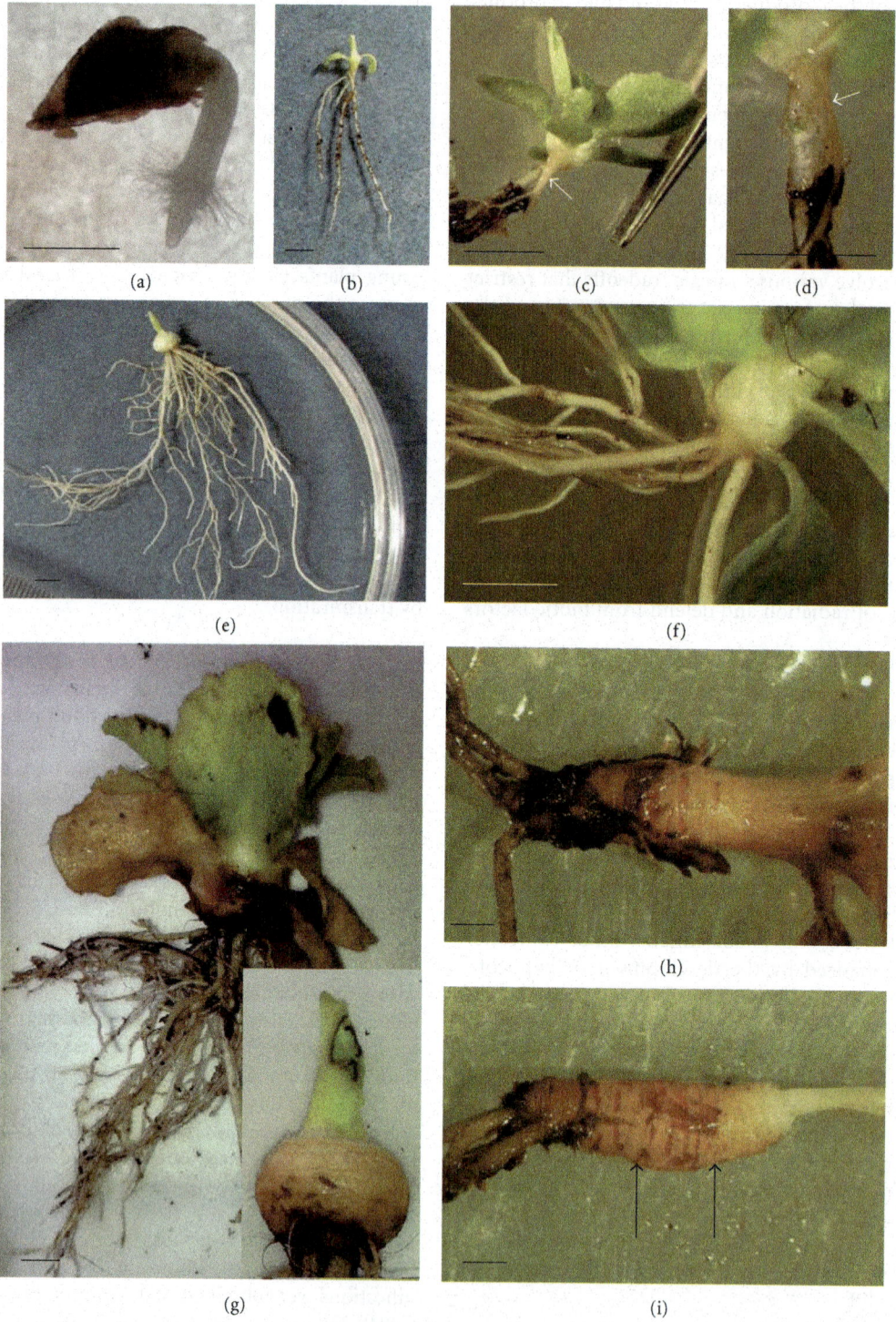

FIGURE 1: Developmental stages of *P. palinuri* plants: (a) germinated seeds with emerging root, hypocotyl ring hairs and hypocotyl; (b)–(d) very young plants at 5–7-leaf stage showing the hypocotyl enlarging into a cone in continuity with the primary root ((c), (d) arrows) and vigorous adventitious roots adjacent to the cone; (e)–(g) very young plants at 10-leaf stage showing the hypocotyl expanded into a round organ with numerous adventitious roots emerging from the base; (h)-(i) young plants with elongated hypocotyl becoming a rhizome; arrows show constrictions marking the probable transition between subsequent years/periods of growth. Bars are 1 mm in a and 0.5 cm in (b)–(i).

Occurrence of Morphological and Anatomical Adaptive Traits in Young and Adult Plants of the Rare Mediterranean
Cliff Species Primula palinuri Petagna

11

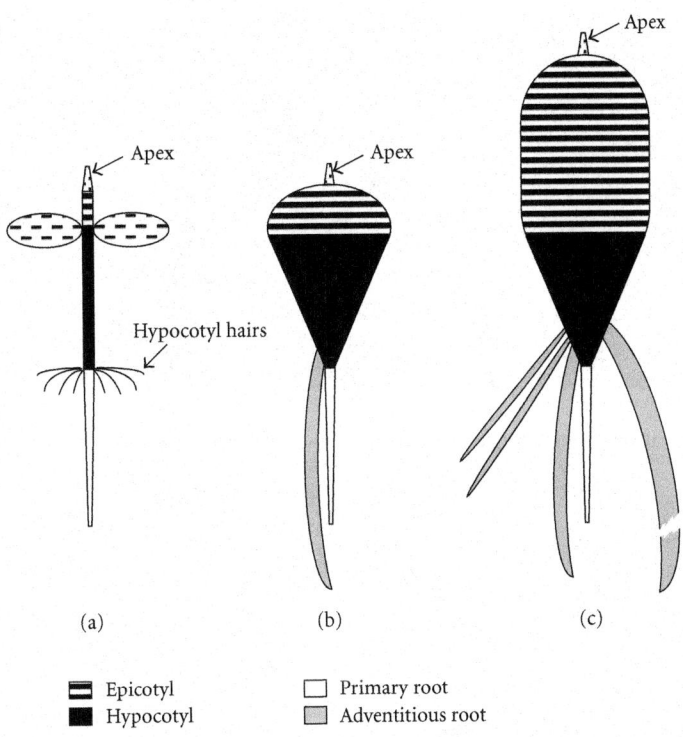

Epicotyl — Primary root
Hypocotyl — Adventitious root

FIGURE 2: Schematic representation of the development of *P. palinuri* plants from seedling stage up to the beginning of rhizome formation:
(a) seedling; (b) very young plants with hypocotyl elongating into a small cone and tuberizing epicotyl; (c) young plant with tuberized
epicotyl elongating to form a rhizome and many adventitious roots emerging at the base of the hypocotyl cone.

through the development of tissues in the epicotyl zone,
while the hypocotyl enlarges into a small cone in continuity
with the primary root (Figures 1(b)–1(d), 2(b)). At this stage
(5–7 leaves, 9–12 weeks) the root system is also made up by
1–3 adventitious roots at times more vigorous than the primary root (Figure 1(d)). At the 10-leaf stage (15 weeks) the
epicotyl expands into a round shape, while many adventitious roots develop from its base (Figures 1(e)– 1(f)). Plant
growth continues with the hypocotyl elongation that marks
the beginning of rhizome formation (Figures 1(g)–1(i), 2(c)).
The rhizomes of two small, young plants found in the field
are reported to show subsequent growth (Figures 1(h), 1(i)).
The rhizome develops indefinitely with a seasonal rhythm:
vegetative growth starts at the beginning of autumn, continues throughout the winter and stops at the end of spring.
New leaves develop together with rhizome elongation; they
all dry up in summer with the exception of a few small apical
leaves. As a consequence, plants assume two different habitus
during the year (Figure 3). Growing rhizomes gradually bend
due to gravity and lie on the substrate. Long rhizomes usually
hang on the steep or vertical cliff-faces and develop upside
down (Figures 3(c)–3(e)). Leaves are oriented independently
of rhizome position. Lateral buds can burst and start the rhizome branching (Figures 3(e)-3(f)). Adventitious roots can
emerge from the rhizome; frequently there are thick roots
acting as tie-rods (Figure 3(g)).

The overall morphological analysis allowed to differen-
tiate four different growth stages in *P. palinuri*: (a) seedling,
immediately after seed germination until the development of
the true leaves; (b) very young plants (about 1–24 months)
with hypocotyl elongating into a small cone and a round
tuberizing epicotyl; (c) young plant (about 2–5 years), with
tuberized epicotyl elongating and enlarging to form a rhizome; (d) adult plant with long, large, and often branched
rhizome.

Long rhizomes are either covered by dry leaves or naked.
In the first case, lower parts of the inflorescence peduncles
persist and mark yearly growth. In the second case a round,
tight scar pinpoints the beginning of the new growth. Bio-
metric measurements of these subsequent growth intervals
in the field showed that rhizomes of adult plants elongate
on average only 0.83 cm (SE ± 0.25) every growing season.
Sometimes they break due to wind, gravity or grazing, and
new rhizomes develop from basal buds. As a consequence,
even if apparently small, most of the plants in the field are
several tens of years old.

In order to explain why the transition from the seedling
to the adult stage is a critical point in the life cycle of
P. palinuri, we finely scanned and compared anatomical characteristics of plant organs of different ages. We analysed seedlings from seeds germinated in the laboratory. For very young
plants and young plants, we used individuals both collected
in the field and developed in the laboratory.

As regards leaves, although their size was dependent
on plant age or time of season, their anatomical structure
was comparable in all of them. Leaf cross-sections showed

FIGURE 3: Adult plants of *P. palinuri*: (a)-(b) view of plants growing on the cliffs; (c) plants in the winter *habitus*; (d) plants in the summer *habitus* with wilting and shedding leaves; (e) branched rhizome of a single plant developing upside down; (f) rhizome with bursting lateral bud (arrow); (g) adventitious roots emerging from the rhizome and acting as tie-rods. Bars are 5 cm in (c)-(d).

a dorsiventral structure, made of abaxial and adaxial epidermis characterised by large cells and covered by a thin cuticle, a mesophyll made of a 1-2 layered palisade parenchyma with not very elongated cells and a thick spongy parenchyma with many wide intercellular spaces (Figure 4(a)). The leaf lamina was flat, and stomata, mainly present on the adaxial surface, were well exposed. The whole structure was not dense, and a few glandular trichomes were present on both surfaces. Epifluorescence microscopy showed the presence of phenolic compounds both linked to the chloroplast membranes and

Occurrence of Morphological and Anatomical Adaptive Traits in Young and Adult Plants of the Rare Mediterranean
Cliff Species Primula palinuri Petagna

13

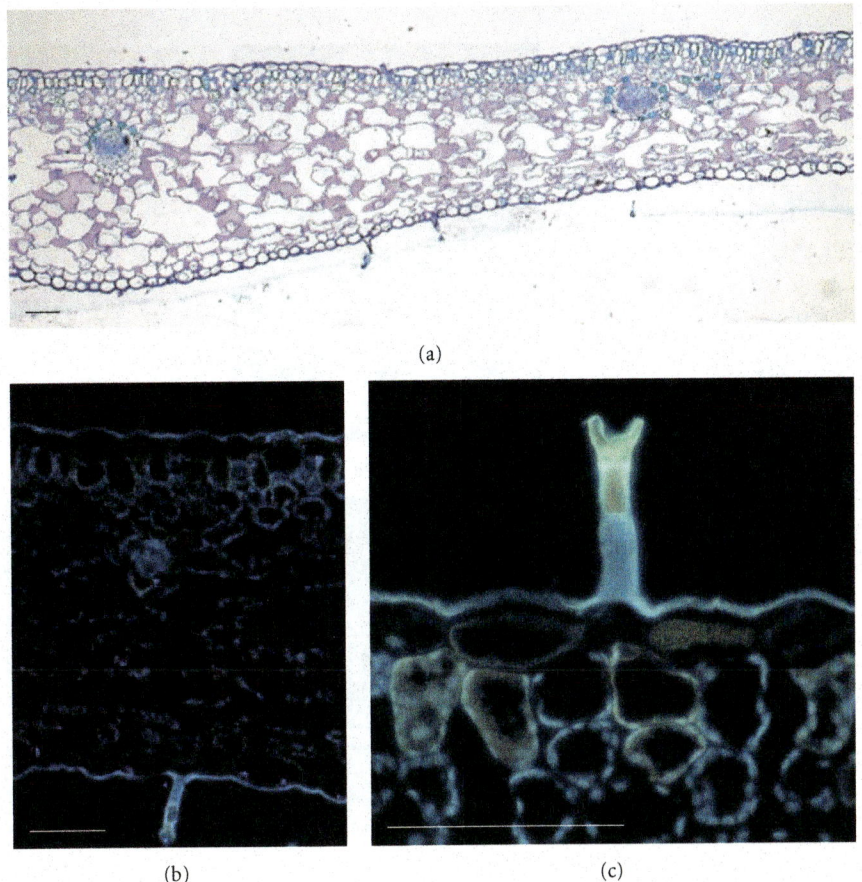

(a)

(b)   (c)

FIGURE 4: Photomicrograph of cross-sections of *P. palinuri* leaves viewed through light (a), epifluorescence (b) and (c) microscopy. UV-microscopy evidenced phenolic compounds linked to the chloroplast membranes (blue fluorescence) and filling some cells in the epidermis, palisade parenchyma and spongy parenchyma around vascular bundles (yellow-orange fluorescence). Bars are 50 μm.

filling some cells in the epidermis, palisade parenchyma, and spongy parenchyma around vascular bundles (Figures 4(b) and 4(c)). In general, leaves from both young and old plants do not survive the dry season and do not develop any anatomical trait of xeromorphy.

Roots with a diameter up to 2 mm, regardless of type (primary or adventitious roots) and growth conditions (field or laboratory), show a similar anatomical structure (Figures 5(a)–5(n)). The stele shows a poorly developed xylem surrounded by a well-developed endodermis with Casparian bands (Figures 5(b), 5(f), 5(i), and 5(n)). The cortical parenchyma consists of many layers of cells with thickened walls; the latter are not autofluorescent under UV-microscopy, suggesting that neither lignin nor suberin is encrusting them (Figures 5(b), 5(e), 5(i), and 5(n)). A suberized exodermis is present under the rhizoderm is also in the absorption zone, as testified by the presence of suberized hairs (Figures 5(b), 5(e), 5(g), 5(l), and 5(n)). The cortical parenchyma is deputed to the accumulation of water and starch.

Tie-rod roots differ from the other root types in having a more developed xylem (evaluated by comparing different root types of similar diameters), and many subepidermal layers of cells with suberized walls (Figures 5(o)–5(r)). Moreover, cortical parenchyma cells have very thickened walls and accumulate yellow-fluorescent phenolic compounds related to membranes (Figure 5(r)). Phenolics also fill the cells around the xylem and those of the subepidermal layers (Figures 5(q) and 5(r)).

Anatomical analysis of entire very young plants through light and epifluorescence microscopy showed that in moving from the above-ground towards the below-ground region there is a gradient of increasing starch accumulation and the gradual appearance of structures favouring water saving from simple cuticles up to many layers of suberized subepidermal cells (Figures 6(a)–6(g)). More specifically, the occurrence of 2-3 layers of suberized cells started at the periphery of the hypocotyl zone (Figures 6(f)–6(h)); the thickness of this suberized layer increases towards the root zone (Figures 6(f), 6(g), and 6(i)).

The rhizome of adult plants is characterised by a homogeneous parenchyma tissue accumulating water and starch; many steles, often interconnected, are dispersed throughout this ground tissue of parenchyma cells (Figures 6(l), and 6(m)). Each stele has a periphloematic vascular bundle and is surrounded by an endodermis with Casparian bands (Figure 6(n)). The periphery of the rhizome contains many layers of subepidermal suberized cells which accumulate phenolic compounds (Figure 6(o)).

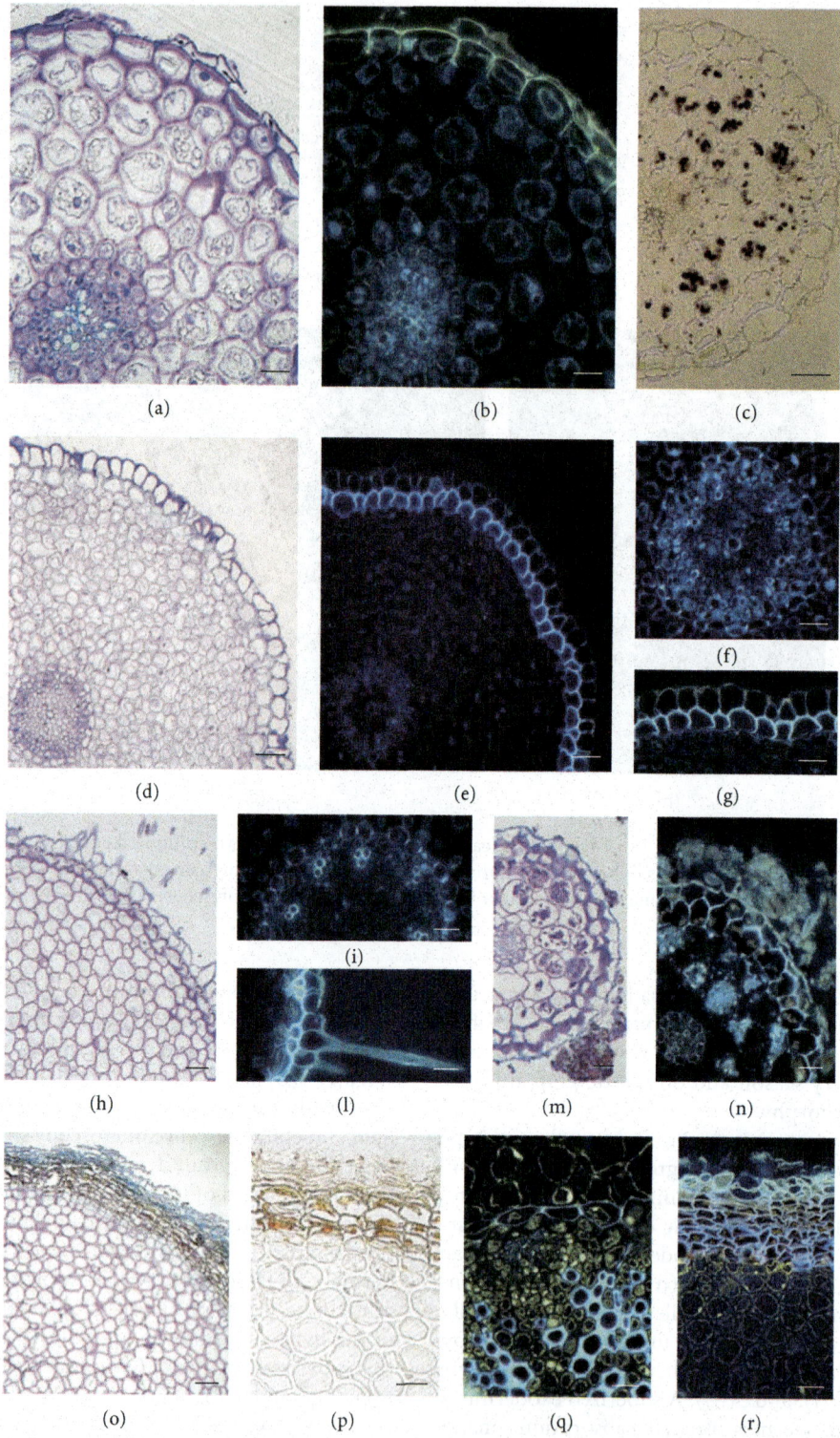

FIGURE 5: Photomicrographs of root cross-sections of *P. palinuri* plants developed under controlled conditions (a)–(g) or field-grown (h)–(r) and viewed under light and epifluorescence microscopy: (a)–(c) primary root of a very young plant; (b)–(g) adventitious root of a very young plant; (h)–(n) lateral roots from rhizomes of adult plants; (o)–(r) tie-rod roots from rhizomes of adult plants. UV-microscopy evidenced the blue fluorescence of endodermal Casparian bands (f), (i) and (q) of exodermal cell walls (e), (g), (n), and of subepidermal suberized layers (r). Phenolic compounds are evidenced by the yellow fluorescence (q) and (r). Bars are 20 μm in (a), (b), (f), (g), (i), (l)–(n), and (q), 50 μm in (c)–(e), (h), (p), (r), and 100 μm in (o).

Occurrence of Morphological and Anatomical Adaptive Traits in Young and Adult Plants of the Rare Mediterranean
Cliff Species Primula palinuri Petagna

15

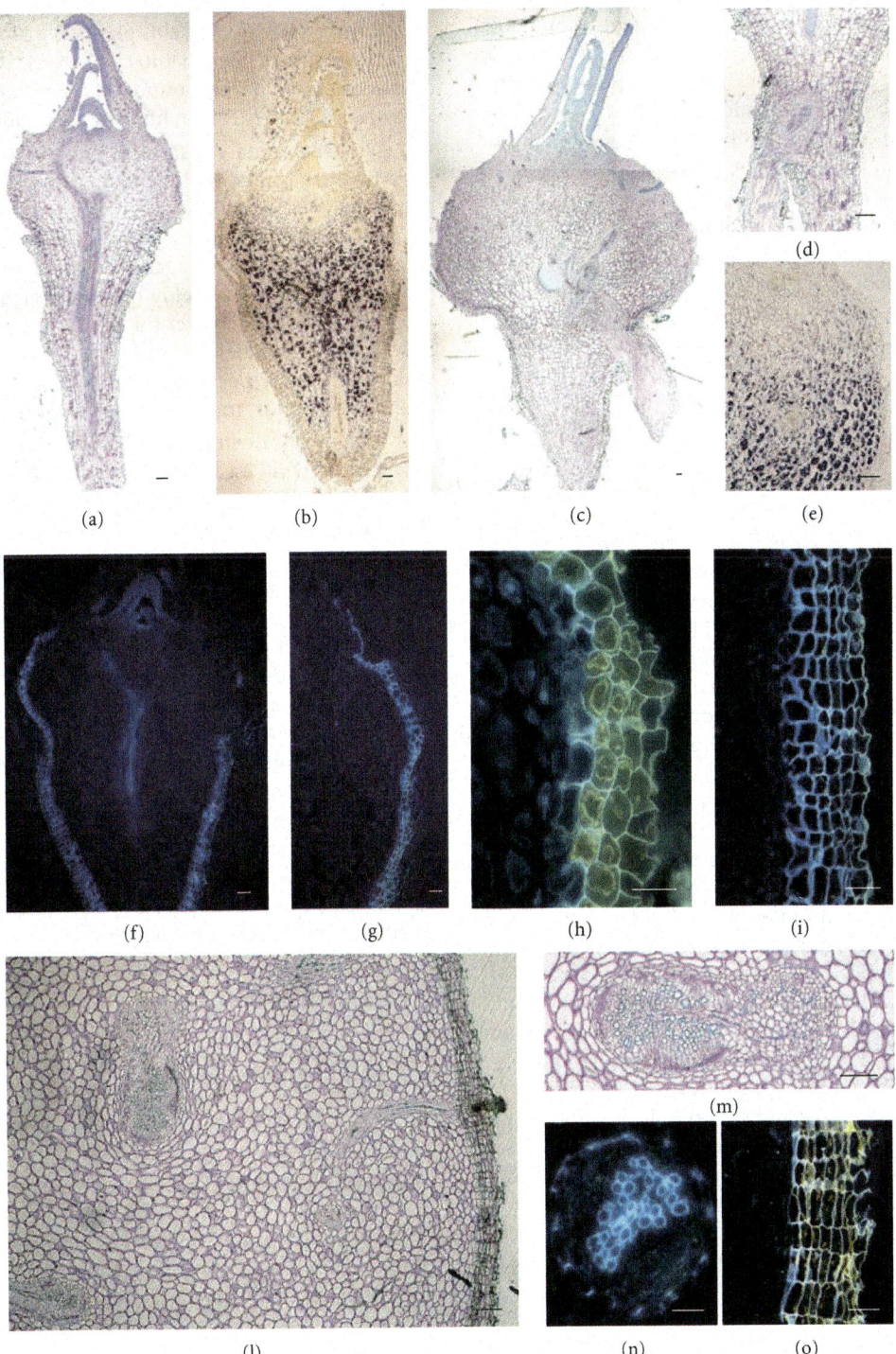

FIGURE 6: Photomicrographs of longitudinal sections of very young *P. palinuri* plants developed in controlled conditions (a)–(i) and of cross-sections of the rhizome of young and adult plants collected in the fields (l)–(o) viewed under light and epifluorescence microscopy. Subepidermal suberized layers (blue fluorescence) increasing in number if moving towards the below-ground region of the very young plants (f) and (g); endodermal Casparian bands in the stele of rhizome (n) and suberized subepidermal layers at the periphery of the rhizome (o). (h) hypocotyl; (i) root; (m) periphloematic vascular bundle. Bars are 10 μm in (a)–(f), (l), (m), and 50 μm in (h), (i), (n), and (o).

4. Discussion

Cliffs are ecosystems which are subjected to harsh physical conditions that limit the normal development of individual plants. Nevertheless, it is reported that these habitats are also characterised by high frequency of species intolerant of competition and biotic disturbance [1].

The success of seedling establishment relies on many factors, including constitutive traits of the seeds, such as size and amount of reserves. It also depends on the morphofunctional features of developing seedlings which allow prompt anchoring and absorption from the substrate [5, 20]. *P. palinuri* has developed a strategy to favour seedling survival based on both the development of hypocotyl hairs, which enhances water absorption and mechanical support, and the accumulation of phenolic compounds acting as feeding deterrents against animal predation [14]. Our field observations showed that seedling survival is a critical phase in the life cycle of this species: the low survival capability of plants at early stages of development may increase the extinction risk of *P. palinuri*.

In Mediterranean environments, species overcome summer drought by being sclerophylls, seasonally dimorphic or summer deciduous, depending on whether they adopt a strategy of tolerance or avoidance [21, 22]. Our microscopy analyses showed that leaves of *P. palinuri* have no xeromorphic traits. Their structure is not designed to accumulate or save water by reducing transpiration (e.g., low mesophyll density, reduced frequency of trichomes, and presence of well-exposed stomata at the lamina surface). Hence, plants dry leaves during periods of summer drought. *P. palinuri* leaves seem to defend themselves better against predators and to be shielded from high levels of radiation due to the presence of phenolic compounds in epidermal cells and along chloroplast membranes. The role of phenolic compounds in protection against predation and in photoprotection has been described for many species populating Mediterranean ecosystems [23–26]. In this scenario, also glandular trichomes on the leaves might be involved in a plant-animal interaction. Further work should be done to verify if they release volatile or toxic compounds.

Unlike leaves, the roots and rhizomes in *P. palinuri* adult plants seem to be specially designed to accumulate and save water as well as store reserves when starch is available after photosynthesis. Tie-rod roots growing upslope are also devised to withstand tension stress thanks to the occurrence of a large amount of lignified tissue in the stele which correlates with breaking strength as reported in other species [27]. Regulation of radial water flux in both rhizome and in roots of *P. palinuri* is exerted by the presence of a suberized exodermis in juvenile structures and of many layers of subepidermal cells with suberized walls in older structures. The occurrence of suberized layers of cells at the periphery of roots is common in Mediterranean species and is considered an effective way for the regulation of the inverse flux of water that, in extreme drought conditions, could pass from the root to the soil [28]. During Mediterranean summer drought, herbivores might be tempted by the presence of authentic "underground water tanks" in the form of rhizomes: the

accumulation of phenolic compounds along membranes in the parenchyma cells of the rhizome in *P. palinuri* can be considered a valuable means to protect against predation. Defence from herbivory of cliff plants is primarily provided by inaccessibility of their habitat [1]. Nevertheless, the occurrence of phenolic compounds in fresh tissues of *P. palinuri* may play a further role in limiting predation.

While perennial organs of *P. palinuri* adult plants appear as a fortress against biotic and abiotic constraints, plants at early stages of development do not seem to be so well protected. In juvenile phases, above-ground organs seem to be inadequately protected against transpiration by a thin and delicate cuticle, while below-ground organs are designed to accumulate and save water thanks to suberized subepidermal layers whose number increases with depth in the soil. However, tuberized hypocotyls and roots do not accumulate phenolic compounds, thus being easy preys of herbivores.

In light of our findings, *P. palinuri* adult plants are well adapted to living under the particular environmental conditions of the cliff microsites in the area of Cilento National Park. By contrast, Morphoanatomical traits of plants at early stages of development do not appear specialized enough to survive the environmental constraints of Mediterranean cliffs.

Transition from seedling to young plant is a critical stage for many species; some of them avoid mortality by passing through the vulnerable stages quickly [29]. This strategy cannot be applied by *P. palinuri* whose hallmark is slow in growth throughout its development. We assume that the reduced resource availability and limited protection against drought and predation of seedlings or very young plants do not permit their survival even to brief periods of unfavourable conditions. Moreover, the limited amount of water and starch stored in the tuberized epicotyl of young plants allow them to survive only to short periods of adverse conditions.

In conclusion, our analysis of the Morphoanatomical traits of juvenile and adult plants of *P. palinuri* has added knowledge about this rare species. Obtained information, concerning critical aspects of its biology, will also be useful to assess the degree of its vulnerability and draw up measures for its conservation.

Aknowledgments

This study was funded by the National Park of Cilento and Diano Valley. The authors are grateful to Maurizio Buonanno for critical contribution to the paper.

References

[1] D. W. Larson, U. Matthes, and P. E. Kelly, *Cliff Ecology*, Cambridge University Press, Cambridge, UK, 2000.

[2] S. Snogerup, "Evolutionary and plant geographical aspects of chasmophytic communities," in *Plant Life of South West Asia*, P. H. Davis, P. C. Harper, and I. C. Hedge, Eds., pp. 157–170, Botanical Society of Edinburgh, Aberdeen, UK, 1971.

[3] P. H. Davis, "Cliff vegetation in the eastern Mediterranean," *Journal of Ecology*, vol. 39, pp. 63–93, 1951.

Occurrence of Morphological and Anatomical Adaptive Traits in Young and Adult Plants of the Rare Mediterranean
Cliff Species Primula palinuri Petagna

17

[4] G. Aronne, V. De Micco, and M. Scala, "Plant reproductive ecology for conservation of mediterranean species," *The Botanica*, vol. 54, pp. 23–29, 2004.

[5] M. Fenner and K. Kitajima, "Seed and seedling ecology," in *Handbook of Functional Plant Ecology*, F. I. Pugnare and F. Valladares, Eds., pp. 589–621, Marcel-Dekker, New York, NY, USA, 1999.

[6] A. Traveset, N. Riera, and R. E. Mas, "Ecology of fruit-colour polymorphism in *Myrtus communis* and differential effects of birds and mammals on seed germination and seedling growth," *Journal of Ecology*, vol. 89, no. 5, pp. 749–760, 2001.

[7] G. Aronne and V. De Micco, "Hypocotyl features of Myrtus communis (Myrtaceae): a many-sided strategy for possible enhancement of seedling establishment in the Mediterranean environment," *Botanical Journal of the Linnean Society*, vol. 145, no. 2, pp. 195–202, 2004.

[8] C. Baraloto, D. E. Goldberg, and D. Bonal, "Performance trade-offs among tropical tree seedlings in contrasting micro-habitats," *Ecology*, vol. 86, no. 9, pp. 2461–2472, 2005.

[9] A. T. Moles and M. Westoby, "What do seedlings die from and what are the implications for evolution of seed size?" *Oikos*, vol. 106, no. 1, pp. 193–199, 2004.

[10] J. J. Stachowicz, "Mutualism, facilitation, and the structure of ecological communities," *BioScience*, vol. 51, no. 3, pp. 235–246, 2001.

[11] F. M. Padilla and F. I. Pugnaire, "Rooting depth and soil moisture control Mediterranean woody seedling survival during drought," *Functional Ecology*, vol. 21, no. 3, pp. 489–495, 2007.

[12] F. M. Padilla, J. D. D. Miranda, and F. I. Pugnaire, "Early root growth plasticity in seedlings of three Mediterranean woody species," *Plant and Soil*, vol. 296, no. 1-2, pp. 103–113, 2007.

[13] D. Uzunov, C. Gangale, and G. Cesca, "*Primula palinuri* Petagna. Flora da conservare," *Informatore Botanico Italiano*, vol. 40, pp. 101–102, 2008.

[14] G. Aronne, V. De Micco, and S. Barbi, "Hypocotyl features of *Primula palinuri* Petagna (Primulaceae) an endemic and rare species of the Southern Tyrrhenian Coast," in *Proceedings of the Ecologia Emergenza Pianificazione, 18th Congresso Nazionale della Società Italiana di Ecologia (SItE '10)*, G. Giordani, V. Rossi, and P. Viaroli, Eds., pp. 113–119, Società Italiana di Ecologia, 2010.

[15] P. Daget, "Le bioclimat mediterraneen: caracteres generaux, modes de caracterisation," *Vegetatio*, vol. 34, no. 1, pp. 1–20, 1977.

[16] I. Nahal, "The mediterranean climate from a biological viewpoint," in *Ecosystems of the World 11, Mediterranean-Type Shrublands*, F. di Castri, D. W. Goodall, and R. L. Specht, Eds., pp. 63–86, Elsevier Scientific Publishing Company, Amsterdam, The Netherlands, 1981.

[17] N. Feder and T. P. O'Brien, "Plant microtechnique: some principles and new methods," *American Journal of Botany*, vol. 55, pp. 123–142, 1968.

[18] K. Fukuzawa, "Ultraviolet microscopy," in *Methods in Lignin Chemistry*, S. Y. Lin and C. W. Dence, Eds., pp. 110–131, Springer, Berlin, Germany, 1992.

[19] S. E. Ruzin, *Plant Microtechnique and Microscopy*, Oxford University Press, New York, NY, USA, 1999.

[20] J. A. Young and E. Martens, "Importance of hypocotyl hairs in germination of Artemisia seeds," *Journal of Range Management*, vol. 44, no. 5, pp. 438–442, 1991.

[21] N. S. Margaris, "Adaptive strategies in plants dominating Mediterranean-type ecosystems," in *Ecosystems of the World 11, Mediterranean-Type Shrublands*, F. di Castri, D. W. Goodall, and R. L. Specht, Eds., pp. 309–315, Elsevier Scientific Publishing Company, Amsterdam, The Netherlands, 1981.

[22] V. de Micco and G. Aronne, "Seasonal dimorphism in wood anatomy of the Mediterranean *Cistus incanus* L. subsp. incanus," *Trees*, vol. 23, no. 5, pp. 981–989, 2009.

[23] G. Agati and M. Tattini, "Multiple functional roles of flavonoids in photoprotection," *New Phytologist*, vol. 186, no. 4, pp. 786–793, 2010.

[24] V. Lattanzio, P. A. Kroon, and S. Quideau, "Plant phenolics—secondary metabolites with diverse functions," in *Recent Advances in Polyphenol Research*, F. Daay and V. Lattanzio, Eds., pp. 1–35, Wiley-Blackwell, Oxford, UK, 2008.

[25] V. De Micco and G. Aronne, "Anatomical features, monomer lignin composition and accumulation of phenolics in 1-year-old branches of the Mediterranean *Cistus ladanifer* L," *Botanical Journal of the Linnean Society*, vol. 155, no. 3, pp. 361–371, 2007.

[26] V. de Micco, C. Arena, L. Vitale, G. Aronne, and A. V. de Santo, "Anatomy and photochemical behaviour of mediterranean *Cistus incanus* winter leaves under natural outdoor and warmer indoor conditions," *Botany*, vol. 89, no. 10, pp. 677–688, 2011.

[27] V. De Micco and G. Aronne, "Root structure of rumex scutatus growing on slopes," *IAWA Journal*, vol. 31, no. 1, pp. 13–28, 2010.

[28] E. Hose, D. T. Clarkson, E. Steudle, L. Schreiber, and W. Hartung, "The exodermis: a variable apoplastic barrier," *Journal of Experimental Botany*, vol. 52, no. 365, pp. 2245–2264, 2001.

[29] M. Fenner and K. Thompson, *The Ecology of Seeds*, Cambridge University Press, Cambridge, UK, 2005.

Antioxidant Potential and Oil Composition of *Callistemon viminalis* Leaves

Muhammad Zubair,[1] **Sadia Hassan,**[1] **Komal Rizwan,**[1] **Nasir Rasool,**[1]
Muhammad Riaz,[1] **M. Zia-Ul-Haq,**[2] **and Vincenzo De Feo**[3]

[1] *Department of Chemistry, Government College University, Faisalabad 38000, Pakistan*
[2] *Department of Pharmacognosy, University of Karachi, Karachi 75270, Pakistan*
[3] *Department of Pharmaceutical and Biomedical Sciences, University of Salerno, 84100 Salerno, Italy*

Correspondence should be addressed to Vincenzo De Feo; defeo@unisa.it

Academic Editors: A. Bosabalidis, J. C. Parajo, and B. Vyskot

The present study was designed to investigate the antioxidant potential and oil composition of *Callistemon viminalis* leaves. GC-MS analysis of the *n*-hexane extract revealed the presence of 40 compounds. Leaves contained appreciable levels of total phenolic contents (0.27–0.85 GAE mg/g) and total flavonoid contents (2.25–7.96 CE mg/g). DPPH radical scavenging IC_{50} and % inhibition of linoleic acid peroxidation were found to be in the ranges of 28.4–56.2 μg/ml and 40.1–70.2%, respectively. The haemolytic effect of the plant leaves was found in the range of 1.79–4.95%. The antioxidant activity of extracts was also studied using sunflower oil as an oxidative substrate and found that it stabilized the oil. The correlation between the results of different antioxidant assays and oxidation parameters of oil indicated that leaves' methanolic extract, exhibiting higher TPC and TFC and scavenging power, was also more potent for enhancing the oxidative stability of sunflower oil.

1. Introduction

The genus *Callistemon* belongs to family Myrtaceae that has a great medicinal importance. The majority of *Callistemon* species are found in east and southeast of Australia. Phytochemical studies of different *callistemon* species revealed the presence of different monoterpenes, sesquiterpenes flavonoids. *Callistemon* species are used for forestry, essential oil production, farm tree/windbreak plantings, degraded land reclamation, and as bioindicators for environmental management and ornamental horticulture among other applications [1, 2]. Earlier phytochemical explorations of members of this genus resulted in the identification of C-methyl flavonoids, triterpenoids, and phloroglucinol derivatives [3–6]. Moreover, some medicinal properties like antimicrobial, antistaphylococcal, antithrombin and nematicidal activities and larvicidal and pupicidal values have been reported for this genus [7]. *Callistemon viminalis* (family: Myrtaceae) is a small tree or shrub with pendulous foliage, although some forms are more pendulous than others. This is an ornamental plant commonly known as bottlebrush that is found in several areas with the exception of localities extremely cold and dry. It is also found along the streets and in the botanical gardens [1–3]. *C. viminalis* is edible, and its leaves are a tea substitute and have a delightfully refreshing flavour and fragrance. Antihelminthic and antibacterial activities of *C. viminalis* various parts have been reported [8]. It has been used to prepare a hot drink locally referred to as "tea" for the treatment of gastroenteritis, diarrhea and skin infections [9]. As part of our studies on exploring medicinal flora of Pakistan for their compositional, nutritional and antioxidant potential [10–16], we studied the plant *C. viminalis* leaves to explore its antioxidant potential and oil composition.

2. Materials and Methods

2.1. Materials. The fresh leaves of the fully matured plant *C. viminalis* were collected on the basis of intensive review and ethnopharmacological information from Botanical Garden, University of Agriculture Faisalabad, Pakistan (A plane

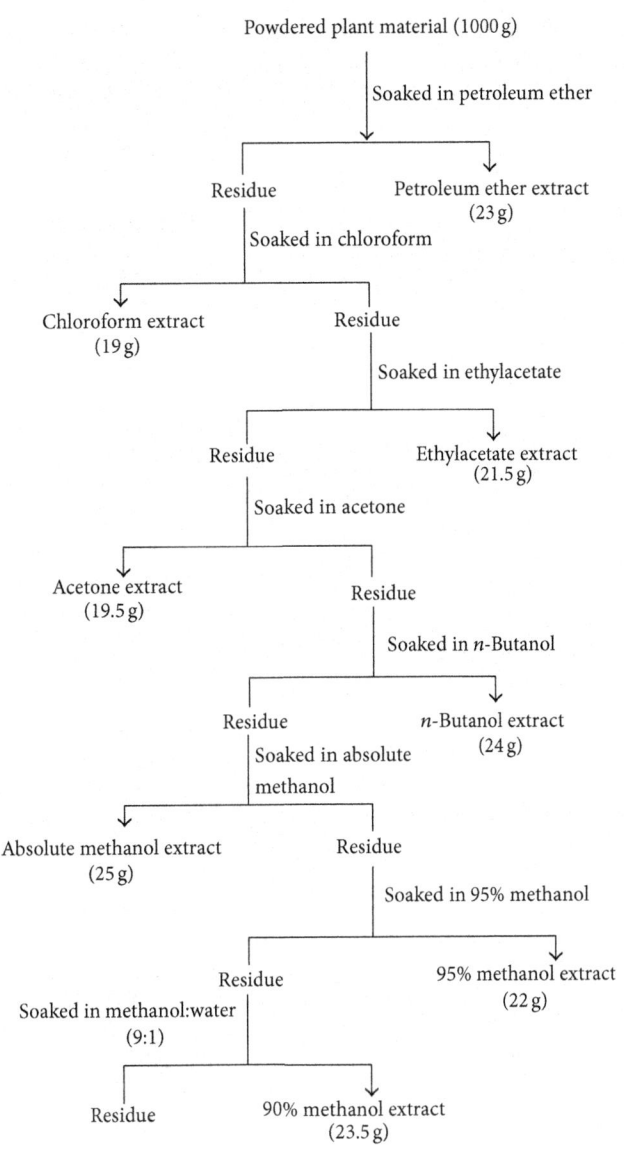

FIGURE 1: Schematic diagram showing preparation of extracts of *C. viminalis* leaves.

region latitude 31°-26′ N, longitude 73°-06′ E, and altitude 184.4 meters above main sea level) and further identified by a Taxonomist, Dr. Mansoor Hameed from Department of Botany, University of Agriculture Faisalabad, Pakistan.

2.2. Sample Preparation. The plant leaves were washed with distilled water and then shade dried. The grinded fine powder of leaves was extracted with petroleum ether (2×2 L) for 6 h at room temperature. After filtering, the extract was concentrated through rotary vacuum evaporator (Eyela, Tokyo Rikakikai Co., Ltd., Japan). This process was repeated thrice to obtain a sufficient quantity of petroleum ether extract. The remaining plant residue was further extracted with other different polarity-based solvents and obtained successively chloroform, ethylacetate, acetone, *n*-butanol, absolute methanol, 95% methanol (95 : 5, methanol : water, v/v), and 90% methanol (90 : 10, methanol : water, v/v) extracts (Figure 1).

All obtained extracts after drying were stored at −4°C till further analysis.

2.3. Preparation of n-Hexane Extract for GC-MS Analysis. The plant material (10 g of powdered leaves) was extracted with 300 mL *n*-hexane by using a Soxhlet apparatus for 6 h. The obtained *n*-hexane extract was filtered and evaporated by using a rotary evaporator and freeze dryer, respectively, to give the crude dried extract. The dried extract was stored at 4°C until used.

2.4. Evaluation of Antioxidant Activity. The plant leaves total phenolic contents (TPCs), total flavonoid contents (TFCs), DPPH radical scavenging IC_{50}, and % inhibition of linoleic acid peroxidation were determined by the following methods described by Rasool et al [17]. Reducing the potential of the plant extracts was also determined [18].

2.5. Haemolytic Activity. Haemolytic activity of the plant was evaluated by following the already reported procedure [19].

2.6. Determination of Antioxidant Efficacy Using Sunflower Oil as Oxidation Substrate

2.6.1. Stabilization of Sunflower Oil. The crude concentrated various extracts of the plant were separately added into the preheated (50°C), refined, bleached, and deodorized sunflower oil (SO) at concentration of 300 ppm (w/w). The oil samples were stirred for 30 minutes at 50°C for uniform dispersion. All oil samples were separately stabilized and stored in 100 mL airtight bottle. A control sample was also prepared (without extract) under the same set of analytical conditions. Samples were stored at room temperature. Synthetic antioxidant (BHT) was employed at its legal limit of 200 ppm to compare the efficacy of extracts. Stabilized and control oil samples (100 mL) were placed in dark brown airtight glass bottles with narrow necks and subjected to accelerated storage in an electric hot air oven (IM-30, Irmeco Gmbh & Co., Germany) at 60°C for 28 days. All oil samples were prepared in triplicate. Oil samples were taken after every 7-day intervals.

2.6.2. Measurement of Oxidation Parameters of Sunflower Oil. The oxidative deterioration level was assessed by the measurement of peroxide value (PV), free fatty acids (FFAs) conjugate diene (CD), conjugate triene (CT), and *p*-anisidine values. Determination of the FFA and PV of stabilized and control sunflower oil samples were made following the AOCS Official methods Cd 8–53 and F 9a-44, respectively [20]. The oxidation products such as conjugated dienes and conjugated trienes were analyzed by following the IUPAC method II D.23 [21]. The absorbance was noted at 232 and 268 nm, respectively. The determination of the *p*-anisidine value was made following an IUPAC method II. D. 26 [21].

2.7. Oil Analysis. The GC-MS analysis of the *n*-hexane extract was performed using GC 6850 network gas chromatographic system equipped with 7683B series auto injector and 5973 i inert mass selective detector (Agilent Technologies USA). Compounds were separated on an HP-5 MS capillary column having 5% phenyl polysiloxane, a stationary phase with the column length 30.0 m, internal diameter 0.25 mm, and film thickness 0.25 μm. The temperature of injector port was 300°C, and 1.0 μL of sample was injected in the split mode with split ratio 30 : 1. The helium was used as the carrier gas at constant flow with the flow rate of 1.5 mL/min. The temperature program used for column oven was an initial temperature 150°C and held for 1 min, then ramped at a rate of 10°C/min up to 290°C, and finally held at this temperature for 5 min. The temperature of MSD transfer line was 300°C. For mass spectra determination, MSD was operated in electrom ionization (EI) mode, with the ionization energy of 70 eV, while the mass range scanned was 3–500 *m/z*. The temperature of ion source was 230°C and that of MS quadrupole 150°C. The identification of components was based on the comparison of their mass spectra with those of NIST mass spectral library [22, 23].

2.8. Statistical Analysis. Each sample was analyzed individually in triplicate, and data were reported as mean ($n = 3 \times 3 \times 1$) ± standard deviation ($n = 3 \times 3 \times 1$). Data were analyzed by analysis of variance (ANOVA) using Minitab 2000 version 13.2 statistical software (Minitab Inc., Pennysylvania, USA).

3. Results and Discussion

3.1. Antioxidant Analysis. The yield (g/100 g) of various extracts from the plant *C. viminalis* leaves using different solvents ranged from 1.90% to 2.5%. The amounts of TPC and TFC (Table 1) from plant leaves in different solvent systems were found to be ranged from 0.27 to 0.85 GAE (mg/g of leaves extracts) and 2.25–7.96 CE (mg/g of leaves extracts), respectively. The ability of different solvents to extract TPC was found asfollows: absolute methanol > 90% methanol > chloroform > acetone > 95% methanol > ethylacetate > *n*-butanol > petroleum ether. The effect of different solvents on TFC values was found in the following order: absolute methanol > 95% methanol > chloroform > acetone > 90% methanol > ethylacetate > *n*-butanol > petroleum ether. These differences in the amount of TPC and TFC may be due to the varied efficiency of the extracting solvents to dissolve endogenous compounds. The leaves extracts exhibited different radical scavenging activity having IC$_{50}$ value 28.4–56.2 μg/mL. Absolute methanolic extract exhibited lowest IC$_{50}$ (28.4 μg/mL) followed by 95% methanol (34.1 μg/mL), chloroform (38.2 μg/mL), acetone (41.4 μg/mL), 90% methanol (45.2 μg/mL), ethylacetate (45.8 μg/mL), *n*-butanol (52.1 μg/mL), and petroleum ether (56.2 μg/mL) extracts. The free radical scavenging activity of absolute methanol and 95% methanol extracts was superior to that of other solvent extracts. However, All extracts offered slightly less scavenging activity as compared to the synthetic antioxidant BHT (19.2 μg/mL). The nature and amount of secondary metabolites of the plant cause the variation in free radical scavenging ability [24]. The free radical scavenging activity depends upon the chemical composition of extracts. Percent inhibition of linoleic acid oxidation ranged from 40.1% to 70.2%. The absolute methanolic extract exhibited the highest inhibition of linoleic acid oxidation (70.2%) and petroleum ether exhibited the lowest inhibition (40.1%). When the results of % inhibition of linoleic acid oxidation were compared with standard BHT (92.8%), all the samples showed significantly ($P < 0.05$) less antioxidant activity (Table 1). The order of inhibition of linoleic acid oxidation offered by various extracts of leaves was as follows: BHT > absolute methanol > 95% methanol > chloroform > acetone > 90% methanol > ethylacetate > *n*-butanol > petroleum ether.

Results of the present study showed that among all the solvent extracts, absolute methanolic extract of plant leaves extracted the highest amount of TPC and TFC, which also demonstrated the highest antioxidant activity as measured by DPPH radical scavenging and inhibition of linoleic acid oxidation. This may be due to the high polarity of methanol, whereas, petroleum ether demonstrated the least antioxidant activity probably because of its low polarity. Previous reports [25, 26] also revealed that the methanolic extracts of plant materials offer more effective antioxidants. Antioxidant

TABLE 1: Antioxidant activity of *C. viminalis* leaves[a].

Extracts/fraction/standard	Yield (g/100 g)	Total phenolic contents[b] (mg/g)	Total flavonoid contents[c] (mg/g)	DPPH, IC_{50} (μg/mL)	Inhibition in linoleic acid system (%)
Petroleum ether	2.30 ± 0.04	0.27 ± 0.001	2.25 ± 0.04	56.2 ± 0.54	40.1 ± 0.52
Chloroform	1.90 ± 0.02	0.71 ± 0.007	6.53 ± 0.06	38.2 ± 0.45	66.4 ± 0.77
Ethylacetate	2.15 ± 0.03	0.48 ± 0.005	5.11 ± 0.04	45.8 ± 0.51	54.1 ± 0.52
n-Butanol	2.40 ± 0.04	0.44 ± 0.004	3.96 ± 0.04	52.1 ± 0.62	48.2 ± 0.52
Acetone	1.95 ± 0.02	0.63 ± 0.007	6.43 ± 0.06	41.4 ± 0.52	56.8 ± 0.65
Absolute methanol	2.50 ± 0.04	0.85 ± 0.009	7.96 ± 0.08	28.4 ± 0.19	70.2 ± 0.77
95% methanol	2.20 ± 0.03	0.52 ± 0.006	6.69 ± 0.07	34.1 ± 0.41	68.2 ± 0.77
90% methanol	2.35 ± 0.03	0.75 ± 0.007	5.95 ± 0.07	45.2 ± 0.51	56.4 ± 0.61
BHT	—	—	—	19.2 ± 0.22	92.8 ± 0.91

[a]Values are mean ± S.D of three separate experiments.
[b]Total phenolic contents expressed as gallic acid equivalent.
[c]Total flavonoid contents are expressed as catechin equivalent.

compounds were extracted from *Catharanthus roseus* shoots and found that methanol gave the maximum antioxidant yield [17]. Similar results were observed in the present investigations as methanol was most effective to extract antioxidative compounds.

3.2. Haemolytic Activity.

Haemolytic activity was analyzed against human red blood cells (RBCs) using Triton X-100 as positive control. The % lysis of RBCs caused by the plant extracts was observed. Ethylacetate extract showed the highest haemolytic effect (4.95%) followed by petroleum ether (4.48%), 90% methanol (3.94%), chloroform (2.61%), 95% methanol (2.49%), acetone (2.33%), absolute methanol (2.03%), and *n*-butanol (1.79%) extracts, respectively. The haemolytic effect of *n*-butanol and absolute methanol extract was less then other extracts (Table 2). The mechanical stability of the erythrocytic membrane is a good indicator of the effect of various *in vitro* studies by various compounds for the screening of cytotoxicity. The percentage lysis of human erythrocytes was below 5.0% for all samples. All these results were in safe range. Though it can be expected that the plant extracts have a minor cytotoxicity [19, 27]. So pharmacologically this kplant may be safe to use for human beings as a source of potential drug.

3.3. Oil Analysis.

The chemical compounds identified by GC-MS analysis of *n*-hexane extract presented in Table 3. The mass spectrum of each compound was compared with that in the NIST 05 library. Almost 40 different compounds were identified in *n*-hexane extract of plant leaves. The major compounds determined in the *n*-hexane extract were 2,5,5,6,8a-pentamethyl-trans-4a,5,6,7,8,8a-hexahydro-gamma-chromene (27.60%), (10E,12E)-10,12-tetradecadienyl acetate (11.62%), Z-7-tetradecenal (4.98%), 1,3-cyclohexadiene (3.97%), respectively. Some of the compounds were present in traces or less concentration as compared to other identified compounds. The *n*-hexane extract may have some fatty acids/methyl esters which may be implicated in some antioxidant and antimicrobial activities. During the literature review, it was found that the activities of some phytocomponents such as flavonoids, palmitic acid (hexadecanoic acid,

TABLE 2: Percentage haemolysis caused by *C. viminalis* leaves different extracts[a].

Extracts	Percentage of haemolysis
Petroleum ether	4.48 ± 0.01
Chloroform	2.61 ± 0.02
Ethylacetate	4.95 ± 0.05
Acetone	2.33 ± 0.02
n-Butanol	1.79 ± 0.01
Absolute methanol	2.03 ± 0.02
95% methanol	2.49 ± 0.04
90% methanol	3.94 ± 0.03
Phosphate Buffer Saline (PBS)	0.00
Triton X-100	99.8 ± 1.01

[a]Values are mean ± S.D of three separate experiments.

ethyl ester and *n*-hexadecanoic acid), unsaturated fatty acid, docosatetraenoic acid, and octadecatrienoic acid as antimicrobial, antiinflammatory, and antioxidant activities [28]. Therefore, the chemical constituents found in may *C. viminalis* leaves play major roles in the biological activities and pharmacological properties.

3.4. Stabilization of Sunflower Oil.

Formation of free fatty acids (FFAs) might be an important measure of rancidity of foods. FFAs are formed due to hydrolysis of triglycerides and may get promoted by the reaction of oil with moisture [29]. FFA content went on increasing with the increase in storage period for all the samples, but no regular pattern of increase could be observed. Control exhibited the highest FFA, while sunflower oil stabilized with BHT exhibited least (Figure 2). Initially, there was no increase in FFA of stabilized oil samples, but after seven days of storage, an increase that was observed showed the free fatty acid (FFA) contents of oil samples stabilized with leaves extract of *C. viminalis* under ambient storage conditions. All the oil samples stabilized

TABLE 3: GC-MS analysis of *n*-hexane extract of *C. viminalis* leaves.

Peak number	Retention time	Compounds	% area
1	5.411	*n*-Octane	0.87
2	6.350	2-Methyloctane	0.62
3	7.754	o-Xylene	0.56
4	7.882	3-Methyloctane	0.68
5	8.964	*n*-Nonane	1.44
6	13.004	1,3-Cyclohexadiene	3.97
7	16.301	Undecane	1.07
8	19.590	*n*-Dodecane	1.02
9	22.354	2-Isopropyl-5-methylphenol	1.64
10	23.309	Exo-2-hydroxy cineole	1.26
11	24.266	Etyhl-5,9-dimethyl-2,4-decadienoate	3.95
12	25.413	3-Methyl-5-(2,6-dimethylheptyl)-1,5-pent-2-enolide	1.20
13	25.497	*n*-Tetradecane	1.27
14	26.880	4-Oxo-β-isodamascol	0.71
15	28.026	*n*-Pentadecane	0.62
16	28.170	2,4-Di-tert-butylphenol	2.75
17	28.391	Durohydroquinone	0.92
18	28.867	(10E,12E)-10,12-Tetradecadienyl acetate	11.62
19	29.027	2,5,5,6,8a-Pentamethyl-trans-4a,5,6,7,8,8a-hexahydro-gamma-chromene	27.60
20	29.783	(−)-Spathulenol	2.07
21	30.117	Hexadecane	2.34
22	31.919	10-Methylicosane	1.25
23	33.537	Heneicosane	3.09
24	33.583	Origanene	0.54
25	35.008	*n*-Nonadecane	0.62
26	35.837	Eicosanoic acid	3.37
27	36.382	*n*-Tetracosane	1.86
28	37.672	1,54-Dibromotetrapentacontane	0.49
29	37.791	Trans-phytol	1.05
30	38.074	Cis-9,cis-12-octadecadienoic acid	2.38
31	38.157	Z-7-Tetradecenal	4.98
32	38.411	Stearic acid	0.60
33	38.892	*n*-Tetratricoaconate	0.81
34	40.175	Ergost-7,22-dien-9,11-epoxy-3-ol, acetate(ester)	0.75
35	40.558	Furostan-12-one	2.37
36	42.576	Mono(2-ethylhexyl)phthalate	2.21
37	44.934	Trans-squalene	1.74
38	45.506	*n*-Hexatriacontane	0.50
39	47.309	α-Tocopherol-β-D-mannoside	0.82
40	49.184	Gamma-sitosterol	2.42

with plant extracts were found to show a slow followed by a gradual increase in free fatty acid contents. The lower values of free fatty acid contents of stabilized oil samples than control indicated the effectiveness of leaves extracts as natural antioxidant in retarding the free fatty acid contents. Of all extracts, methanol extract was most efficient after BHT to inhibit the formation of FFA contents.

Peroxide value (PV) is usually used to evaluate the extent of primary oxidation products in oils. The highest PV was observed for control sample followed by petroleum ether > chloroform > ethylacetate > *n*-butanol > acetone > 90% methanol > 95% methanol > absolute methanol > BHT, respectively. All used extracts of the plant controlled peroxide value appreciably, revealing good antioxidant efficacy of extracts in stabilization of oil. A regular increase in PV as a function of storage time was observed for all the samples at all intervals. Initially, the difference in peroxide content of control and stabilized oil samples was not noticeable; it became significant ($P < 0.05$) just after heating up to one day (Figure 3).

The results for para-anisidine values (PAVs) which usually determine the amount of aldehyde in oils presented in Figure 4. The control sample showed the maximum increase in para-anisidine values indicating a higher rate of secondary product formation. A slow increase in PAV of stabilized sunflower oil as compared with the control indicating the

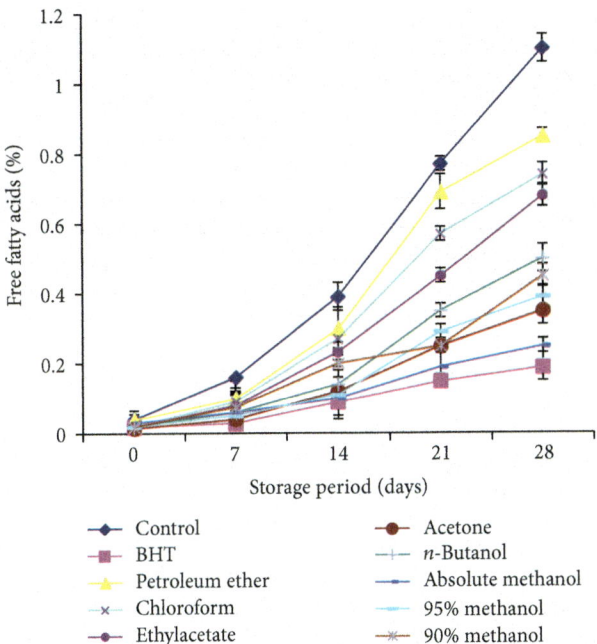

FIGURE 2: Free fatty acid contents (%) of sunflower oil stabilized with *C. viminalis* leaves extracts.

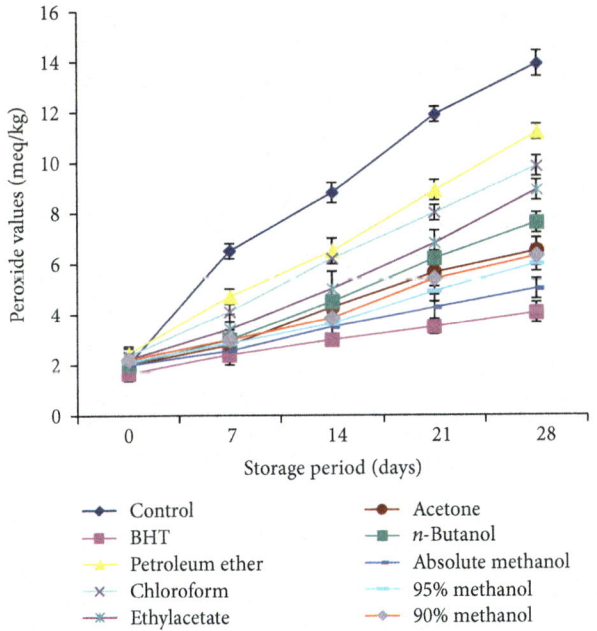

FIGURE 3: Peroxide values of sunflower oil stabilized with *C. viminalis* leaves extracts.

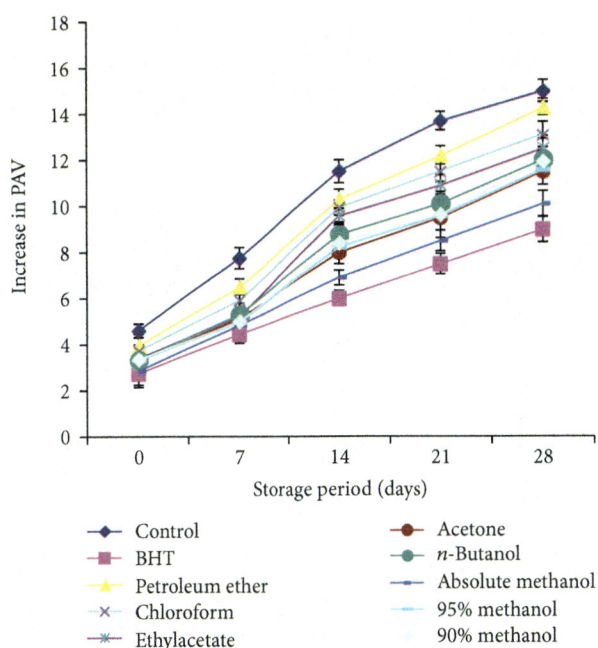

FIGURE 4: Relative increase in *p*-ansidine values of sunflower oil stabilized with *C. viminalis* leaves extracts.

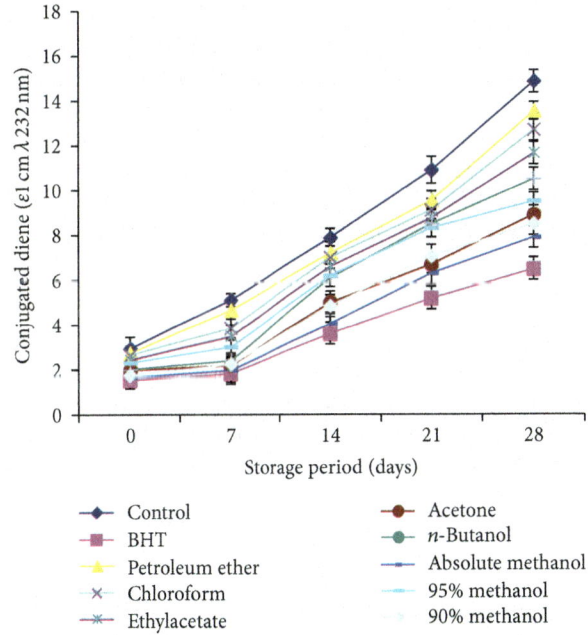

FIGURE 5: Relative increase in conjugated dienes content (CD) of sunflower oil stabilized with *C. viminalis* leaves extracts.

antioxidant potential of the plant leaves. A decreasing order of stability of oil treated with different extracts of plant regarding para-anisidine values was found to be BHT > absolute methanol > acetone > 90% methanol > 95% methanol > *n*-butanol > ethylacetate > chloroform > petroleum ether > control.

The formation of conjugated diene (Figure 5) and triene (Figure 6) analyzed for the control and stabilized sunflower oil, respectively. Highest contents were observed for control, indicating greater intensity of oxidation.

The determination of CD and CT is a good measure of the oxidative state of oils [30] and thus a good indicator of effectiveness of antioxidants. CD and CT contents went on increasing with the increase in storage time. A slow increase in CD and CT of the stabilized sunflower oil as compared with those of the control indicated the antioxidant potential

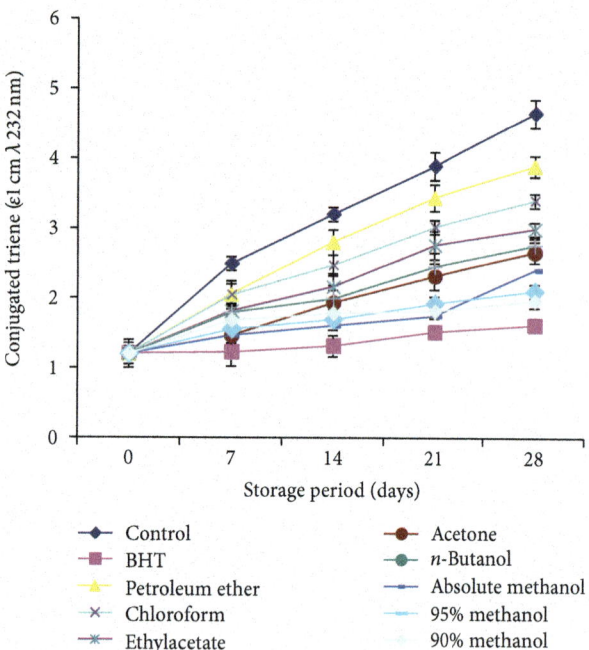

FIGURE 6: Relative increase in conjugated trienes content (CT) of sunflower oil stabilized with *C. viminalis* leaves extracts.

of the *C. viminalis* leaves. Stabilized sunflower oil conjugated diene and triene contens values were found to be in range of 1.66–13.54 and 1.2–3.89 (ɛ1 cm 232 nm), respectively. Absolute methanolic extract showed the lowest values and petroleum ether exhibited highest. All extracts played a prominent role for stabilization of sunflower oil but after standard BHT, methanol extract was most efficient to stabilize the oil.

4. Conclusions

The results of the present study concluded that the plant possessed considerable antioxidant potential, and it may also be used to stabilize the edible sunflower oil. Hence, the plant leaves investigated can be explored as a potential antioxidant source of natural origin. Cytotoxicity of plant extracts against human erythrocytes was checked, and it was in safe range, so the investigated plant may be safe use for pharmaceutical and natural therapies.

Conflict of Interests

The authors do not have a direct financial relation with the commercial identities mentioned in this paper that might lead to a conflict of interests.

References

[1] R. D. Spencer and P. F. Lumley, "Callistemon," in *Flora of New South Wales*, G. J. Harden, Ed., vol. 2, pp. 168–173, New South Wales University Press, Sydney, Australia, 1991.

[2] G. S. Wheeler, "Maintenance of a narrow host range by Oxyops vitiosa; a biological control agent of Melaleuca quinquenervia," *Biochemical Systematics and Ecology*, vol. 33, no. 4, pp. 365–383, 2005.

[3] J. W. Wrigley and M. Fagg, *Bottlebrushes, Paperbarks and Tea Trees and All Other Plants in the Leptospermum Alliance*, Angus and Rovertson, Sydney, Australia, 1993.

[4] E. Wollenweber, R. Wehde, M. Dörr, G. Lang, and J. F. Stevens, "C-Methyl-flavonoids from the leaf waxes of some Myrtaceae," *Phytochemistry*, vol. 55, no. 8, pp. 965–970, 2000.

[5] F. Huq and L. N. Misra, "An alkenol and C-methylated flavones from *Callistemon lanceolatus* leaves," *Planta Medica*, vol. 63, no. 4, pp. 369–370, 1997.

[6] R. S. Varma and M. R. Parthasarathy, "Triterpenoids of *Callistemon lanceolatus* leaves," *Phytochemistry*, vol. 14, no. 7, pp. 1675–1676, 1975.

[7] C. Gomber and S. Saxena, "Anti-staphylococcal potential of *Callistemon rigidus*," *Central European Journal of Medicine*, vol. 2, no. 1, pp. 79–88, 2007.

[8] S. K. Srivastava, A. Ahmad, K. V. Syamsunder, K. K. Aggarwal, and S. P. S. Khanuja, "Essential oil composition of *Callistemon viminalis* leaves from India," *Flavour and Fragrance Journal*, vol. 18, no. 5, pp. 361–363, 2003.

[9] M. M. Cowan, "Plant products as antimicrobial agents," *Clinical Microbial Research*, vol. 12, pp. 564–582, 1999.

[10] K. Rizwan, M. Zubair, N. Rasool, M. Riaz, M. Zia-Ul-Haq, and V. de Feo, "Phytochemical and Biological Studies of *Agave attenuata*," *International Journal of Molecular Sciences*, vol. 13, pp. 6440–6451, 2012.

[11] M. N. Bari, M. Zubair, K. Rizwan et al., "Biological activities of *Opuntia Monacantha* cladodes," *Journal of Chemical Society Pakistan*, vol. 34, pp. 990–995, 2012.

[12] M. Zubair, K. Rizwan, N. Rasool, N. Afshan, M. Shahid, and V. Ahmed, "Antimicrobial potential of various extract and fractions of leaves of *Solanum nigrum*," *International Journal of Phytomedicine*, vol. 3, no. 1, pp. 63–67, 2011.

[13] N. Rasool, K. Rizwan, M. Zubair et al., "Antioxidant activity of various extracts and organic fractions of *Ziziphus jujuba*," *International Journal of Phytomedicine*, vol. 3, pp. 346–352, 2011.

[14] M. Zia-Ul-Haq, S. Cavar, M. Qayum, I. Imran, and V. Defeo, "Compositional studies: antioxidant and antidiabetic activities of *Capparis decidua* (Forsk.) Edgew," *International Journal of Molecular Sciences*, vol. 12, pp. 8846–8861, 2011.

[15] M. Zia-Ul-Haq, S. Ahmad, S. Iqbal, D. L. Luthria, and R. Amarowicz, "Antioxidant potential of lentil cultivars commonly consumed in Pakistan," *Oxidation Communication*, vol. 34, pp. 819–831, 2011.

[16] M. Zia-Ul-Haq, S. Ahmad, L. Calani et al., "Compositional study and antioxidant potential of *Ipomoea hederacea* Jacq. and *Lepidium sativum* L. sedes," *Molecules*, vol. 17, pp. 10306–10321, 2012.

[17] N. Rasool, K. Rizwan, M. Zubair, K. U. R. Naveed, I. Imran, and V. U. Ahmed, "Antioxidant potential of different extracts and fractions of *Catharanthus roseus* shoots," *International Journal of Phytomedicine*, vol. 3, no. 1, pp. 108–114, 2011.

[18] B. Sultana, F. Anwar, and R. Przybylski, "Antioxidant activity of phenolic components present in barks of *Azadirachta indica*, *Terminalia arjuna*, *Acacia nilotica*, and *Eugenia jambolana* Lam. trees," *Food Chemistry*, vol. 104, no. 3, pp. 1106–1114, 2007.

[19] W. A. Powell, C. M. Catranis, and C. A. Maynarrd, "Design of self processing antibacterial peptide of plant protection," *Letters in Applied Microbiology*, vol. 31, pp. 163–165, 2000.

[20] American Oil Chemists Society (AOCS), *Official and Recommended Practices of the American Oil Chemists Society*, AOCS Press, Champaign, Ill, USA, 5th edition, 1997.

[21] International Union of Pure And Applied Chemistry (IUPAC), *Standard Methods For the Analysis of Oils, Fats and Derivatives*, C. Paquot and A. Hautfenne, Eds., Blackwell Scientific, London, UK, 7th edition, 1987.

[22] Y. Massada, *Analysis of Essential Oils by Gas Chromatography and Mass Spectrometry*, John Wiley & Sons, New York, NY, USA, 1976.

[23] Mass Spectral Library. NIST/EPA/NIH: USA, 2002, http://www.nist.gov/srd/nistla.htm.

[24] Y. Sudjaroen, R. Haubner, G. Würtele et al., "Isolation and structure elucidation of phenolic antioxidants from Tamarind (*Tamarindus indica* L.) seeds and pericarp," *Food and Chemical Toxicology*, vol. 43, no. 11, pp. 1673–1682, 2005.

[25] P. Siddhuraju and K. Becker, "Antioxidant properties of various solvent extracts of total phenolic constituents from three different agroclimatic origins of drumstick tree (*Moringa oleifera Lam.*) leaves," *Journal of Agricultural and Food Chemistry*, vol. 51, no. 8, pp. 2144–2155, 2003.

[26] S. A. S. Chatha, F. Anwar, M. Manzoor, and J. U. R. Bajwa, "Evaluation of the antioxidant activity of rice bran extracts using different antioxidant assays," *Grasas y Aceites*, vol. 57, no. 3, pp. 328–335, 2006.

[27] P. Sharma and J. D. Sharma, "*In vitro* hemolysis of human erythrocytes—by plant extracts with antiplasmodial activity," *Journal of Ethnopharmacology*, vol. 74, no. 3, pp. 239–243, 2001.

[28] P. Kumar, S. Kumaravel, and C. Lalitha, "Screening of antioxidant activity, total phenolics and GC-MS study of *Vitex negundo*," *African Journal of Biotechnology*, vol. 4, pp. 191–195, 2010.

[29] N. Frega, M. Mozzon, and G. Lercker, "Effects of free fatty acids on oxidative stability of vegetable oil," *Journal of the American Oil Chemists' Society*, vol. 76, no. 3, pp. 325–329, 1999.

[30] S. H. Yoon, S. K. Kim, M. G. Shin, and K. H. Kim, "Comparative study of physical methods for lipid oxidation measurement in oils," *Journal of the American Oil Chemists' Society*, vol. 62, no. 10, pp. 1487–1489, 1985.

4

Influence of Abscisic Acid and Sucrose on Somatic Embryogenesis in Cactus *Copiapoa tenuissima* Ritt. forma *mostruosa*

J. Lema-Rumińska, K. Goncerzewicz, and M. Gabriel

Laboratory of Biotechnology, Department of Ornamental Plants and Vegetable Crops,
University of Technology and Life Sciences in Bydgoszcz, Bernardyńska 6, 85-029 Bydgoszcz, Poland

Correspondence should be addressed to J. Lema-Rumińska; lem-rum@utp.edu.pl

Academic Editors: U. Feller, A. Roldán Garrigós, and H. Verhoeven

Having produced the embryos of cactus *Copiapoa tenuissima* Ritt. forma *mostruosa* at the globular stage and callus, we investigated the effect of abscisic acid (ABA) in the following concentrations: 0, 0.1, 1, 10, and 100 μM on successive stages of direct (DSE) and indirect somatic embryogenesis (ISE). In the indirect somatic embryogenesis process we also investigated a combined effect of ABA (0, 0.1, 1 μM) and sucrose (1, 3, 5%). The results showed that a low concentration of ABA (0-1 μM) stimulates the elongation of embryos at the globular stage and the number of correct embryos in direct somatic embryogenesis, while a high ABA concentration (10–100 μM) results in growth inhibition and turgor pressure loss of somatic embryos. The indirect somatic embryogenesis study in this cactus suggests that lower ABA concentrations enhance the increase in calli fresh weight, while a high concentration of 10 μM ABA or more changes calli color and decreases its proliferation rate. However, in the case of indirect somatic embryogenesis, ABA had no effect on the number of somatic embryos and their maturation. Nevertheless, we found a positive effect of sucrose concentration for both the number of somatic embryos and the increase in calli fresh weight.

1. Introduction

Copiapoa tenuissima Ritt. forma *mostruosa* is a member of the Cactaceae family, and it is a spontaneous mutant derived from *Copiapoa tenuissima* Ritt. (syn. *Neochilenia wageringeliana*) species [1]. This species originated in extremely dry desert areas of Chile, and its name has been linked to the town of Copiapo where the small number of its natural habitats exists [2]. Currently, this species is strictly protected (entered into Appendix II Cactaceae species of the CITES 1997). *C. tenuissima* Ritt. forma *mostruosa*, which differs from the original species in the lack of thorns and rare, almost black epidermis color, is one of the most interesting cactus species most desired by collectors globally. Its attractive appearance is additionally enhanced by white wool-like areoles. To prove an effective protection to the natural habitats of this cactus, it is necessary to develop effective propagation methods under artificial conditions. Under natural conditions the formation of new axillary shoots is slow; shoot growth by the activation of meristems in areoles can be much faster [3]. But the most effective methods of micropropagation are based on somatic embryogenesis which allows for producing valuable genotypes on a large scale in a very short time. Somatic embryogenesis may be induced via a direct or indirect pathway. For direct somatic embryogenesis, embryos develop directly on the surface of organized tissue. Alternatively, indirect somatic embryogenesis may occur via an intermediate step involving callus formation. Both the direct and indirect somatic embryogenesis make the regeneration of plants from single somatic cells possible [4]. Minocha and Mehra [5] reported the first regeneration of somatic embryos in cactus *Neomammillaria prolifera*. Since then, many applicable reports on cacti have been published [6–10], but only one on *Copiapoa* genus [11]. A critical stage of somatic embryogenesis is the maturation stage when embryos accumulate up storage materials [12, 13]. This stage depends on the presence of specific plant growth regulators (PGRs), mostly abscisic acid (ABA) and sucrose [14–16]. ABA increases the level of storage proteins and fatty acids in somatic embryos [15–17]. Abscisic acid plays a significant role in the regulation of many physiological processes of plants. It is often used in tissue culture systems to promote somatic embryogenesis and enhance somatic embryo quality by increasing desiccation tolerance and preventing precocious germination [18]. Sucrose, as a source of energy and carbon skeletons, determines the growth potential of the plant [19] and also affects the quality of embryos [15].

The aim of the present study was to determine the effect of ABA and sucrose on direct and indirect somatic embryogenesis in cactus *Copiapoa tenuissima* Ritt. f. *mostruosa*.

2. Materials and Methods

Plant materials were mammillae of cacti *Copiapoa tenuissima* Ritt. forma *mostruosa*. The cactus was grafted onto the pad (stem) from the genus *Cereus*. The initial explants (400 mammillae with areoles) were taken from the central zones of donor plants (average height: 6 cm) from the collection of Licznerski (Jarużyn Kolonia near Bydgoszcz, Poland).

2.1. Direct Somatic Embryogenesis (DSE)

2.1.1. Induction Stage. The explants were surface disinfected with 70% ethanol for 1-2 s and then with 0.79% hypochloride solution for 15 min, followed by three rinses with distilled sterilized water (all steps in laminar flow cabinet). Then they were cultured (one explant per jar) on MS [20] basal salts medium with additional 1506.2 μM CaCl$_2$·6H$_2$O, 50.0 μM FeSO$_4$·7H$_2$O, and 55.3 μM Na$_2$EDTA·2H$_2$O. The medium contained 3% sucrose, solidified with 1.2% Purified Lab Agar (Biocorp); the media pH was adjusted to 5.7 prior to autoclaving. The explants were cultured on MS medium with 9.05 μM auxin 2,4-D (2,4-dichlorophenoxyacetic acid) or MS medium without PGRs (as control). The cultures were kept in a growth room at 24 ± 2°C and exposed to 16 h photoperiod. Daylight was maintained by using Philips TLD 54/34 W lamps with a photon flux density of 40.4 μmol · m^{-2} · s^{-1}. Induction stage developed during 8 weeks.

2.1.2. Influence of ABA on DSE. At the successive stage of the experiment, randomly selected mammillae with somatic embryos were transferred onto 5 types of modified MS media containing ABA at different concentrations: 0; 0.1; 1; 10; 100 μM. Each type of media was represented by 12 explants with four replications (3 explants per jar). Other medium components and *in vitro* culture conditions were identical to the ones described previously in induction stage. After 6 weeks of culture the analysis of somatic embryos was made using the stereomicroscope.

2.2. Indirect Somatic Embryogenesis (ISE)

2.2.1. Induction Calli Stage. Calli were obtained on a modified MS medium containing 9.05 μM 2,4-D from initial explants (mammillae with areoles). The successive calli-proliferating transfers were made regularly every 3 weeks on a modified MS medium supplemented with 13.32 μM BA, 16.11 μM NAA, and 0.57 μM IAA added to produce an adequate amount of callus for further research. The other medium components and culture conditions were the same as described for the induction stage of the direct somatic embryogenesis induction. The calli were yellow-green in color and demonstrated strong proliferation properties.

2.2.2. Effect of ABA and ABA and Sucrose. The calli were divided into fragments and the initial fresh weight was registered. Next, they were cultured onto modified MS media containing ABA at different concentrations (0; 0.1; 1; 10; 100 μM) or onto modified MS media with ABA (0; 0.1; 1 μM) and sucrose (1; 3; 5% w/v). Despite different concentrations of ABA or ABA and sucrose, the MS media contained a fixed number of PGRs facilitating calli proliferation (13.32 μM BA, 16.11 μM NAA, and 0.57 μM IAA). After 5 weeks the calli fresh weight was registered again, and the embryo structures were analyzed under the stereomicroscope.

2.3. Statistical Analysis. The experiments were arranged in a completely randomized design with four replicates per treatment. Each type of media was represented by 12 explants each; four replicates (3 explants per jar). The data were evaluated by analysis of variance, and comparisons between the mean values were made by the *t*-Student test at $\alpha = 0.05$.

3. Results

3.1. Direct Somatic Embryogenesis. During induction stage, we observed the regeneration of somatic embryos at the globular stage on 20.67% of mammillae cultured in media containing 9.05 μM 2,4-D. The embryos were cream yellow in color. Bacterial and fungal contamination accounted for 12.67%. Next, at the stage 2, we transferred the mammillae with somatic embryos at the globular stage onto media supplemented with different ABA concentrations, and we found that further elongation growth of embryos occurred only on the media with a low ABA content (Table 1). The media with a high concentration of this PGR (10 and 100 μM) inhibited the growth of embryos and resulted in an evident color change in embryos from cream-yellow to brown (Figures 1(a) and 1(b)). On media with a low content of ABA (0-1 μM) the embryos were round in shape and their epidermis was shiny, while mammillae remained green. Also, we observed three developmental stages: globular, torpedo, and shoot (Figures 2(a), 2(b), and 2(c)). The globular stage that predominated on all the media was evaluated accounting for 84.6 to 91.7%. Only on MS ABA0.1 and MS ABA1 torpedo stages were able to develop. However, on MSABA0.1 the embryo developed into a spherical shaped shoot white in color with maroon points being the developing areoles (Figure 2(c)). The proximal part

TABLE 1: The influence of ABA on direct somatic embryogenesis in cactus *C. tenuissima* Ritt. forma *monstruosa*.

| Medium | ABA (μM) | Length of embryo | Color of embryo | Somatic embryo | | | | Stage of somatic embryo | | | | | |
| | | | | Correct | | Formed wrinkles | | Globular | | Torpedo | | Shoot | |
				Number	%	Number	%	Number	%	Number	%	Number	%
MS 0ABA	0	0.95 a*	Cream-yellow	11 a	91.7	1 b	8.3	11 a	91.7	1 a	8.3	0 a	0.0
MS 0.1ABA	0.1	0.97 a	Cream-yellow	20 a	100.0	0 b	0.0	17 a	85.0	2 a	10.0	1 a	5.0
MS 1ABA	1	0.90 a	Cream	11 a	91.7	1 b	8.3	11 a	84.6	2 a	15.4	0 a	0.0
MS 10ABA	10	0.80 b	Brown	3 b	25.0	9 a	75.0	11 a	91.7	1 a	8.3	0 a	0.0
MS 100ABA	100	0.80 b	Brown	3 b	25.0	9 a	75.0	11 a	91.7	1 a	8.3	0 a	0.0

*Data in columns marked with the same letter do not differ significantly at $\alpha = 0.05$.

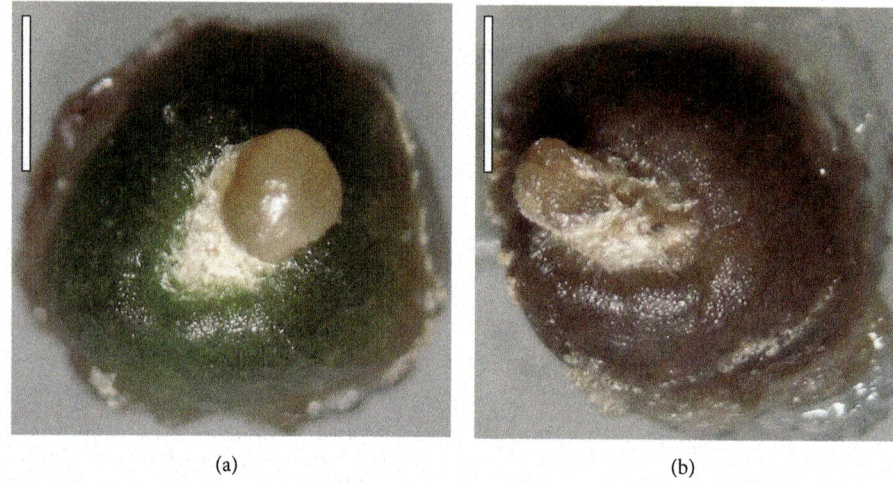

(a)   (b)

FIGURE 1: Somatic embryos on mammillae of the cactus *C. tenuissima* Ritt. forma *mostruosa* developed on the medium (bar = 1 mm): (a) without ABA and (b) with a high concentration of ABA (10 μM).

of the shoot was visibly becoming green. In this medium we also observed secondary somatic embryogenesis.

3.2. Indirect Somatic Embryogenesis.

In experiment 1, media with low ABA concentration (0-1 μM), we noted a more-intensive growth of the fresh weight of callus than on the media of a high ABA concentration (10–100 μM) (Table 2). Additionally, a high concentration of ABA was noted to coincide with changes in calli structure and color from yellow into cream and cream-brown (Figures 3(a) and 3(b)). However, the ABA concentration did not have a significant effect on the number of somatic embryos produced and their development stages (Table 3). Three development stages were found: globular, heart, and torpedo; most embryos, however, were isolated at the globular stage on the media with 0 and 1 μM ABA, where they accounted for 100% of all the embryos. Further development stages, namely, heart, occurred at 10 and 100 μM ABA, while torpedo stage occurred at 0.1, 10, and 100 μM ABA. The results observed in experiment 2, investigating the effect of ABA and sucrose, coincided with the results of experiment 1. Here, we identified no effect of ABA on the number of somatic embryos regenerated and their degree of maturity (Table 4). However, we noted a significant effect of the sucrose concentration on the number of somatic embryos. The lowest number of embryos

was produced at 1% concentration, while at 3 and 5% a significantly higher number of somatic embryos resulted (Table 4). Similarly an increase in the calli fresh weight was greater at sucrose concentrations of 3 and 5%.

4. Discussion

In the cactus *Copiapoa tenuissima* Ritt. forma *mostruosa* auxin 2,4-D plays a decisive role in the induction of somatic embryogenesis both directly and indirectly. Similar to many other plant species, this process is induced by exogenous 2,4-D [10, 21–23]. In the cactus investigated, 9.05 μM 2,4-D resulted in the production of 20.67% explants regenerating embryos at the globular stage and 5.67% explants regenerating calli. Also, secondary somatic embryogenesis was observed on young shoots that were developed on mammillae of the studied cactus. Secondary embryogenesis has been reported for several plant species as coronation [24], peanut [25], *Medicago truncatula* [26], and *Helianthus maximiliani* [27] cultured on media containing variation types of auxins, particularly 2,4-D. Further somatic embryo development stages occur very often after the elimination of auxin from the medium and adding PGRs favourable to somatic embryo maturation. Among the PGRs most frequently applied in the process of somatic embryo maturation

| (a) | (b) | (c) |

FIGURE 2: Developmental stages observed in somatic embryos obtained in direct somatic embryogenesis in the cactus *C. tenuissima* Ritt. forma *mostruosa* (bar = 1 mm): (a) globular, (b) torpedo, and (c) shoot developed from somatic embryo.

| (a) | (b) |

FIGURE 3: Callus of the cactus *C. tenuissima* Ritt. forma *mostruosa* on the media (bar = 0.1 mm): (a) without ABA (magnification 1.5×10) and (b) with a high concentration of ABA (10 μM; magnification 1.4×10).

TABLE 2: Effect of ABA on calli fresh weight increase and percentage of somatic embryos regenerated from calli during indirect somatic embryogenesis of *C. tenuissima* Ritt. forma *monstruosa*.

Medium	ABA (μM)	Calli fresh weight increase (g)	Calli with somatic embryos (%)
MS ABA0[1]	0	8.36 a*	66.7
MS ABA0.1	0.1	8.75 a	83.3
MS ABA1	1	7.29 a	66.7
MS ABA10	10	4.02 a	83.3
MS ABA100	100	0.50 b	66.7

[1]Except ABA, all media contained constant concentration of the following PGRs: 13.32 μM BA; 16.11 μM NAA; 0.57 μM IAA.
*Data in columns marked with the same letter do not differ significantly at $\alpha = 0.05$.

TABLE 3: Effect of ABA concentration on the number and the developmental stage observed in somatic embryos obtained during indirect somatic embryogenesis of *C. tenuissima* Ritt. forma *monstruosa*.

| Medium | ABA (μM) | Number of somatic embryos | | Stage of somatic embryo | | | | | |
| | | | | Globular | | Heart | | Torpedo | |
		Total	On one cultured callus	Total	On one cultured callus	Total	On one cultured callus	Total	On one cultured callus
MS ABA0[1]	0	11	0.9 a	11	0.9 a	0	0.0 a	0	0.0 a
MS ABA0.1	0.1	21	1.8 a	20	1.7 a	0	0.0 a	1	0.1 a
MS ABA1	1	12	1.0 a	12	1.0 a	0	0.0 a	0	0.0 a
MS ABA10	10	35	2.9 a	33	2.8 a	1	0.1 a	1	0.1 a
MS ABA100	100	13	1.1 a	10	0.8 a	1	0.1 a	2	0.2 a

Explanation as in Table 2.

TABLE 4: Influence of ABA and sucrose contents on indirect somatic embryogenesis of *C. tenuissima* Ritt. forma *monstruosa*.

Medium	ABA (μM)	Sucrose (%)	Calli fresh weight increase (g)	Calli with somatic embryo (%)	Number of somatic embryos	
					Total	On one cultured callus
MS 0ABA S1[1]	0	1	2.12 aB*	75	25	2.3 aB
MS 0ABA S3	0	3	8.94 aA	100	195	17.7 aA
MS 0ABA S5	0	5	10.57 aA	100	202	18.4 aA
MS 0.1ABA S1	0.1	1	1.45 aB	92	49	4.5 aB
MS 0.1ABA S3	0.1	3	10.70 aA	100	173	15.7 aA
MS 0.1ABA S5	0.1	5	10.76 aA	100	250	22.7 aA
MS 1ABA S1	1	1	0.96 aB	100	94	8.5 aB
MS 1ABA S3	1	3	6.48 aA	100	237	21.5 aA
MS 1ABA S5	1	5	10.14 aA	100	258	23.5 aA

[1]Except ABA and sucrose (S), all media contained constant concentration of the following PGRs: 13.32 μM BA; 16.11 μM NAA; 0.57 μM IAA.
*a: data in columns marked the influence of ABA with the same lowercase letter do not differ significantly at $\alpha = 0.05$.
A: data in columns marked the influence of sucrose with the same uppercase letter do not differ significantly at $\alpha = 0.05$.

is ABA [28–30]. Cardoza and D'Souza [29] reported the maturation of somatic embryos from the globular to heart and cotyledon stages in *Anacardium occidental* L. on the MS medium containing 2 μM ABA. In the present study a low ABA concentration (0–1 μM) stimulated the elongation of embryos at the globular stage and the number of adequate embryos in direct somatic embryogenesis, while a high ABA concentration (10–100 μM) resulted in growth inhibition, and on somatic embryos wrinkles formed as well as the primary explants, mammillae turning brown. A low concentration of ABA (2.5–7.5 μM) also enhanced the development of adequately developed somatic embryos in *Cocos nucifera* L. [21]. However, as reported by Cailloux et al. [14] in *Hevea brasiliensis*, a high ABA concentration (10 μM) combined with a high concentration of sucrose enhances the maturation of embryos. Baskaran and Van Staden [31] showed that the addition of abscisic acid (1.9 μM) to the medium significantly improved the development of somatic embryos and their conversion to plantlets in *Merwilla plumbea* (Lindl.) Speta. Our results also suggest that the application of low ABA concentrations (0.1–1 μM) is favourable to the maturation of somatic embryos in *Copiapoa*. Similarly, the study of ISE in this cactus suggests that lower ABA concentrations enhance the development of calli fresh weight, while at higher concentrations (10 μM ABA or more) calli color changes and its proliferation rate decreases. Nevertheless, in the ISE no effect of ABA on the number of somatic embryos and their maturation was shown. Similarly, Agarwal et al. [32] reported a negative effect of ABA with an increase in its concentration on *Morus alba* L. and a total inhibition of embryogenesis at the concentration of 10 μM. Another essential factor which facilitates the maturation of somatic embryos is sucrose. We identified a positive effect of sucrose (3 and 5%) both on the number of somatic embryos and on an increased calli fresh weight of *Copiapoa*. The application of a high concentration of sucrose (6%) definitely increases the size of somatic embryos in *Juglans regia* L. [30]. Similarly the reports by Agarwal et al. [32] point to a considerable role of sucrose (6%) in the process of somatic embryogenesis in *Morus alba* L, while Nakagawa et al. [16] reported that sucrose

induces the somatic embryogenesis in melon (*Cucumis melo* L.). A negative effect on the development of cotyledonary in *Merwilla plumbea* was observed by Baskaran and Van Staden [31] with reduction of sucrose (below 3%) in the medium. On the other hand, Charrière and Hahne [33] found that the concentration of sucrose (3 or 12%) had a significant effect on the pattern of organogenesis at a low concentration of stimulated somatic embryogenesis at a higher concentration in sunflower (*Helianthus annuus* L.), whereas Sghaier et al. [17] showed that both a high sucrose concentration (9%) and a high ABA concentration (20 μM) affect the morphology, rate of germination, and the content of storage protein in somatic embryos in date palm (*Phoenix dactylifera* L.).

5. Conclusions

We investigated the effect of abscisic acid (ABA) and sucrose on successive stages of DSE and ISE in cactus *Copiapoa tenuissima* Ritt. forma *mostruosa*. The results showed that a low concentration of ABA (0-1 μM) stimulates the elongation of embryos, while the high ABA concentration (10–100 μM) results in growth inhibition. The ISE study suggests that the lower ABA concentration enhances the increase in calli fresh weight, while the high concentration changes calli color and decreases its proliferation rate. The positive effect of sucrose concentration (3 and 5%) for both the number of somatic embryos and the increase in calli fresh weight was also observed.

Abbreviations

ABA: Abscisic acid
BA: 6-Benzylaminopurine
2,4-D: 2,4-Dichlorophenoxyacetic acid
DSE: Direct somatic embryogenesis
IAA: 3-Indolylacetic acid
ISE: Indirect somatic embryogenesis
MS: Murashige and Skoog medium
NAA: 1-Naphthylacetic acid.

Acknowledgment

The authors are grateful to Professor Dr. Małgorzata Zalewska for critically reading this paper.

References

[1] V. Dornig, "Ist *Neochilenia wageringeliana* eine Standorthybride?" *Kakteen Sukkulenten*, vol. 11, no. 1, pp. 5–6, 1976.

[2] C. Graham, *The Cactus File Handbook*, Cirio, Southampton UK, 1998.

[3] J. Lema-Rumi?ska and I. Licznerska, "Influence of growth regulators on the regeneration of cacti *Copiapoa tenuissima* Ritt *f. monstruosa* from meristematic explant," *Folia Universitatis Agriculturae Stetinensis, Agricultura*, vol. 236, no. 94, pp. 109–114, 2004.

[4] G. H. McGranahan, C. A. Leslie, S. L. Uratsu, and A. M. Dandekar, "Improved efficiency of the walnut somatic embryo gene transfer system," *Plant Cell Reports*, vol. 8, no. 9, pp. 512–516, 1990.

[5] S. C. Minocha and P. N. Mehra, "Nutritional and morphogenetic investigations on callus cultures of *Neomamillaria prolifera* Miller (Cactaceae)," *American Journal of Botany*, vol. 6, pp. 168–173, 1974.

[6] R. Infante, "*In vitro* axillary shoot proliferation and somatic embryogenesis of yellow pitaya *Mediocactus coccineus* (Salm-Dyck)," *Plant Cell, Tissue and Organ Culture*, vol. 31, no. 2, pp. 155–159, 1992.

[7] F. Santacruz-Ruvalcaba, A. Gutierrez-Mora, and B. Rodriguez-Garay, "Somatic embryogenesis in some cactus and agave species," *Journal of the Professional Association for Cactus Development*, vol. 3, pp. 15–26, 1998.

[8] S. P. Da Costa, A. A. Soares, and B. Arnholdt-Schmitt, "Studies on the induction of embryogenic globular structures in *Opuntia ficus-indica*," *Journal of the Professional Association for Cactus Development*, vol. 4, pp. 66–74, 2001.

[9] K. G. Moebius-Goldammer, M. Mata-Rosas, and V. M. Chávez-Avila, "Organogenesis and somatic embryogenesis in *Ariocarpus kotschoubeyanus* (Lem.) K. Schum. (Cactaceae), an endemic and endangered Mexican species," *In vitro Cellular and Developmental Biology*, vol. 39, no. 4, pp. 388–393, 2003.

[10] F. L. A. F. Gomes, F. F. Heredia, P. B. E. Silva, O. Facó, and F. D. A. D. P. Campos, "Somatic embryogenesis and plant regeneration in *Opuntia ficus-indica* (L.) Mill. (Cactaceae)," *Scientia Horticulturae*, vol. 108, no. 1, pp. 15–21, 2006.

[11] J. Lema-Rumińska, "Flow cytometric analysis of somatic embryos, shoots, and calli of the cactus *Copiapoa tenuissima* Ritt. forma monstruosa," *Plant Cell, Tissue and Organ Culture*, vol. 106, no. 3, pp. 531–535, 2011.

[12] S. Misra, "Conifer zygotic embryogenesis, somatic embryogenesis, and seed germination: biochemical and molecular advances," *Seed Science Research*, vol. 4, no. 4, pp. 357–384, 1994.

[13] T. K. Mondal, A. Bhattacharya, A. Sood, and P. S. Ahuja, "Factors affecting germination and conversion frequency of somatic embryos of tea [*Camellia sinensis* (L.) O. Kuntze]," *Journal of Plant Physiology*, vol. 159, no. 12, pp. 1317–1321, 2002.

[14] F. Cailloux, J. Julien-Guerrier, L. Linossier, and A. Coudret, "Long-term somatic embryogenesis and maturation of somatic embryos in *Hevea brasiliensis*," *Plant Science*, vol. 120, no. 2, pp. 185–196, 1996.

[15] F. Pliego-Alfaro, M. J. R. Monsalud, R. E. Litz, D. J. Gray, and P. A. Moon, "Effect of abscisic acid, osmolarity and partial desiccation on the development of recalcitrant mango somatic embryos," *Plant Cell, Tissue and Organ Culture*, vol. 44, no. 1, pp. 63–70, 1996.

[16] H. Nakagawa, T. Saijyo, N. Yamauchi, M. Shigyo, S. Kako, and A. Ito, "Effects of sugars and abscisic acid on somatic embryogenesis from melon (*Cucumis melo* L.) expanded cotyledon," *Scientia Horticulturae*, vol. 90, no. 1-2, pp. 85–92, 2001.

[17] B. Sghaier, W. Kriaa, M. Bahloul, J. V. J. Novo, and N. Drira, "Effect of ABA, arginine and sucrose on protein content of date palm somatic embryos," *Scientia Horticulturae*, vol. 120, no. 3, pp. 379–385, 2009.

[18] M. K. Rai, N. S. Shekhawat, H. Harish et al., "The role of abscisic acid in plant tissue culture: a review of recent progress," *Plant Cell, Tissue and Organ Culture*, vol. 106, no. 2, pp. 179–190, 2011.

[19] R. R. Finkelstein and S. I. Gibson, "ABA and sugar interactions regulating development: cross-talk or voices in a crowd?" *Current Opinion in Plant Biology*, vol. 5, no. 1, pp. 26–32, 2002.

[20] T. Murashige and F. Skoog, "A revised medium for rapid growth and bioassays with tobacco tissue cultures," *Physiologia Plantarum*, vol. 15, pp. 473–497, 1962.

[21] S. C. Fernando and C. K. A. Gamage, "Abscisic acid induced somatic embryogenesis in immature embryo explants of coconut (*Cocos nucifera* L.)," *Plant Science*, vol. 151, no. 2, pp. 193–198, 2000.

[22] A. Martínez-Palacios, M. P. Ortega-Larrocea, V. M. Chávez, and R. Bye, "Somatic embryogenesis and organogenesis of *Agave victoriae-reginae*: considerations for its conservation," *Plant Cell, Tissue and Organ Culture*, vol. 74, no. 2, pp. 135–142, 2003.

[23] A. Pareek and S. L. Kothari, "Direct somatic embryogenesis and plant regeneration from leaf cultures of ornamental species of *Dianthus*," *Scientia Horticulturae*, vol. 98, no. 4, pp. 449–459, 2003.

[24] O. Karami, A. Deljou, and G. K. Kordestani, "Secondary somatic embryogenesis of carnation (*Dianthus caryophyllus* L.)," *Plant Cell, Tissue and Organ Culture*, vol. 92, no. 3, pp. 273–280, 2008.

[25] E. L. Little, Z. V. Magbanua, and W. A. Parrott, "A protocol for repetitive somatic embryogenesis from mature peanut epicotyls," *Plant Cell Reports*, vol. 19, no. 4, pp. 351–357, 2000.

[26] L. O. Das Neves, S. R. L. Duque, J. S. De Almeida, and P. S. Fevereiro, "Repetitive somatic embryogenesis in *Medicago truncatula* ssp. Narbonensis and *M. truncatula* Gaertn cv. Jemalong," *Plant Cell Reports*, vol. 18, no. 5, pp. 398–405, 1999.

[27] D. Vasic, G. Alibert, and D. Skoric, "Protocols for efficient repetitive and secondary somatic embryogenesis in *Helianthus maximiliani* (Schrader)," *Plant Cell Reports*, vol. 20, no. 2, pp. 121–125, 2001.

[28] M. A. Lelu and P. Label, "Changes in the levels of abscisic acid and its glucose ester conjugate during maturation of hybrid larch (*Larix x leptoeuropaea*) somatic embryos, in relation to germination and plantlet recovery," *Physiologia Plantarum*, vol. 92, no. 1, pp. 53–60, 1994.

[29] V. Cardoza and L. D'Souza, "Induction, development and germination of somatic embryos from nucellar tissues of cashew (*Anacardium occidentale* L.)," *Scientia Horticulturae*, vol. 93, no. 3-4, pp. 367–372, 2002.

[30] K. Vahdati, S. Bayat, H. Ebrahimzadeh, M. Jariteh, and M. Mirmasoumi, "Effect of exogenous ABA on somatic embryo maturation and germination in Persian walnut (*Juglans regia* L.)," *Plant Cell, Tissue and Organ Culture*, vol. 93, no. 2, pp. 163–171, 2008.

[31] P. Baskaran and J. Van Staden, "Somatic embryogenesis of *Merwilla plumbea* (Lindl.) Speta," *Plant Cell, Tissue and Organ Culture*, pp. 1–8, 2012.

[32] S. Agarwal, K. Kanwar, and D. R. Sharma, "Factors affecting secondary somatic embryogenesis and embryo maturation in *Morus alba* L," *Scientia Horticulturae*, vol. 102, no. 3, pp. 359–368, 2004.

[33] F. Charrière and G. Hahne, "Induction of embryogenesis versus caulogenesis on *in vitro* cultured sunflower (*Helianthus annuus* L.) immature zygotic embryos: role of plant growth regulators," *Plant Science*, vol. 137, no. 1, pp. 63–71, 1998.

5

Induction of MAP Kinase Homologues during Growth and Morphogenetic Development of Karnal Bunt (*Tilletia indica*) under the Influence of Host Factor(s) from Wheat Spikes

Atul K. Gupta,[1] J. M. Seneviratne,[1] G. K. Joshi,[2] and Anil Kumar[1]

[1] Department of Molecular Biology and Genetic Engineering, College of Basic Sciences and Humanities, GBPant University of Agriculture and Technology, Pantnagar 263 145, India
[2] Department of Biotechnology, HNB Garhwal University, Srinagar, Uttarakhand 246174, India

Correspondence should be addressed to Anil Kumar, anilkumar.mbge@gmail.com

Academic Editor: Laszlo Bogre

Signaling pathways that activate different mitogen-activated protein kinases (MAPKs) in response to certain environmental conditions, play important role in mating type switching (Fus3) and pathogenicity (Pmk1) in many fungi. In order to determine the roles of such regulatory genes in *Tilletia indica*, the causal pathogen of Karnal bunt (KB) of wheat, semi-quantitative and quantitative RT-PCR was carried out to isolate and determine the expression of MAP kinase homologues during fungal growth and development under *in vitro* culture. Maximum expression of TiFus3 and TiPmk1 genes were observed at 14th and 21st days of culture and decreased thereafter. To investigate whether the fungus alters the expression levels of same kinases upon interaction with plants, cultures were treated with 1% of host factors (extracted from S-2 stage of wheat spikes). Such treatment induced the expression of MAPks in time dependent manner compared to the absence of host factors. These results suggest that host factor(s) provide certain signal(s) which activate TiFus3 and TiPmk1 during morphogenetic development of *T. indica*. The results also provides a clue about the role of host factors in enhancing the disease potential due to induction of MAP kinases involved in fungal development and pathogenecity.

1. Introduction

Karnal bunt (KB) caused by *Tilletia indica* (*Neovosia indica Syn.*) is a fungal disease that affects wheat, durum wheat and triticale. KB is an emerging infectious disease (EID) with profound socioeconomic implications [1]. It is a seed-borne disease which typically causes partial conversion of individual kernels into sori filled with fetid teliospores thus affecting yield and quality [2]. *T. indica* is a dimorphic pathogen occurring in haploid (mycelia and sporidia) and diploid (teliospore) stages of its life cycle. It is of particular interest because it is a representative of smut fungi that only cause infection on florets during anthesis. Infection occurs after heading when sporidia produced from teliospores at the soil surface are dispersed to the glumes of the wheat spike. Fungus threads (hyphae) from sporidia penetrate stomata and grow intercellularly to the base of the developing kernel.

Commonly, only some kernels in a spike are affected. The fungus infects one or more developing seed on a head, but usually not all the seeds [3]. As so far, none of the control measures had been proven to be satisfactory for the disease management, hence it becomes inevitable to understand the biology of the pathogenic fungus through dissecting the complex signaling pathways involved in fungal development and pahogenecity [4].

The disease attack at different plant developmental stages has been reported during specific growth stages ranging from boot swelling to partial emergence of spike [5, 6], boot stage to complete emergence of spike [7], early boot stage to anthesis [8] between spike emergence and anthesis [9, 10], or during anthesis [6, 11, 12]. The disease presence is clearly noticed in spike when ear head is just peeping out ($Z = 58$, S2) from the tip compared to the leaves ($Z = 16$, Sv), boot stage ($Z = 46$, S1), ear head completely out from the boot leaves/seed formation stage ($Z = 77$, S3) [13]. Thus, plant development stages reported susceptible to infection vary considerably. Since there is a close association between the host developmental stages and infection of Karnal bunt, it is important to study the development of disease and the role of host factors in processes ranging from flowering to seed development and subsequently their effect on the developmental stages of *T. indica*. In our laboratory, elaborative efforts have been made to study the morphogenetic development of the fungus in the presence of host extracts prepared from different parts of wheat plant. The aqueous, salt, methanol, and acetone extracts of stem, leaf, and S-1, S-2, and S-3 stages of inflorescence exhibited differential growth in terms of radial growth and mycelial biomass. Maximum growth was seen in acetone extracts of S-2 stage of inflorescence, which induced mycelial growth due to involvement of MAP kinase machinery which is required for controlling processes critical for development of disease [14, 15], and key stages of morphogenetic development after the perception of an external stimulus [16].

A fundamental property of living cells is the ability to sense and respond appropriately to changing environmental conditions and various other stimuli. One frequently utilized molecular device for eliciting these responses is the three-tiered cascade of protein kinases known as the mitogen-activateed protein kinase (MAPK) module [17]. Mitogen-activated protein (MAP) kinase signaling pathways are ubiquitous and evolutionarily conserved in eukaryotic organisms connecting cell surface receptors to critical regulatory targets within cells that result in various morphogenetic processes [18]. MAP kinase activity is regulated through a tiered cascade composed of a MAP kinase, a MAP kinase kinase (MEK), and a MAP kinase kinase kinase (MEKK), first recognized in *Saccharomyces cerevisiae*. In *Saccharomyces cerevisiae*, 5 MAP kinase pathways have been identified [19]. These enzymes are regulated by characteristic phosphorelay system in which a series of three protein kinases phosphorylate activating one and another. Intercellular targets are subsequently regulated by phosphorylation and include transcription factors and cytoskeletal proteins [20].

In other fungus *Magnaporthe grisea*, a well-conserved MAP kinase gene Pmk1 is essential for fungal pathogenesis and for production of female reproductive structures [21]. Unlike the situation in fungal-plant pathogens, the Pmk1-like MAPK pathway is not required for virulence in the fungal-fungal interaction [22]. Further studies have revealed how protein orthologous to Pmk1p/Fus3p/Kss1p are required for pathogenecity in many other phyto-pathogenic fungi [23]. The Fus3 MAP kinase pathway controls the transduction of the pheromone signal and is activated in response to the binding of a peptide-mating pheromone to cell type-specific pheromone receptors. In addition to activating transcription, transduction of the mating response results in reorientation of the cytoskeleton and secretary apparatus to polarize toward a mating partner [24]. Cyclic AMP and MAPK signaling are involved in this process. In the phytopathogenic fungus *Ustilago maydis*, pheromonemediated cell fusion is a prerequisite for the

generation of the infectious dikaryon. The pheromone signal elevates transcription of the pheromone genes and elicits the formation of conjugation hyphae.

Dibutyryl c-AMP, an analogue of c-AMP, induces sporidia formation in *T. indica*. The fungus on exposure to dbc-AMP experienced morphological differentiation from vegetative mycelial phase to sporogenous mycelial phase and was induced to produce filiform sporidia [25] through involvement of MAP kinase module(s) in such morphogenetic transition and development. However, so far no orthologous genes/proteins of mitogen signaling pathway have been identified in KB pathogen of wheat which enter the plant through stomata and grow through intercallinary division unlike other *Magnaportha grisea* which enter the plant by direct penetration through formation of specialized infection structures in the form of appressoria. Taking the advantages of evolutionarily conserved MAP kinase signal transduction pathways for regulating critical processes of disease development in diverse pathogenic fungi even distantly related with very different modes of infection, two homologues of MAP kinase (Fus3 and Pmk1) from KB were identified in our laboratory by PCR-based approaches. In order to establish the molecular basis of induced mycelination in presence of host factor(s) derived from wheat spikes and pathogenecity, in the present study, attempts were made to study the expression of TiFus3 and TiPmk1 MAP kinase genes under the influence of host factors and its relation with morphogenetic development and pathogenesis in *T. indica*.

2. Materials and Methods

2.1. Collection of Seeds. The resistant (HD29) and susceptible wheat varieties (WH542) were collected from Punjab Agriculture University, Ludhiana. The resistant genotype (HD29) was developed through conventional breeding approach and showed resistance against Karnal bunt as evident by pathogen inoculation studies.

2.2. Preparation of Host Factor(s). Acetone extracts were prepared from spike tissues collected from susceptible (WH542) wheat varieties in boot emergence stage (S2). Using pestle and mortar, wheat spikes (50 g) were ground in liquid nitrogen to a fine powder. Finely ground plant tissues suspended in cold acetone, in the ratio of 1 g of sample in 10 mL of acetone. The suspension was then agitated in cold condition for 5 hours and filtered through muslin cloth to remove larger debris and stored at 4°C in tightly capped bottles. Before starting the experiments, acetone was evaporated at room temperature using flash evaporator or blowing hot air over the solution. Dried material obtained was resuspended in 1/10th of the volume of sterilized distilled water and filtered through 0.22 μ filter before incorporation into the culture media as host factors [14].

2.3. Fungal Culture and Harvesting of Mycelium. The fungus *T. indica* was cultured in modified potato dextrose liquid media. All the constituents of the potato dextrose media were dissolved in distilled water. 100 mL of liquid media was transferred to 250 mL conical flasks and autoclaved.

Then these flasks were inoculated with mycelial discs or loop-full of inoculums from the slants prepared earlier. The cultures were incubated in BOD incubator at $22 \pm 2°C$ under light and dark conditions. The growing liquid cultures of *T. indica* were harvested at the 7th, 14th, 21st, and 30th days with and without host factor(s). The media containing the mycelial mat of *T. indica* was filtered through a folded muslin cloth and washed several times in PBS (0.05 M, pH 7.2) followed by sterilized distilled water. The wet mycelial biomass and total soluble protein extracted from fungal cultures (grown in presence and absence of host factors) at different intervals were determined for plotting the growth curves. The concentration of soluble protein of mycelia was determined by Bradford method [26]. The wet mycelia were lyophilized for 5 hours to obtain the dry weight. Dry mycelial masses were stored in $-80°C$ for subsequent expression studies.

2.4. Morphological Observation and Sporidial Count.

The fungal cultures were stained with cotton blue and observed under light microscopy. The formation of sporidia in fungal cultures grown on solid PDA medium at different time intervals of growth and development of *T. indica* was calculated using haemocytometer. 10 mL of sterile distilled water was added to petri plate and gently moved back and forth and water containing sporidia was decanted to sterilized oak-ridge tubes and centrifuged at 4000 rpm for 10 min to get sporidial pellet. Supernatant was discarded and the pellet was again washed in 1 mL sterile distilled water and centrifuged for 15 minutes at 3000 rpm. The pelleted sporidia were finally dissolved in 1 mL sterile distilled water. The sporidia collected were enumerated with the help of hamocytometer under microscope.

2.5. Pathogenicity Testing.

Disease scoring is primarily based on the percentage of infected kernels. Surface-sterilized wheat seeds (*Triticum aestivum* cv. WH542 and HD29 susceptible and resistant, resp.) were germinated on wet paper and were planted on commercial soil mix. Plants were grown at $22°C/18°C$ (12 h light/12 h dark) in a glass house. The pot experiment was laid out in a randomized block design having different treatment (control (C), pathogen inoculation (P), and host factor(s) treated pathogen inoculation (HFP)) in both susceptible and resistant wheat cultivars having five replicates and designated as SC, SP, SHFP and RC, RP, RHFP, respectively. The injection technique was adopted in which the inoculums are injected with a hypodermic syringe into the boot just as awns emerged [27, 28]. High percentage of infection can be obtained with this technique. Five ear heads were artificially inoculated using hypodermic syringe with the 21-day-old sporidial cultures (10^6 sporidia/mL) in the month of January. Inoculated ear heads were covered with butter paper to prevent natural infection as well as to maintain moisture. Average percent-infected grains were calculated after harvesting in three successive years.

2.6. Disease Scoring.

Disease scoring was done on the basis of the percentage of infected kernels. Average percent-infected grains were calculated after harvesting. As most of the bunted grains were partially infected, numerical values, depending upon the extent of damage to the grains, were given for calculating coefficient of infection. Number of grains showing incipient infection, blackening extended up to half of the grain, 3/4 of grain and infected grains were multiplied with the numerical values 0.25, 0.5, 0.75, and 1.0, respectively, and then divided by 100 to obtain percent coefficient of infection [29].

2.7. Preparation of Total RNA and cDNA Synthesis.

Total RNA was isolated from each sample by using one-step RNA isolation reagent (Trizole) from Bio Basic Inc., according to the manufacturer's instructions. RNA preparations were subjected to DNase digestion according to manufacturer's instruction (Fermentas International Inc., Canada). Total RNA (5 μg) of each sample was used to synthesize first-strand cDNA by using oligo(dT)$_{18}$ primer with RevertAid H Minus M-MuLV Reverse Transcriptase (RT) (Fermentas International Inc., Canada) according to the manufacture's instruction. The efficiency of cDNA synthesis was assessed by reverse transcriptase PCR amplification of a basal/housekeeping transcript, for example, ribosomal protein S17 (RPS17).

2.8. In Silico Sequence Analysis.

In silico analysis were performed in order to confirm the MAP kinase sequences from *T. indica* and their relatedness. The homology search of the Fus3 and Pmk1 was done through Blast search tool of NCBI (http://www.ncbi.nlm.nih.gov) using Blastn and tBlastx algorithm. All sequences (Figure 1) from different fungi were aligned using ClustalW method after that phylogenetic tree was constructed using UPGMA method of MEGA version 4.0.02 [30]. Each node was tested using the bootstrap approach by taking 1,000 replications and a random seeding of 24,054 to ascertain the reliability of nodes. The number is indicated in percentages against each node. The branch lengths are drawn to scale indicated.

2.9. Semiquantitative RT PCR Analysis.

To analyze the MAPKs (TiFus3 and TiPmk1) transcript levels in *T. indica* culture grown in presence and absence of host factors at different time intervals, MAPKs (TiFus3 and TiPmk1) specific primers were designed from first-time cloned, sequenced and nucleotide sequences of TiFus3 and TiPmk1 that were submitted as partial DNA fragments of 300 kb and 223 kb from our laboratory to NCBI database with accession numbers as HQ268553 and FJ571362.1, respectively. The gene-specific primers for the expression analysis by RT-PCR and qRT-PCR were used TiFus3 Fwd ACAATTCA-GAGCCCACAGGT & Rev ATCTCTGCCAGGGAAGATTG and TiPmk1 Fwd CCGATGACCACTGTCAGTACTTT & Rev CAACGTATTCGGTCATGAAACC which yield the product size of 180 and 210 bp, respectively. Ribosomal protein S17 (RPS17) gene was selected as endogenous internal standard, because it is a house-keeping gene and expressed constantly. The RPS17 primer used as internal control forward 5′-CGA ACC AAG ACG GTG AAG AAG-3′ Reverse 5′-CCT GCA ACT TGA TGG AGA TAC C-3′. cDNAs were exponentially amplified using (Fermentas International Inc., Canada) Taq Polymerase. PCR was performed in 25 μL of 1x KCl buffer

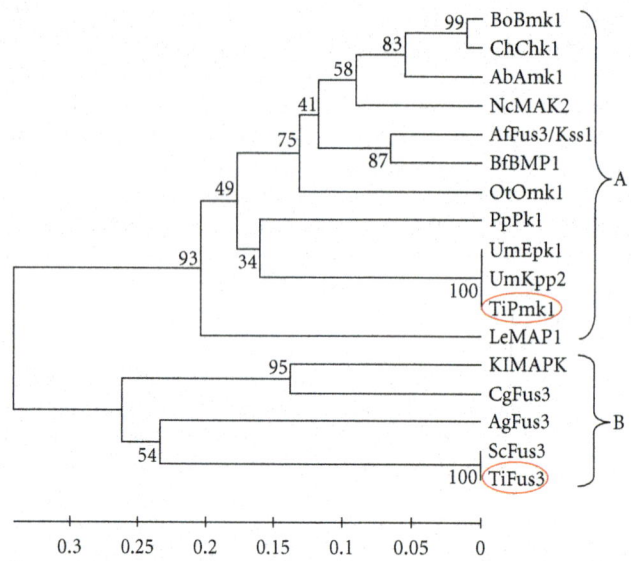

FIGURE 1: Phylogenetic analysis with TiFus3, TiPmk1, and selected fungal MAP kinases (GenBank accession numbers in parenthesis). *Bipolaris oryzae*, BoBmk1 (AB180104.1); *Cochliobolus heterostrophus*, ChChk1 (AF178977.1); *Alternaria brassicicola*, AbAmk1 (AY515257.1); *Neurospora crassa*, NcMAK2 (AF348490.1); *Aspergillus flavus*, AfFus3/Kss1 (XM002374886.1); *Botryotinia fuckeliana*, BfBMP1 (AF205375.1); *Olpitrichum tenellum*, OtOmk1 (EU479712.1); *Postia placenta*, PpPk1 (XM002471663.1); *Ustilago maydis*, UmEpk1 (XM754359.1), UmKpp2 (AF193614.1); *Tilletia indica*, TiPmk1 (FJ571362.1), TiFus3 (HQ268553); *Lentinula edodes*, LeMAP1 (AB446447); *Kluyveromyces lactis*, KlMAPK (XM454426.1); *Candida glabrata*, CgFus3 (XM447892.1); *Ashbya goesypii* AgFus3 (NM210920.1); *Saccharomyces cerevisiae* ScFus3 (NM001178256.1). The phylogram was constructed by UPGMA method of MEGA version 4.0.02. Bootstrap values are indicated against each branch.

(Fermentas International Inc., Canada) containing 0.2 mM dNTPs, 30 ng of each primer, 1.5 mM MgCl$_2$, 0.8 U Taq DNA polymerase (Fermentas International Inc., Canada), and 100 ng of cDNA. Amplification was carried out according to the following temperature profile: 4 min initial denaturation at 94°C; 30 cycles of 94°C for 30 second; 57°C for 20–30 second, 72°C for 45 second; final extension of 5 min at 72°C; final hold at 4°C. Densitometry analysis was done with the help of Gene Profiler software, Alpha Innotech Corporation USA. Briefly, individual gels were scored by placing the curser over individual band and recording the relative densitometry values of three independent gels used for expression analysis.

2.10. Quantitative Real-Time PCR. Real-time PCR was done in triplicate reaction of three different cDNAs prepared from three RNA samples isolated separately at different time intervals in presence and absence of host factor(s) using the 5 Prime Real Master Mix SYBR ROX (Eppendorf India Ltd.) according to manufacturer's instructions. The 5 Prime uses the fluorescent dye, SYBR green, to detect PCR products. The thermocycler used was eppendorf thermocycler ep realplex. Two-step real time PCR was carried out using cDNA prepared as mentioned earlier from different time interval in presence and absence of host factors. The primers for MAPKs and RPS17 genes used were the same as used for semiquantitative RT-PCR analysis. The reverse transcription efficiencies of MAPKs and RPS17 genes were almost equal as analyzed by comparing the C_T values at different dilutions of cDNA [31]. Final concentrations, in a total volume of 20 μL, were 2.5x Real Master Mix SYBR ROX/20x SYBR solution,

100 nM of each forward and reverse primers, and 100 ng of cDNA. The following amplification program was used: 95°C for 2 min, 40 cycles of 95°C for 15 sec, followed by 1 minute at 60°C. All samples were amplified in triplicate and the mean value was considered. Completely randomized design (CRD) was used for analyzing the gel data and real-time data. The cycle threshold (Ct) value is the number of cycles required to accumulate enough SYBR green fluorescent signal to exceed the threshold (background) level. The Ct value is proportional to the amount of RT-PCR product and was used for quantification. The relative value obtained for quantitation was expressed at $2_T^{-\Delta\Delta C}$, where ΔC_T represents the difference between the C_T value of the sample and that of RPS17 (endogenous control) in the same sample and $\Delta\Delta C_T$ is the difference between the ΔC_T value of a sample and that of its respective control.

2.11. Statistical Analysis. Three independent determinations for disease scoring, time-dependent expression of MAPKs homologues (TiFus3 and TiPmk1 genes) in absence and presence of host factors, were taken and mean ± SE values were calculated for statistically analysis using paired *t*-test and GraphPad Prism 5.04 software.

3. Results and Discussion

Mitogen-activated signal transduction pathways play a crucial role [32, 33] in development of virulence levels in pathogens. Through MAPK pathways, pathogens respond to external stimuli and alter their own features such as

Induction of MAP Kinase Homologues during Growth and Morphogenetic Development of Karnal Bunt (Tilletia indica) under
the Influence of Host Factor(s) from Wheat Spikes

37

cell wall integrity, mating, morphological transition, and adaptation to stress factors and this modification leads to generate different virulence levels in phyto pathogens [24]. Homologous of several MAPKKK, MAPKK, and MAPK have been characterized in several pathogenic filamentous ascomycetes including wheat pathogen and play key roles in infection structure (appressorium) formation and host colonization [15, 34]. The KB is slow growing fungal pathogen and its development is concomitantly dependent on the host's flowering to grain filling stages. The boot emergence stage of developing spikes at anthesis is the most susceptible stage and disease infectivity significantly drops when ear head is completely out from boot leaf and at the postanthesis stage. Hence, it is quite worthwhile to study the influence of host factor(s) on the expression of fungal MAPKs genes in order to simulate the fungal growth and morphogenetic development under in vitro cell culture by mimicking some of the microenvironment of anthesis (S-2) stage of developing spikes.

3.1. Pathogenicity Testing of Cultivars after Host Factor(s) Treated Sporidial Suspension Inoculation.

The most favorable weather for KB infection coincides with wheat heading. It was demonstrated that infection occurred most reliably after hypodermically injecting a suspension of sporidia into the wheat boot at the awns emerging stage when the ear head just peeps out at the tip or from center [35]. Percentage disease severity of this stage is 22.2 as compared to 6.97 of boot stage and 6.9 of ear head half outside boot leaf [36]. Therefore, in order to check whether host factor(s) induce disease severity, host factor(s) pretreatment was done at the time of KB culturing prior to artificial inoculation of the KB pathogen in wheat spikes of both varieties when the ear head just peeped out of the boot. Results of infectivity tests carried out by boot injection are presented in Table 1.

On comparing the pathogen-inoculated plants of both varieties (RC and SC), the values of % infection, coefficient of infection, and overall response of susceptible variety were found to be greater than the resistant variety. However, on host factor(s), application values of % infections were increased in both susceptible and resistant varieties, showing induction of infection. Host factor(s) mediated induction was found to be more pronounced in susceptible variety than the resistant variety. It was observed that host factor(s) significantly ($P < 0.05$) changed the overall response value toward pathogen from 34.69 to 136.51 in susceptible (SP versus SHFP) and from 10.59 to 20.51 in resistant variety (RP versus RHFP). The coefficient of infection (CI) was also high in the presence of host factor(s) in pathogen-inoculated varieties from 0.07 to 0.12 in case of susceptible variety and from 0.04 to 0.05 in case of resistant variety. The role of host factor(s) in increasing disease incidence also became evident from observation of seeds harvested from both varieties. It was found that host factor(s) considerably increase the amount of seed blackening in wheat seeds after pathogen inoculation in both varieties. This clearly suggests that host factor(s) can elicit the signal responses in favor of the pathogen to enable it to increase KB pathogenesis. This indicates the causal link between host factor(s) and disease

progression only through modulation of fungal development probably through alteration of MAP kinase machinery responsible for fungal development and virulence.

3.2. Sequence Analysis.

Phylogenetic tree of TiPmk1 and TiFus3 was constructed taking genes from Bipolaris oryzae, Cochliobolus heterostrophus, Alternaria brassicicola, Neurospora crassa, Gibberella zeae, Aspergillus flavus, Botryotinia fuckeliana, Olpitrichum tenellum, Postia placenta, Ustilago maydis, Lentinula edodes, Kluyveromyces lactis, Candida glabrata, Ashbya goesypii, and Saccharomyces cerevisiae by UPGMA method of MEGA version 4.0.02 which showed two major clusters A and B as shown in Figure 1. Cluster-A contains both Pmk1 and Fus3 genes of different fungi while Cluster-B contains Fus3 gene of different fungi. Pmk1 genes of T. indica and U. maydis—both belonging to Basidomyceates class—are present in same Cluster-A while Fus3 genes of T. indica and S. cerevisiae are present in same Cluster-B. The in silico phylogenetic study clearly reveals that both amplified products (TiPmk1 and TiFus3) are MAP kinase homologues isolated from T. indica and shows the homology with MAPKs of other fungi.

3.3. Influence of Host Factor(s) on Fungal Growth and Development.

The vegetative mycelium increased exponentially (logarithmic growth phase) by lateral intercalary division up to 14 days followed by a decrease in the rate of multiplication (stationary/decline growth phase; Figure 2). The mass of mycelial mat at exponential stage was approximately 1.2 g/100 mL, and total soluble protein 3.1 mg/100 mL of culture at the 7th day rose to 5.6 g/100 mL mass of mycelial mat, and total soluble protein was 14.7 mg/100 mL of culture at the 14th day of growth cycle in absence of host factor(s) (Figures 2(a) and 2(b)); however, in presence of host factor(s), the mass of mycelia mat was 2.6 g/100 mL, and total soluble protein 8.6 mg/100 mL of culture at the 7th day rose to 6.3 g/100 mL mass of mycelia mat, and total soluble protein was 22.2 mg/100 mL of culture at the 14th day of growth cycle. After that, the decrease in growth in terms of wet mycelia weight and protein contents at the 30th day of culture both in presence and absence of host factors may be due to mycelial death on account of exhaustion of nutrients in media or transition of mycelial phase to sporogeneous phase. It was reported earlier that during sexual development of bunt fungi, fusion of compatible sporidia leads to conversion of haploid mycelial or sporidial phase to diploid teliosporic phase [37, 38].

These transition from mycelial to sporogeneous phase was examined at different time intervals by microscopic and haemocytometer. A clear-cut variation was observed in sporidial count as well as morphological features in fungal cultures grown at different time intervals in presence and absence of host factor(s). As shown in Table 2, host factor(s) induces the formation of mycelination which prolongs up to 21 days, while in absence of host factors, the fungal cultures undergo transition from mycelia to sporogenous phase. The mycelial growth in presence of host factor(s) was pronounced up to 21 days with intercalary division that led to thickening with multiple nuclei. In contrast, thin, long, less septet,

TABLE 1: Pathogenicity testing of two varieties under different treatments on the basis of disease scoring after the crop harvest.

Treatments	Total number of seeds	Number of seeds with susceptible reaction	Percentage infection	Coefficient of infection in %	Overall response value***
(1)	(2)	(3)	(4)	(5)	(6)
Resistant					
RC	116	0	0	0	0
RP	85	15 (14* + 1**)	17.65	0.04	10.59#
RHFP	88	19 (17* + 2**)	21.59	0.05	20.51#
Susceptible					
SC	109	0	0	0	0
SP	89	21 (11* + 10**)	23.6	0.07	34.69#
SHFP	90	32 (17* + 15**)	35.55	0.12	136.51#

$t = 0.3172$.
#$P < 0.05$.
*Incipient infection.
**Blackening of seed upto 1/2.
***Products of numerical values in column 3, 4, 5.

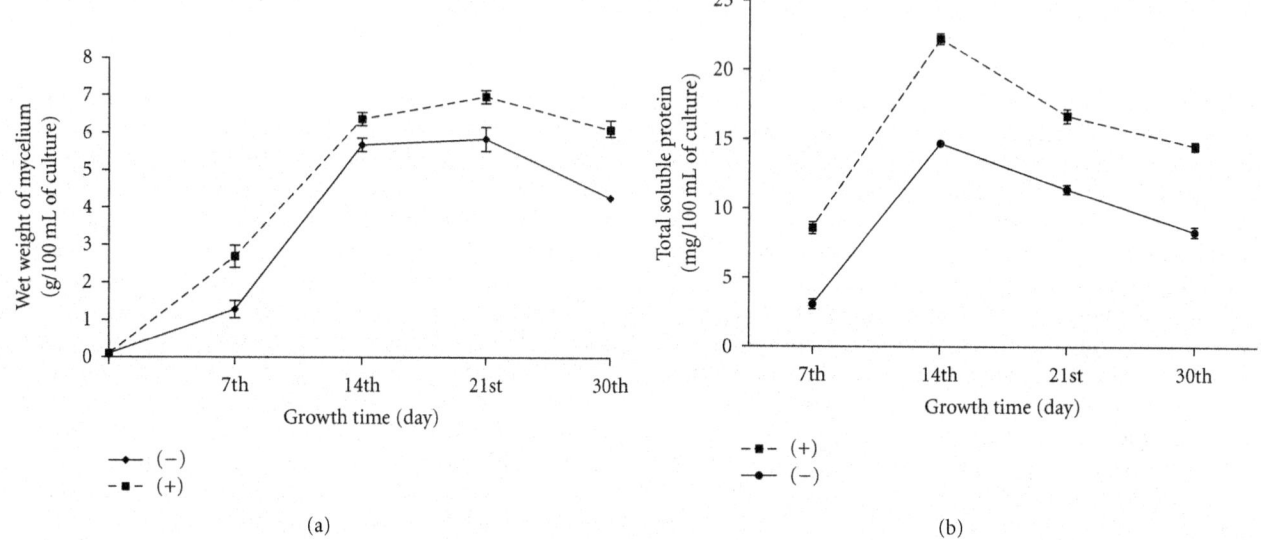

(a)

(b)

FIGURE 2: (a) Growth kinetics of *T. indica* isolate grown in presence (+) and absence (−) of host factor(s) in terms of total biomass production (g/100 mL on wet basis) in different time intervals. (b) Total soluble protein content of *T. indica* isolate grown in presence (+) and absence (−) of host factor(s) (mg/100 mL of culture) at different time intervals.

and nucleic formation was observed in mycelial growth in absence of growth factor(s). At the 21st day of growth, there are appearances of banana-shaped sporidia which are less in the presence of host factor(s) when compared in absence of host factor(s). At 30 days, both cultures grown in presence and absence of host factors showed the transition in the developmental stages after sensing the nutritional status. At the tip of mycelia, crimpled sporogenous mycelia with formation of more banana-shaped sporidia at the tip of hyphae in absence of growth factor(s); however, in presence of growth factor(s), and the enlarged sporogenous mycelia with less formation of such sporidias were observed. The formation of few chlamydospore (immature teliospores like entities) and high number of banana-shaped allantoid sporidia (8.2×10^9) were observed in fungal cultures grown

in absence of host factors while no chlamydospore and comparatively less allantoid sporidia (1.3×10^8) formation was observed in fungal cultures grown in presence of host factors. The influence of host factor(s) on fungal growth and development was subsequently related with stage-dependent expression of MAP kinase genes keeping in the view of investigating their role in pathogenecity and sexual development of *T. indica*.

3.4. Expression Analysis of MAP Kinase Genes under the Influence of Host Factor(s). Over the past decade, it has emerged that the signal transduction pathways which regulate key virulence functions are highly conserved across a wide range of plant pathogenic fungi. In the present study, the expression of MAP kinase genes (responsible

Induction of MAP Kinase Homologues during Growth and Morphogenetic Development of Karnal Bunt (Tilletia indica) under the Influence of Host Factor(s) from Wheat Spikes

39

TABLE 2: Determination of morphological variation and sporidial count of *T. indica* isolate grown in presence and absence of host factor(s).

Days of culture	In absence of host factor(s)		In presence of host factor(s)	
	Morphological variation	Sporidial count (Number of sporidia/mL)	Morphological variation	Sporidial count (number of sporidia/mL)
7th	Thin long, interwoven vegetative mycelia with less septation and nuclei	Nil	Thick, comparatively small vegetative mycelia with extensive septation and multiple nuclei	Nil
14th	Thickened, branched mycelia which produces secondary sporidia at the tip of lateral small branches	1.8×10^3	Thickened, small vegetative mycelia	Nil
21st	Acicular, tapered, banana-shaped high number sporidia and sporogenous mycelia seen occasionally	3.9×10^9	Thickened mycelia which produce few banana-shaped allantoid sporidia	4.1×10^7
30th	Crumpled sporogenous mycelia with short sporogenous hypal branches, terminal/intercalary rounding up of hypal cells, sometimes in chain. Few small chlamydospore like structures	8.2×10^9	Enlarged and many sporogenous mycelia and less hyphal branches and not much chlamydospore formation	1.3×10^8

for pathogenicity in fungal systems) with respect to fungal growth and morphology was studied. Total RNA was isolated from *T. indica* cultures grown in absence and presence of host factor(s) at different time intervals (7th, 14th, 21st, and 30 day after inoculation) and isolated RNAs were subjected for two-step RT-PCR and real-time PCR using the gene-specific primers for MAPKs designated as TiFus3 and TiPmk1. As the difference in melting temperature of forward and reverse primers used in the second strand synthesis, a gradient PCR was performed for each and every combination of primers in order to decide the ideal annealing temperature for primer combinations and optimum number of PCR cycles required for discrimination of expression. It was observed that TiFus3 and TiPmk1 genes were amplified from cDNAs derived from fungal samples grown in absence and presence of host factor(s) in time-dependent manner as given below. The RT-PCR amplicons were cloned and nucleotide sequence of independent clones was determined with the dye terminator kit (ABI Prism, Perkin Elmer, NJ) and analyzed on Applied Biosystems 370 at University of Delhi. The obtained sequences are similar to the CDS sequence that we had submitted earlier in NCBI (FJ571362.1 and HQ268553).

3.5. Study of Expression TiFus3 under the Influence of Host Factor(s).
For the second strand synthesis of TiFus3 gene, annealing temperature for PCR amplification was optimal to 57°C as determined by the gradient PCR. A 180 bp band was detected after the amplification of cDNA by RT-PCR analysis which showed higher expression of TiFus3 transcripts in fungal cultures grown in presence of host factors while comparing with control (absence of host factors). In the case of positive control, a 210 bp band of RPS17 was amplified. No amplification was detected in negative control which was carried out without cDNA indicating no DNA or mRNA

contamination in reagents used in the amplification. For the time-dependent expression of Fus3 transcripts, KB fungal cultures grown in absence and presence of host factors at the 7th, 14th, 21st, and 30th day were used in the study. For semiquantitative RT-PCR, densitometry (relative values) and real-time PCR clearly showed that RPS17 gene expression was constitutive, that is, similar expression of RPS17 gene, whereas Fus3 mRNA transcripts were differentially expressed at all time intervals in both host factors treated and untreated fungal cultures (Figures 3(a) and 3(b)). The expression profiling and densitometry analysis of TiFus3 gene showed gradual increase in amplification up to 14 days and henceforth a reduction was noticed with no amplification at 30th day. The results of RT-PCR and real-time PCR was almost parallel to each other.

For reverse transcription efficiency in real-time PCR, reaction conditions were optimized with endogenous control (RPS17) and TiFus3 genes. Different dilutions of cDNA were used and based on C_T values, these efficiencies were almost equal. This showed that RPS17 gene can be used as endogenous control to analyze the relative expression of TiFus3 gene at different time interval. Quantitation of the TiFus3 transcripts was done in fungal cultures grown in absence and presence of host factors at the 7th, 14th, 21st, and 30th days Relative expression of Fus3 was calculated (Figure 3(c)) and it was expressed at significantly higher level (1.5 folds) at 14th day, followed by 21st day (0.8 fold), and 30th day (0.0, no expression) as compared to 7th day of cultures grown in absence of host factors. However, comparing the fungal cultures grown in presence of host factors, it was found that TiFus3 gene was expressed at significantly higher level at the 7th day (2.3 folds), 14th day (3.9 folds), 21st day (1.8 folds), and 30th day no expression was detected. The expression of TiFus3 gene was positively

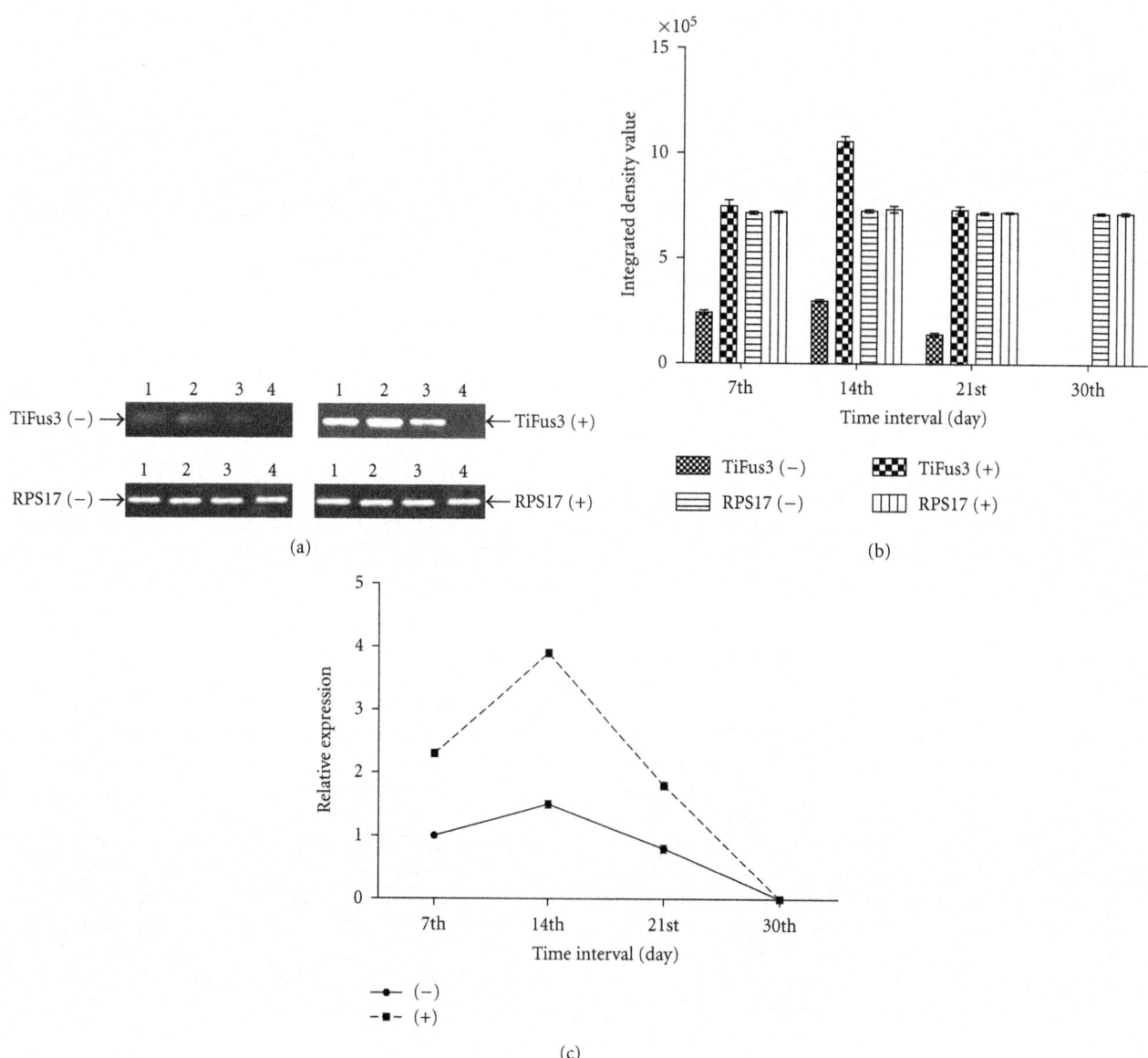

FIGURE 3: (a, b and c) Expression analysis of TiFus3 gene in *T. indica* isolate in absence of host factor (−) and in presence of host factor (+) at different days of growth and development in liquid culture. (a) Semiquantitative RT-PCR analysis—RT-PCR of expressed messenger RNA at 7th, 14th, 21st, and 30th day (lane 1 to 4, resp.) with RPS17 rRNA transcript used as internal control. (b) Densitometry analysis-Integrated density value based on densitometry analysis was done with the help of gene profiler software, Alpha Innotech Corporation USA. (c) Quantitative real-time PCR analysis—relative expression using real-time PCR.

correlated with wet mycelia weight and total soluble protein in the presence ($r = 0.0713$ and 0.4581) and in the absence of host factor(s) ($r = 0.0991$ and 0.3610), respectively.

Components of three MAP kinase pathways have been identified by genome sequence analysis in the filamentous fungus *Neurosphora crassa*. One of the predicted MAPK in *N. crassa*, MAK-2, shows similarity to Fus3p and Kss1p of *Saccharomyces cereviceae*, which are involved in sexual reproduction (mating type switching) and filamentation, respectively [39]. Participation in multiple MAPKs is a characteristic of the MAPK signaling pathways of *S. cereviceae*, where Fus3p is specific to the pheromone-induced mating MAPK pathway, while other components of this pathway also

participate in the MAPK pathway which controls filamentous growth [24]. MAP kinase signaling components (Fus3) might regulate downstream transcription factors gene and transcription factor gene shows the transcriptional changes with disease development or penetration, dimorphic switch in the presence of pheromone. In the present study, Fus3 gene expression has been noticed in inducible manner by host factor(s) and the gene expression is very negligible and almost constitutive in control (absence of host factors). Host factor(s) treatment tends to increase the mycelination in fungal cultures by lowering the sporidial production. Hence, it can be concluded that increase in mycelination of fungi leads to impose more pathogenicity levels in the

Induction of MAP Kinase Homologues during Growth and Morphogenetic Development of Karnal Bunt (Tilletia indica) under the Influence of Host Factor(s) from Wheat Spikes

41

host and prolific multiplication of pathogen inside host leading to more damage to developing grains. These findings reveal that the host factor(s) provide an environment for induction of TiFus3 gene expression and in turn might promote morphogenetic development through mating type switching and induced mycelination under the influence of host factors.

3.6. Study of Expression TiPmk1 under the Influence of Host Factor(s). In *Magnaporthe grisea*, a well-conserved mitogen-activated protein (MAP) kinase gene Pmk1 is essential for fungal pathogenesis. Zheng et al. [21] tested whether the same MAP kinase is essential for plant infection in the gray mold fungus *Botrytis cinerea*, a necrotrophic pathogen that employs infection mechanisms different from those of *M. grisea*. They used a polymerase chain reaction-based approach to isolate MAP kinase homologues from *B. cinerea*, the *Botrytis* MAP kinase required for pathogenesis (BMP). MAP kinase gene was highly homologous to the *M. grisea* Pmk1, Bmp1 is a single-copy gene. Bmp1 gene replacement mutants produced normal conidia and mycelia but were reduced in growth rate on nutrient-rich medium. Bmp1 mutants were nonpathogenic on carnation flowers and tomato leaves. Reintroduction of the wild-type Bmp1 allele into the Bmp1 mutant restored both normal growth rate and pathogenicity. Further studies indicated that conidia from Bmp1 mutants germinated on plant surfaces but failed to penetrate and macerate plant tissues, Bmp1 mutants also appeared to be defective in infecting through wounds. These results indicated that Bmp1 is essential for plant infection in *B. cinerea*, and this MAP kinase pathway may be widely conserved in pathogenic fungi for regulating infection processes. However, under *in vitro* cell culture system, host factors might regulate MAP kinase signaling components (Pmk1) and its downstream transcription factor genes which led the transcriptional changes with invasive growth, cell wall integrity, and spore wall assembly.

With respect to the cDNA amplification of TiPmk1 gene, 57°C was used as the annealing temperature. Samples of cDNA prepared from fungal culture grown in absence and presence of host factors were subjected for second strand synthesis of semiquantitative RT-PCR and quantitative real-time PCR. RT-PCR results showed amplification of an amplicon 210 bp in size in both types of samples. However, the expression of Pmk1 was higher in fungal cultures grown in presence of host factors when compared with its control (absence of host factors). In this case positive control (RPS17 gene) amplified 210 bp band from cDNA isolated from both types of fungal cultures. In order to determine the change in time-dependent expression of Pmk1, fungal cultures grown in absence and presence of host factors were harvested at the 7th, 14th, 21st, and 30th day in cultures. Gradual increase in amplification was detected up to the 21st day and thereafter, a decline was observed and with no amplification at the 30th day. The expression profiling and densitometry analysis of TiPMK1 gene in absence and presence of host factors are given in Figures 4(a) and 4(b). No amplification was detected in negative control which was carried out without cDNA

indicating no DNA or mRNA contamination in reagents used in the amplification.

The results of reverse transcription and real-time PCR were almost parallel to each other and also observed in case of TiFus3 gene. Quantitation of the TiPmk1 transcripts expressed in fungal cultures grown in absence and presence of host factors at different days (7th, 14th, 21st, and 30th day) was noticed. Relative expression of TiPmk1 was calculated in both types of fungal cultures (Figure 4(c)) and it was expressed at significantly higher level in the 21st day (1.4 folds) followed by 14th (1.3 folds), 30th (0.0, no expression) as compared to the 7th day of cultures grown in absence of host factors. However, comparing the fungal cultures grown in presence of host factors, it was found that TiPMK1 gene was expressed at significantly higher level at the 7th day (1.8 folds), 14th day (2.9 folds), 21st day (3.1 folds), and 30th day (0.0 fold, no expression was detected). The expression of TiPmk1 gene was positively correlated with wet mycelia weight and total soluble protein in the presence ($r = 0.2179$ and 0.4477) and in the absence of host factor(s) ($r = 0.2868$ and 0.3740), respectively.

Such results clearly revealed the higher expression of TiPmk1 gene in presence of host factors and in turn induction of pathogenecity. It is also evident from the results of pathogenecity testing under the influence of host factors in two varieties of wheat differing in their resistance to KB. The percentage of disease scoring was increased (percentage of infection from 23.6 to 35.5 in susceptible and from 17.7 to 21.6 in resistant varieties in Table 1) by the fungal culture grown in presence of host factors. It might be possible that the host-pathogen interaction requires mechanism to detect suitable signals from host that triggers a signal transduction cascade in fungal pathogen and that induce the expression of appropriate virulence factors, as has been well characterized in numerous other fungal-plant-pathogen interaction [22].

The above findings related to the expression of MAP kinase genes indicate the clear involvement of MAPK signal transduction cascade in development of pathogenesis in *T. indica*. In these perspectives, it is appropriate to study the expression of genes of MAPK module which will provide strong clues in order to unveil the molecular mechanism associated with fungal development and pathogenesis of *T. indica*. The expression of two genes belongs to MAPK (TiFus3 and TiPmk1) were detected by RT-PCR and real-time PCR in *T. indica*. In the present study, we found the amplification of TiFus3 and TiPmk1 gene by gene-specific primer at different time interval and also found expression increase in presence of host factor(s) in *T. indica* when compared in absence of host factor(s). It supports a causal link between host factors and MAP kinase function in influencing the disease progression and pathogenesis during host-pathogen interaction. With the present observations, it can be indirectly interpreted the involvement of mating type switching (TiFus3) as well as pathogenicity (TiPmk1) subpathways of fungal MAPK signal transduction cascade in the morphogenetic development and pathogenesis of *T.indica*. Remarkably, we have further shown the role of such orthologues of TiFus3 and TiPmk1 in *T. indica* that are also present in *Magnaporthe grisea*

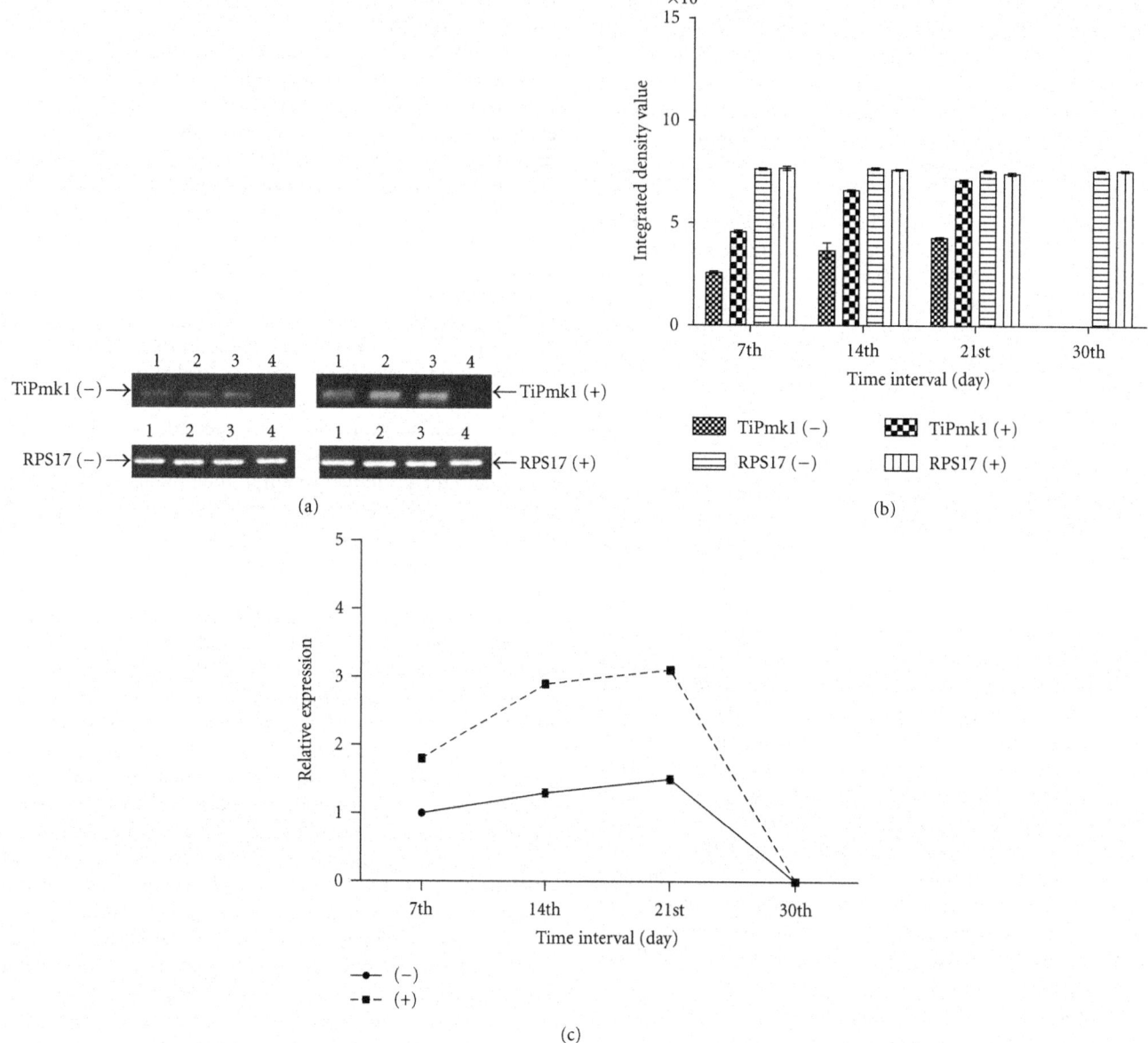

FIGURE 4: (a, b and c) Expression analysis of TiPmk1 gene in *T. indica* isolate in absence of host factor (−) and in presence of host factor (+) at different days of growth and development in liquid culture. (a) Semiquantitative RT-PCR analysis—RT-PCR of expressed messenger RNA at the 7th, 14th, 21st, and 30th day (lane 1 to 4, resp.) with RPS17 rRNA transcript used as internal control. (b) Densitometry analysis-integrated density value based on densitometry analysis was done with the help of gene profiler software, Alpha Innotech Corporation USA. (c) Quantitative real-time PCR analysis—relative expression using real-time PCR.

and *S. cereviceae*. Identification of downstream MAP kinase regulated transcription factors, most importantly those that regulate the critical penetration and colonization, would open the door to a broader understanding of the nature of the regulatory networks which govern pathogenic development and why certain signaling pathways are universally essential for pathogenicity. Ultimately, this may will lead to the identification of potential novel fungicides hitting such key targets of MAP kinase machinery which contribute significantly to the fungal growth and pathogenecity. Hence, delineation of signal transduction machinery is very much

crucial in framing biotechnological control measures against such pathogens.

Acknowledgments

The authors wish to acknowledge the Department of Science and Technology, Government. of India for providing financial support at GB Pant University of Agriculture and Technology, Pantnagar (Grant no. SR/SO/PS-83/2005 dated 20/12/2007). The support provided by the Director, Experiment Station, and Dean, College of Basic Sciences and Humanities, GB Pant University of Agriculture and Technology, Pantnagar, is also greatly acknowledged.

References

[1] P. K. Anderson, A. A. Cunningham, N. G. Patel, F. J. Morales, P. R. Epstein, and P. Daszak, "Emerging infectious diseases of plants: pathogen pollution, climate change and agrotechnology drivers," *Trends in Ecology and Evolution*, vol. 19, no. 10, pp. 535–544, 2004.

[2] K. S. Sekhon, A. K. Saxena, S. K. Randhawa, and S. S. Gill, "Effect of Karnal bunt disease on quality characteristics of wheat," *Bulletin of Grain Technology*, vol. 18, no. 3, pp. 208–212, 1980.

[3] B. J. Goates, "Histology of ionfection of wheat by *T. indica*. The Karnal bunt pathogen," *Phytopathology*, vol. 77, pp. 371–375, 1988.

[4] A. Kumar, U. S. Singh, A. Singh, V. S. Malik, and G. K. Garg, "Molecular signaling in pathogenicity and host recognition in smut fungi taking Karnal bunt as a model system," *Indian Journal of Experimental Biology*, vol. 38, no. 6, pp. 525–539, 2000.

[5] J. Kumar and S. Nagarajan, "Role of flag leaf and spike emergence stage on the incidence of Karnal bunt in wheat," *Plant Disease*, vol. 82, no. 12, pp. 1368–1370, 1998.

[6] A. K. Sharma, G. P. Singh, J. Rane, and S. Nagarajan, "Alternate selection approaches for targeted cropping sequences," in *Wheat Research Needs Beyond 2000*, pp. 227–235, AD. Narosa, New Delhi, India, 1998.

[7] S. S. Aujla, A. S. Grewal, K. S. Gill, and I. Sharma, "A screening technique for Karnal bunt disease of wheat," *Crop Improvement*, vol. 7, pp. 145–146, 1989.

[8] C. M. Rush, J. M. Stein, R. L. Bowden, R. Riemenschneider, T. Boratynski, and M. H. Royer, "Status of Karnal bunt of wheat in the United States 1996 to 2004," *Plant Disease*, vol. 89, no. 3, pp. 212–223, 2005.

[9] G. Fuentes-Davila and S. Rajaram, "Sources of resistance to *Tilletia indica* in wheat," *Crop Protection*, vol. 13, no. 1, pp. 20–24, 1994.

[10] G. Singh, S. Rajaram, J. Montoya, and G. Fuentes-Davila, "Genetic analysis of Karnal bunt resistance in 14 Mexican breadwheat genotypes," *Plant Breeding*, vol. 114, pp. 439–441, 1995.

[11] M. R. Bonde, G. L. Peterson, N. W. Schaad, and J. L. Smilanick, "Karnal bunt of wheat," *Plant Disease*, vol. 81, no. 12, pp. 1370–1377, 1997.

[12] M. R. Bonde, G. L. Peterson, G. Fuentes-Davila, S. S. Aujla, G. S. Nanda, and J. G. Phillips, "Comparison of the virulence of isolates of Tilletia indica, causal agent of Karnal bunt of wheat, from India, Pakistan, and Mexico," *Plant Disease*, vol. 80, no. 9, pp. 1071–1074, 1996.

[13] K. S. Gill, A. S. Randhawa, S. S. Aujla, H. S. Dhaliwal, A. S. Grewal, and I. Sharma, "Breeding wheat varieties resistant to Karnal bunt," *Crop Improvement*, vol. 8, pp. 73–80, 1981.

[14] R. Manish, C. Arora, B. Ram, and A. Kumar, "Floral specificity of Karnal bunt infection due to presence of fungal growth-promotory activity in wheat spikes," *Journal of Plant Biology*, vol. 28, no. 3, pp. 283–290, 2001.

[15] J. R. Xu, "MAP kinases in fungal pathogens," *Fungal Genetics and Biology*, vol. 31, no. 3, pp. 137–152, 2000.

[16] D'Souza C. A. and J. Heitman, "Conserved cAMP signaling cascades regulate fungal development and virulence," *FEMS Microbiol*, vol. 25, no. 3, pp. 349–364, 2001.

[17] C. Widmann, S. Gibson, M. B. Jarpe, and G. L. Johnson, "Mitogen-activated protein kinase: conservation of a three-kinase module from yeast to human," *Physiological Reviews*, vol. 79, no. 1, pp. 143–180, 1999.

[18] L. Chang and M. Karin, "Mammalian MAP kinase signalling cascades," *Nature*, vol. 410, no. 6824, pp. 37–40, 2001.

[19] I. Herskowitz, "Map kinase pathways in yeast: for mating and more," *Cell*, vol. 80, no. 2, pp. 187–197, 1995.

[20] R. Treisman, "Regulation of transcription by MAP kinase cascades," *Current Opinion in Cell Biology*, vol. 8, no. 2, pp. 205–215, 1996.

[21] L. Zheng, M. Campbell, J. Murphy, S. Lam, and J. R. Xu, "The BMP11 gene is essential for pathogenicity in the gray mold fungus Botrytis cinerea," *Molecular Plant-Microbe Interactions*, vol. 13, no. 7, pp. 724–732, 2000.

[22] P. D. Collopy, R. C. Amey, M. J. Sergeant et al., "The *pmk1*-like mitogen-activated protein kinase from *Lecanicillium* (*Verticillium*) *fungicola* is not required for virulence on *Agaricus bisporus*," *Microbiology*, vol. 156, no. 5, pp. 1439–1447, 2010.

[23] X. Zhao, R. Mehrabi, and J. R. Xu, "Mitogen-activated protein kinase pathways and fungal pathogenesis," *Eukaryotic Cell*, vol. 6, no. 10, pp. 1701–1714, 2007.

[24] R. E. Chen and J. Thorner, "Function and regulation in MAPK signaling pathways: lessons learned from the yeast Saccharomyces cerevisiae," *Biochimica et Biophysica Acta*, vol. 1773, no. 8, pp. 1311–1340, 2007.

[25] A. Kumar, K. Tripathi, M. Rana, S. Purwar, and G. K. Garg, "Dibutyryl c-AMP as an inducer of sporidia formation: biochemical and antigenic changes during morphological differentiation of Karnal bunt (*Tilletia indica*) pathogen in axenic culture," *Journal of Biosciences*, vol. 29, no. 1, pp. 23–31, 2004.

[26] M. M. Bradford, "A rapid and sensitive method for the quantitation of microgram quantities of protein utilizing the principle of protein dye binding," *Analytical Biochemistry*, vol. 72, no. 1-2, pp. 248–254, 1976.

[27] R. Duran and R. Cromarty, "*Tilletia indica*: a heterothallic wheat bunt fungus with multiple alleles controlling compatibility," *Phytopathology*, vol. 67, no. 7, pp. 812–815, 1977.

[28] R. A. Singh and A. Krishna, "Susceptible stage for inoculation and effect of Karnal bunt on viability of wheat seed," *Indian Phytopathology*, vol. 35, pp. 54–56, 1982.

[29] S. Dutt, D. Pandey, and A. Kumar, "Jasmonate signal induced expression of cystatin genes for providing resistance against Karnal bunt in wheat," *Plant Signaling and Behavior*, vol. 6, no. 6, pp. 821–830, 2011.

[30] K. Tamura, J. Dudley, M. Nei, and S. Kumar, "MEGA4: molecular evolutionary genetics analysis (MEGA) software version 4.0," *Molecular Biology and Evolution*, vol. 24, no. 8, pp. 1596–1599, 2007.

[31] K. J. Livak and T. D. Schmittgen, "Analysis of relative gene expression data using real-time quantitative PCR and the 2-ΔΔCT method," *Methods*, vol. 25, no. 4, pp. 402–408, 2001.

[32] S. Lev, A. Sharon, R. Hadar, H. Ma, and B. A. Horwitz, "A mitogen-activated protein kinase of the corn leaf pathogen *Cochliobolus heterostrophus* is involved in conidiation, appressorium formation, and pathogenicity: diverse roles for mitogen-activated protein kinase homologs in foliar pathogens," *Proceedings of the National Academy of Sciences of the United States of America*, vol. 96, no. 23, pp. 13542–13547, 1999.

[33] M. C. Gustin, J. Albertyn, M. Alexander, and K. Davenport, "Map kinase pathways in the yeast *Saccharomyces cerevisiae*,"

Microbiology and Molecular Biology Reviews, vol. 62, no. 4, pp. 1264–1300, 1998.

[34] B. Kramer, E. Thines, and A. J. Foster, "MAP kinase signalling pathway components and targets conserved between the distantly related plant pathogenic fungi Mycosphaerella graminicola and Magnaporthe grisea," *Fungal Genetics and Biology*, vol. 46, no. 9, pp. 667–681, 2009.

[35] S. S. Aujla, I. Sharma, and B. B. Singh, "Method of teliospore germination and breaking of dormancy in *Neovossia indica*," *Indian Phytopathology*, vol. 39, pp. 574–577, 1986.

[36] E. J. Warham and N. L. Cashion, " Evaluation of inoculation methods in the green house and field," in *Proceedings of International Maize and Wheat Improvement Center (CIMMYT '84)*, pp. 11–13, Cuidad Obregon, Mexico, April 1984.

[37] A. Krishna and R. A. Singh, "Multiple alleles controlling the incompatibility in *N. indica*," *Indian Phytopathology*, vol. 36, p. 746, 1983.

[38] G. Rai, A. Kumar, A. Singh, and G. K. Garg, "Modulation of antigenicity of mycelial antigens during developmental cycle of Karnal bunt (*Tilletia indica*) of wheat," *Indian Journal of Experimental Biology*, vol. 38, no. 5, pp. 488–492, 2000.

[39] A. Pandey, M. G. Roca, N. D. Read, and N. L. Glass, "Role of a mitogen-activated protein kinase pathway during conidial germination and hyphal fusion in *Neurospora crassa*," *Eukaryotic Cell*, vol. 3, no. 2, pp. 348–358, 2004.

Effects of CO_2 Enrichment on Growth and Development of *Impatiens hawkeri*

Fan-Fan Zhang, Yan-Li Wang, Zhi-Zhe Huang, Xiao-Chen Zhu, Feng-Jiao Zhang, Fa-Di Chen, Wei-Min Fang, and Nian-Jun Teng

College of Horticulture, Nanjing Agricultural University, Nanjing 210095, China

Correspondence should be addressed to Nian-Jun Teng, njteng@njau.edu.cn

Academic Editors: Z. Nishio and C. Varotsos

The effects of CO_2 enrichment on growth and development of *Impatiens hawkeri*, an important greenhouse flower, were investigated for the purpose of providing scientific basis for CO_2 enrichment to this species in greenhouse. The plants were grown in CO_2-controlled growth chambers with 380 (the control) and 760 (CO_2 enrichment) μmol $\cdot$ mol^{-1}, respectively. The changes in morphology, physiology, biochemistry, and leaf ultrastructure of *Impatiens* were examined. Results showed that CO_2 enrichment increased flower number and relative leaf area compared with the control. In addition, CO_2 enrichment significantly enhanced photosynthetic rate, contents of soluble sugars and starch, activities of peroxidase (POD), superoxide dismutase (SOD), and ascorbate peroxidase (APX), but reduced chlorophyll content and malondialdehyde (MDA) content. Furthermore, significant changes in chloroplast ultrastructure were observed at CO_2 enrichment: an increased number of starch grains with an expanded size, and an increased ratio of stroma thylakoid to grana thylakoid. These results suggest that CO_2 enrichment had positive effects on *Impatiens*, that is, it can improve the visual value, promote growth and development, and enhance antioxidant capacity.

1. Introduction

Atmospheric concentrations of greenhouse gases such as CO_2, N_2O, and CH_4 are increasing quickly since the beginning of the industrial revolution, which results in a rise in ground-level air temperatures [1–4]. These global climate changes will have produced profound effects on plant physiology and growth, structure and function of plant populations, and species distributions [2–4]. Among these factors, CO_2 is one of the raw materials of photosynthesis and has great influences on plant growth and development. The current atmospheric CO_2 concentration is about 380 μmol$\cdot$mol^{-1}, which is far below the optimum concentration of plant photosynthesis [5], especially for those plants such as ornamentals grown in greenhouse where ventilation is so limited to supplement the CO_2 consumed by plant photosynthesis, thus seriously affecting the growth, development, yield, and visual value of greenhouse-grown ornamentals [6, 7].

Over the past three decades, a large number of studies have focused on the effects of CO_2 enrichment (elevated CO_2) on the growth and development of plants. Generally, plants grown at elevated CO_2 relative to those grown at ambient CO_2 often exhibit increased growth and photosynthesis, lower transpiration, inhibited respiration, improved water use efficiency, decreased mineral nutrient concentrations, increased plant hormones contents, reduced stomatal density and conductance, and so forth [8–11]. However, most of these studies are focused on trees, steppe plants, crop plants, and greenhouse-grown vegetables, but substantial knowledge about potential influences of CO_2 enrichment on greenhouse-grown ornamentals is lacking [12]. Besides, available studies on greenhouse-grown ornamentals are usually focused on the morphology, photosynthesis, yield, and visual value of plants [7], few of them have investigated the impacts of CO_2 enrichment on antioxidant enzyme system and leaf ultrastructure, which are very important for an integrative understanding of plant responses to elevated CO_2.

Impatiens hawkeri or *Impatiens* New Guinea is a perennial species with rich colors and long flowering period, which has become a popular and important potted flowering plant throughout the world in recent years [13]. According to our knowledge, a study concentrating on the responses of *Impatiens* New Guinea to CO_2 enrichment has not been reported until now. Therefore, to better understand the effects of CO_2 enrichment on growth and development of *Impatiens* New Guinea, we investigated the influences of CO_2 enrichment on its morphological characters, photosynthetic rate, chlorophyll content, nonstructural carbohydrates content, in particular antioxidant enzyme activity, and leaf ultrastructure. Although the results in the artificial greenhouse environment are possibly different from the results expected in the real world because of the nonlinear nature of the CO_2 concentrations and the temporal and spatial variability of the CO_2 concentration in the filed observations [1–3], our study will provide useful knowledge for supplementing CO_2 during growth of ornamentals and even other horticultural crops in greenhouse.

2. Materials and Methods

2.1. Plant Materials and Growth Conditions. Seedlings of *Impatiens* were bought from a local company (Changshu Agricultural Technology Development Co., Ltd.) and rooted in 200 cm³ plastic pots filled with medium consisting of a 2 : 1 [volume/volume (v/v)] mixture of peat and vermiculite. After that, plants were cultivated for ten weeks in CO_2-controlled growth chambers which can automatically and accurately control environmental factors including temperature, light, relative humidity, photosynthetically active radiation (PAR), and CO_2 concentration according to preset data. The plants were alternately watered to saturation with Murashige-Skoog (MS) solution or deionized water. During the first four weeks, plants were fertilized to saturation with 1/3 MS solution every two weeks and with 1/2 MS solution every two weeks in the following six weeks. Plants were grown under a 12 h photoperiod with $300\,\mu mol\cdot m^{-2}\cdot s^{-1}$ PAR and day : night temperatures of 25 : 18°C. The relative humidities during daytime and at time were maintained at 70–80% and over 90%, respectively. Control plants were grown in one CO_2-controlled growth chamber with CO_2 concentration of $380 \pm 30\,\mu mol\cdot mol^{-1}$, while the treated plants were grown in another one with CO_2 concentration of $760 \pm 50\,\mu mol\cdot mol^{-1}$ during daytime and $380 \pm 30\,\mu mol\cdot mol^{-1}$ at night. The purity of CO_2 applied in this study was 99.999%. Except for CO_2 concentration of CO_2, other environmental conditions in two growth chambers were common. In order to keep the potential for interactive effects between the chambers and the developmental stage of the plants to a minimum, the CO_2 concentrations of the two chambers were swapped, and the pots were moved between chambers and randomly rearranged weekly. This purpose is to average out any possible effects from the chambers and pot positions within the chambers.

2.2. Growth and Development Analysis. After ten weeks' cultivation under either $380\,\mu mol\cdot mol^{-1}$ or $760\,\mu mol\cdot mol^{-1}$ CO_2 concentration, plant growth (three plants for each treatment) was assessed by regular (at 9:00 am every day) and nondestructive measurements of leaf length and leaf width (the third or fourth round of leaves), flower diameter, flower number, flower bud number, and lateral branch number for one month. Half of the product of leaf length multiplied by leaf width was regarded as the relative leaf area, and the mean of three times measurements of a flower's diameter from different angles was defined as the flower diameter. Flowers with fully unfolded corolla and buds with initial appearance of petals were included in the number of flower and flower bud, respectively. Leaf shape was quantified by means of bivariate allometric relationships between log-transferred leaf length and leaf width ($n = 34$), and the relationship between leaf length and leaf width determined the leaf shape of *Impatiens* New Guinea [12].

2.3. Determination of Photosynthetic Rate. Leaf Photosynthetic rate was measured with a Portable Photosynthesis system LI-6400 (LI-COR Inc., Lincoln, Neb, USA) after 45 days with three fully expanded leaves from each of five plants randomly selected from each treatment. The measurements for ambient CO_2-grown plants were carried out at $1500\,\mu mol\cdot m^{-2}\cdot s^{-1}$ PAR, 2.0–2.5 KPa vapour pressure deficit (VPD), $23 \pm 1°C$ and $380\,\mu mol\cdot mol^{-1}$ CO_2, and for elevated CO_2-grown plants at $1500\,\mu mol\cdot m^{-2}\cdot s^{-1}$ PAR, 2.0–2.5 KPa VPD, $23 \pm 1°C$ and $760\,\mu mol\cdot mol^{-1}$ CO_2.

2.4. Determination of Physiological and Biochemical Indexes. Leaves were obtained at the end of ten weeks' treatment from five plants of the control and elevated CO_2-grown *Impatiens* New Guinea. Parts of the leaves were oven-dried at 60°C for the determination of starch and soluble sugars, and the other parts were stored in $-80°C$ refrigerator following freezing in liquid nitrogen for the determination of contents of chlorophyll, MDA and activities of POD, SOD, APX. The content of starch was measured with the iodine colorimety method [14]. Chlorophyll, soluble sugars, MDA contents, and activities of POD, SOD were measured according to the methods of Wang [15]. The activity of APX was measured by method of Chen and Wang [16].

2.5. Leaf Ultrastructure Analysis. Areas beside the primary veins of fully expanded leaves were dissected into 1-2 mm² squares and immediately fixed in 2.5% (v/v) glutaraldehyde (in $0.1\,mol\cdot L^{-1}$ phosphate buffer, pH 7.0) for 24 h. Then the samples were washed 5 times with the same buffer and postfixed in 1% osmium tetroxide for 3 h. After being washed with the same buffer for 3 times, leaf tissues were immediately passed through an ethanol dehydration series and then infiltrated and embedded in epoxy resin Epon-812. An ultramicrotome LKB-5 was used to cut sections. Thin sections were stained with uranyl acetate and lead citrate, and finally observations were carried out with a transmission electron microscope JEM-1200EX [9].

2.6. Statistical Analysis. The data are shown as mean ± standard deviation to indicate significant differences. Data were subjected to one-way analysis of variance and *t*-test using the SPSS software 16.0 (SPSS Inc, Chicago, Ill, USA).

TABLE 1: Leaf parameters and flower diameter of *Impatiens* New Guinea grown under two CO_2 concentrations: ambient CO_2 ($380 \mu mol \cdot mol^{-1}$) and elevated CO_2 ($760 \mu mol \cdot mol^{-1}$).

	Ambient CO_2 ($380 \mu mol \cdot mol^{-1}$)	Elevated CO_2 ($760 \mu mol \cdot mol^{-1}$)	% increase	P value
Log leaf length	0.863 ± 0.004	0.893 ± 0.004	3.5	<0.001
Log leaf width	0.392 ± 0.003	0.401 ± 0.003	2.3	0.009
Relative leaf area (cm^2)	9.016 ± 0.137	9.864 ± 0.140	9.4	<0.001
Flower diameter (cm)	6.615 ± 0.028	6.656 ± 0.023	0.6	0.131

Values given are mean ± standard deviation. Mean values (n = 34 samples, with 9 leaves or flowers from 3 plants per sample) were compared by Student's t-test.

TABLE 2: Chlorophyll content and photosynthetic rate of *Impatiens* New Guinea grown under two CO_2 concentrations: ambient CO_2 ($380 \mu mol \cdot mol^{-1}$) and elevated CO_2 ($760 \mu mol \cdot mol^{-1}$).

	Ambient CO_2 ($380 \mu mol \cdot mol^{-1}$)	Elevated CO_2 ($760 \mu mol \cdot mol^{-1}$)	% increase	P value
Chlorophyll a ($\mu g \cdot mg^{-1}$)	1.03 ± 0.02	0.76 ± 0.01	-26.2	<0.001
Chlorophyll b ($\mu g \cdot mg^{-1}$)	0.45 ± 0.01	0.39 ± 0.01	-13.3	0.15
Total chlorophyll ($\mu g \cdot mg^{-1}$)	1.81 ± 0.03	1.48 ± 0.02	-18.2	<0.001
Chlorophyll a/b radio	2.28 ± 0.02	1.97 ± 0.06	-13.6	0.004
Photosynthetic rate ($\mu mol \cdot m^{-2} \cdot s^{-1}$)	11.09 ± 0.30	13.96 ± 0.30	25.9	<0.001

Values given are mean ± standard deviation. Mean values (n = 3 samples, with 5 plants per sample) were compared by Student's t-test. Total chlorophyll: Chlorophyll a + Chlorophyll b. FW: fresh weight.

Morphological characteristics were determined based on 3 plants per sample, whereas physiological, chemical, and cellular characteristics were determined based on 5 plants per sample. All the analyses were repeated three times.

3. Results

3.1. Vegetative and Developmental Responses to Elevated CO_2.
The *Impatiens* New Guinea used in this study displayed altered external features when subjected to elevated CO_2 (Figures 1, 2, 3 and Table 1). The most conspicuous change of *Impatiens* New Guinea was the increased flower number per individual plant (Figures 1, 2), which was significantly enhanced by 72.18% ($P < 0.0001$), leading to higher visual value. Elevated CO_2 also increased flower bud number per individual plant by 14.97% (Figure 2, $P = 0.003$). However, elevated CO_2 had no significant effects on flower diameter (Table 1, $P = 0.131$). Besides, lateral branch number per individual plant was significantly reduced by 29.39% (Figure 2, $P < 0.0001$), thus inducing a relatively well-proportioned plant shape compared with plants in ambient CO_2 which had more little lateral branches to occupy more space but seldom flowered (Figure 1).

Elevated CO_2 also increased relative leaf area by 9.4% (Table 1, $P < 0.001$). The allometric relationships between leaf length and leaf width were analyzed to predict the changes of leaf shape (Figure 3). Log leaf length showed a 3.5% (Table 1, $P < 0.0001$) increase and log leaf width showed a 2.3% (Table 1, $P = 0.009$) increase, indicating that leaf length had a relatively bigger increase. Analyzing Table 1 and Figure 3 together, mean value of (log leaf length, log leaf width) of plants in ambient CO_2 was (0.863, 0.392) and that

of plants in elevated CO_2 was (0.893, 0.401), showing that leaves in elevated CO_2 appeared to be longer and wider than leaves in ambient CO_2. Although the change of leaf shape seemed to be small, it was noticeable and measurable.

3.2. Responses of Photosynthetic Rate and Chlorophyll Content to Elevated CO_2.
Photosynthetic rate of *Impatiens* New Guinea grown in elevated CO_2 was significantly accelerated compared with that of *Impatiens* New Guinea grown in ambient CO_2 (Table 2), showing a 25.9% ($P < 0.001$) increase. However, total chlorophyll (chlorophyll a and chlorophyll b) content per unit leaf fresh weight of *Impatiens* New Guinea in elevated CO_2 was significantly reduced (Table 2), showing an 18.2% ($P < 0.001$) decrease from that of *Impatiens* New Guinea in ambient CO_2. Both chlorophyll a and chlorophyll b contributed to that response, which have a 26.2% and a 13.3% decrease, respectively, in spite of the indistinctive decrease of chlorophyll b ($P = 0.15$). The radio of chlorophyll a/b also dropped by 13.6% (Table 2, $P = 0.004$), indicating that the significant decrease of total chlorophyll mainly resulted from the decrease of chlorophyll a rather than chlorophyll b.

3.3. Responses of Nonstructural Carbohydrates to Elevated CO_2.
Main nonstructural carbohydrates found in leaves are total soluble sugars and starch. Figure 4 shows that elevation of CO_2 had significant effect on nonstructural carbohydrates content per unit leaf dry weight of *Impatiens* New Guinea plants, which dramatically improved contents of soluble sugars and starch, showing a 77.81% ($P < 0.001$) and a 122.39% ($P < 0.001$) increase, respectively. Hence, the total nonstructural carbohydrates content of plants grown

(a)   (b)

FIGURE 1: Growth situation of *Impatiens* New Guinea grown under two CO_2 concentrations: ambient CO_2 ($380\,\mu\text{mol}\cdot\text{mol}^{-1}$) and elevated CO_2 ($760\,\mu\text{mol}\cdot\text{mol}^{-1}$) for 10 weeks.

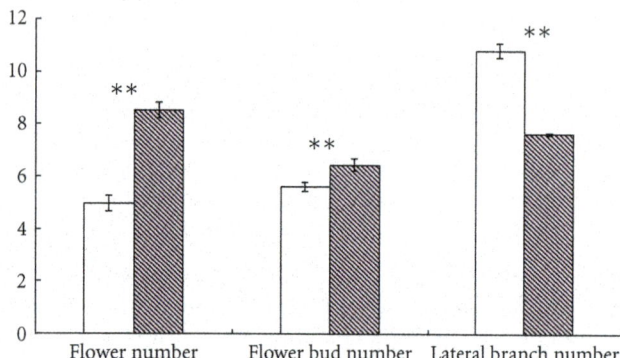

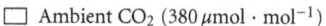
□ Ambient CO_2 ($380\,\mu\text{mol}\cdot\text{mol}^{-1}$)
▨ Elevated CO_2 ($760\,\mu\text{mol}\cdot\text{mol}^{-1}$)

FIGURE 2: The numbers of flower, flower bud, and lateral branch of *Impatiens* New Guinea grown under two CO_2 concentrations: ambient CO_2 ($380\,\mu\text{mol}\cdot\text{mol}^{-1}$) and elevated CO_2 ($760\,\mu\text{mol}\cdot\text{mol}^{-1}$). Asterisks show statistically significant differences ($^*P < 0.05$; $^{**}P < 0.01$; Student's t-test; $n = 34$ samples, with 3 plants per sample) and the bar is the standard deviation.

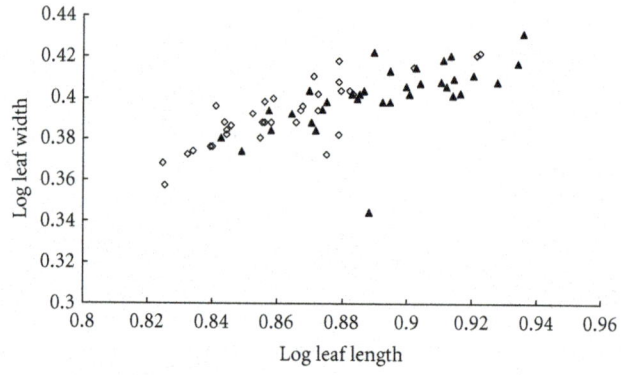

◇ Ambient CO_2 ($380\,\mu\text{mol}\cdot\text{mol}^{-1}$)
▲ Elevated CO_2 ($760\,\mu\text{mol}\cdot\text{mol}^{-1}$)

FIGURE 3: Allometric relationships between leaf length and leaf width for *Impatiens* New Guinea grown under two CO_2 concentrations: ambient CO_2 ($380\,\mu\text{mol}\cdot\text{mol}^{-1}$) and elevated CO_2 ($760\,\mu\text{mol}\cdot\text{mol}^{-1}$). $n = 34$ samples, with 9 leaves from 3 plants per sample.

in elevated CO_2 was doubled (a 103.5% increase, $P < 0.001$) compared with that of plants grown under ambient CO_2.

3.4. Responses of Antioxidant Enzyme Activity and MDA Content to Elevated CO_2.
Elevated CO_2 stimulated the activity of antioxidant enzymes, including peroxidase (POD), superoxide dismutase (SOD), and ascorbate peroxidase (APX) in these experiments of *Impatiens* New Guinea leaves (Table 3). The activity of POD, SOD, and APX increased by 119.78%, 11.01%, 73.26%, respectively, among which the increases of POD activity and APX activity reached significance level

($P = 0.05$), whereas the increase of SOD activity did not ($P > 0.05$). Besides, MDA content of *Impatiens* New Guinea leaves was remarkably reduced by 61.13% ($P = 0.006$) due to CO_2 elevation (Table 3).

3.5. Responses of Leaf Ultrastructure to Elevated CO_2.
Figures 5(a) and 5(b) show cross-sections through typical cells of *Impatiens* New Guinea leaves in ambient CO_2 and elevated CO_2, respectively. Chloroplasts were located peripherally and almost occupied half of the volume in the typical cell in elevated CO_2 (Figure 5(b)). Starch grains were also observed, most of which are located at lateral sides of chloroplasts (Figures 5(a) and 5(b), arrows), leading to an ear-shaped

TABLE 3: Antioxidant enzyme activity and MDA content of *Impatiens* New Guinea grown under two CO_2 concentration: ambient CO_2 $(380\,\mu mol \cdot mol^{-1})$ and elevated CO_2 $(760\,\mu mol \cdot mol^{-1})$.

	Ambient CO_2 $(380\,\mu mol \cdot mol^{-1})$	Elevated CO_2 $(760\,\mu mol \cdot mol^{-1})$	% increase	P-value
POD (U/(g∗min))	10.77 ± 2.83	23.67 ± 2.22	119.78	0.011
SOD (U/g)	527.75 ± 4.47	585.83 ± 39.51	11.01	0.141
APX ((U/(g∗min))	20.64 ± 0.88	35.76 ± 4.41	73.26	0.039
MDA ($\mu mol/g$)	0.01114 ± 0.00142	0.00433 ± 0.00062	−61.13	0.006

Values given are mean ± standard deviation. Mean values ($n = 3$ samples, with 5 plants per sample) were compared by Student's t-test.

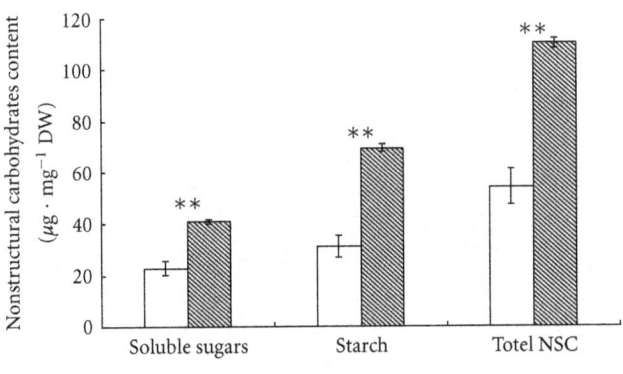

□ Ambient CO_2 $(380\,\mu mol \cdot mol^{-1})$
▨ Elevated CO_2 $(760\,\mu mol \cdot mol^{-1})$

FIGURE 4: Nonstructural carbohydrates content of *Impatiens* New Guinea grown under two CO_2 concentration: ambient CO_2 $(380\,\mu mol \cdot mol^{-1})$ and elevated CO_2 $(760\,\mu mol \cdot mol^{-1})$. Asterisks show statistically significant differences ($^*P < 0.05$; $^{**}P < 0.01$; Student's t-test; $n = 3$ samples, with 5 plants per sample) and the bar is the standard deviation. Total NSC= soluble sugars + starch.

protuberance arising from the surface. Figures 5(c) and 5(d) show that grana can be found throughout the chloroplast other than the volume occupied by starch grains. Figures 5(e) and 5(f) clearly show that grana were composed of numbers of grana thylakoids which were stacked orderly, making grana regular shape of cylinders. Moreover, stroma thylakoids were seen between grana (Figures 5(e) and 5(f), arrows), and plastoglobules were observed to spread in chloroplast around grana (Figure 5(f), arrows).

As a consequence of elevated CO_2, striking changes of the leaf ultrastructure were seen in the chloroplast. Firstly, chloroplasts showed an increased number of starch grains with expanded size (Figures 5(a), 5(b), 5(c), and 5(d)). Secondly, the number of grana in chloroplasts with elevated CO_2 treatment seemed to be smaller than that of grana in ambient CO_2 (Figures 5(c) and 5(d)), because of the greater number and size of starch grains. Furthermore, grana seemed to be reduced in size (Figures 5(e) and 5(f)), namely, the number of thylakoids making-up grana was reduced. However, elevated CO_2 increased the number of stroma thylakoids between grana (Figures 5(e) and 5(f)), leading to an increased ratio of stroma thylakoid to grana thylakoid combining the opposite responses of stroma thylakoids

and grana thylakoid numbers to CO_2 elevation. Grana thylakoids in ambient CO_2 were closely and orderly aligned (Figure 5(e)), whereas grana thylakoids in elevated CO_2 were relatively loosely aligned (Figure 5(f)). This phenomenon was significantly observed in grana next to starch grains (Figures 5(c) and 5(d)). Finally, more plastoglobules seemed to be observed in chloroplasts with elevated CO_2 treatment compared with those in ambient CO_2 (Figures 5(e) and 5(f)).

4. Discussion

4.1. Morphological Characters. Flower number and flower bud number of *Impatiens* New Guinea were both significantly increased by elevated CO_2. Similar results were found in other species such as Miniature rose, *Alstroemeria,* and *Lilium* [17–19]. Jablonski et al. [20] who used meta-analysis to integrate data from 159 enrichment papers providing information on 79 species found that growth at elevated CO_2 resulted in the production of more (+19%) flowers. The significant stimulation of flower number and flower bud number of plants may primarily be a consequence of increased relative growth rate and accelerated developmental process which make plants reach the minimum size required for flowering earlier and have more resources available for reproduction, and finally increase flower number and flower bud number under elevated CO_2 [9]. As flower number per individual plant is an important indicator for quality of ornamental plants, the remarkable enhancement of flower number and flower bud number by CO_2 enrichment will be of great importance to improving the ornamental quality and market competitiveness of *Impatiens* New Guinea.

Besides, elevated CO_2 also noteworthily increased relative leaf area of *Impatiens* New Guinea, that is, both leaf length and leaf width showed measurable increase, but leaf length had a relatively bigger increase. This is in accordance with many early reports [21, 22]. This increase of plant leaf area under elevated CO_2 is probably an outcome of the well documented issue that elevated CO_2 increased photosynthesis and carbohydrate production (discussed below). Ainsworth et al. [23] further demonstrated that at the transcript and metabolite level, CO_2 enrichment stimulate the respiratory breakdown of carbohydrates, which provides increased energy and biochemical precursors for leaf expansion and growth, thus leading to increased leaf area of plants under elevated CO_2. However, in a recent study, elevated CO_2 had distinct effects on two *Aechmea* hybrids:

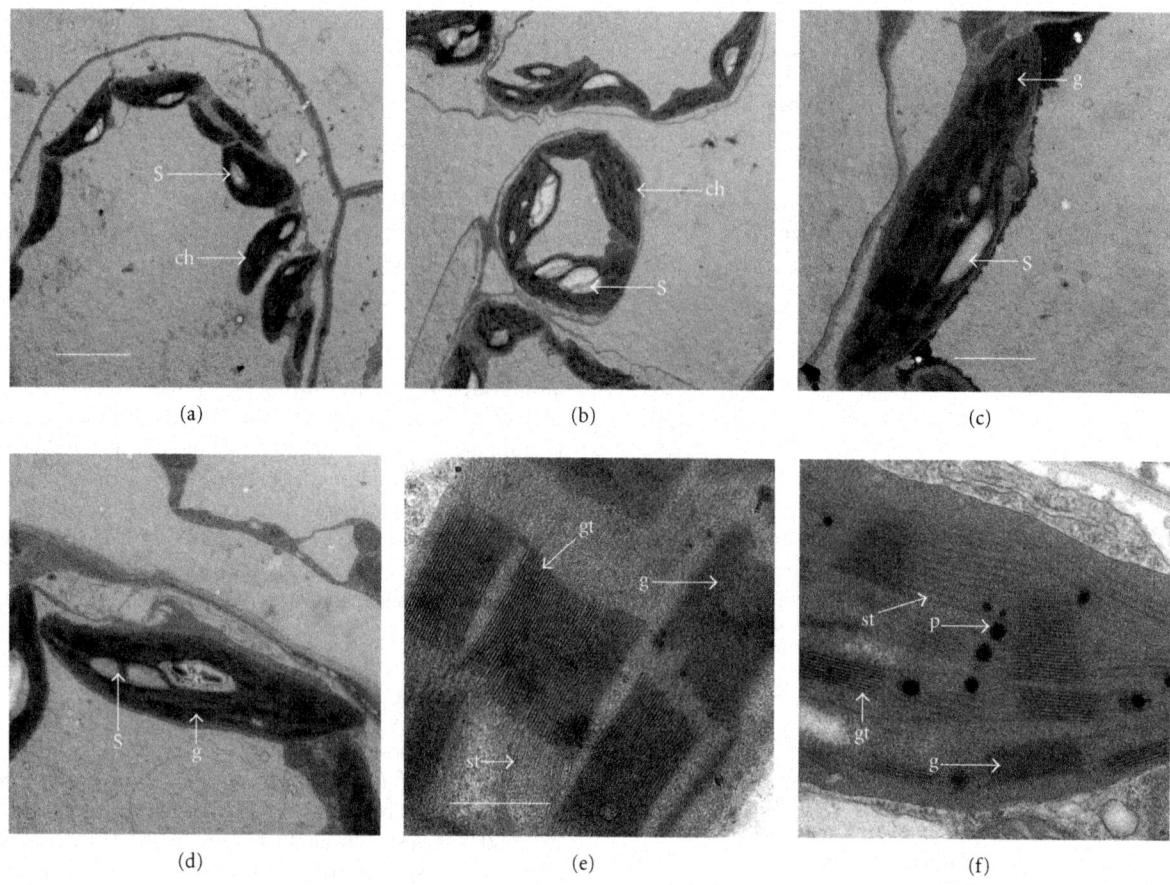

FIGURE 5: Transmission electron micrographs showing leaf ultrastructure of *Impatiens* New Guinea grown under two CO_2 concentrations: ambient CO_2 (a, c, e, $380\,\mu\text{mol}\cdot\text{mol}^{-1}$) and elevated CO_2 (b, d, f, $760\,\mu\text{mol}\cdot\text{mol}^{-1}$). s: starch; ch: chloroplast; g: grana; gt: grana thylakoid; st: stroma thylakoids; p: plastoglobuli. Bars: $5\,\mu$m (a and b); $2\,\mu$m (c and d); $0.5\,\mu$m (e and f).

A. fasciata "Maya" showed for both CO_2 concentrations ($380\,\mu\text{mol}\cdot\text{mol}^{-1}$ and $750\,\mu\text{mol}\cdot\text{mol}^{-1}$) an equal leaf area enhancement throughout the experimental period, whereas *A. fasciata* "Primera" showed a reduction of total leaf area by 41% [12]. A possible reason for this discrepancy is that the period of elevated CO_2 treatment to this two *Aechmea* hybrids was too long (34 weeks), leading to the occurrence of CO_2 acclimation, that is, the positive effects of long-term and high concentration CO_2 treatments on plants will disappear gradually over time [6, 7].

4.2. Photosynthetic Rate, Chlorophyll Content, and Nonstructural Carbohydrates. In our study, photosynthetic rate of *Impatiens* New Guinea grown in elevated CO_2 was significantly accelerated. This result was in accordance with many previous studies [8, 19, 21, 22], thus giving more evidence to the publicly accepted conclusion that CO_2 enrichment is able to promote plant photosynthesis to some extent. However, in our study, total chlorophyll and chlorophyll per unit leaf fresh weight of *Impatiens* New Guinea in elevated CO_2 was significantly reduced. We suppose that there is at least two reasons responsible for this result: firstly, the reduction of chlorophyll content may be caused by the dilution from

excess accumulation of carbohydrates (discussed below); secondly, as Teng et al. [9] have proved of elevated CO_2-induced reductions of mineral nutrient concentrations in plant leaves, including some of the basic components of chlorophyll such as N and Mg, so we hypothesize that the reductions of some necessary mineral nutrient may eventually affect the synthesis of chlorophyll.

A large number of studies have shown significant enhancement of carbohydrates (sugars and starches) in plant leaves by elevated CO_2 [24–26]. It was reported that soluble sugars and starch contents increased by 50% and 160% on average, respectively [27]. We verified the previous studies with the result that contents of soluble sugars and starch of *Impatiens* New Guinea showed a 77.81% and a 122.39% increase, respectively, causing total nonstructural carbohydrates content doubled. We have documented above that elevated CO_2 enhanced photosynthesis, which is in favor of the assimilation of nonstructural carbohydrates; in addition, limited sink capacity as well as some functional restriction such as limited phloem loading capacity and low efficiency of assimilate transport are likely to aggravate the accumulation of nonstructural carbohydrates in plant leaves under elevated CO_2 [12].

However, with the increase of CO_2 concentration or application time, plants tended to bring about the phenomenon of CO_2 acclimation, which have been proved by many past studies to have a close relationship with these excess carbohydrates in plant leaves. Given the links between the phenomenon of CO_2 acclimation and carbohydrates, it can be concluded as two aspects, one was that the excess accumulation of starch in plant leaves caused some physical damage to chloroplast ultrastructure (discussed below); the other was that soluble sugars suppressed photosynthesis by feedback inhibition [28].

4.3. Antioxidant Enzyme System. Antioxidant enzymes including peroxidase (POD), superoxide dismutase (SOD), and ascorbate peroxidase (APX) can neutralize free radicals, ridding the plant body of their harmful effects, therefore the activity of antioxidant enzymes is an important indicator of plant stress resistance [29]. Malonaldehyde (MDA) is a decomposition of peroxidized membrane lipid, so its content, which is closely related to the activity of antioxidant enzymes, can indicate the extent of membrane lipid peroxidation [30]. In our study, CO_2 enrichment significantly improved activities of POD, SOD, and APX, but reduced MDA content, illustrating that antioxidant capacity of *Impatiens* New Guinea leaves was enhanced by elevated CO_2. This phenomenon may be resulted from the significant stimulation of nonstructural carbohydrates in *Impatiens* New Guinea leaves (Figure 4), which can provide more energy substances for antioxidative metabolism, leading to improved activities of antioxidant enzymes and inhibited production of MDA, and eventually enhanced antioxidant capacity. Another possible interpretation for this is that elevated CO_2 causes oxidative stress, thus signaling the need to increase the activity of antioxidant enzymes.

Because of the well-established link between antioxidant enzyme system and plant stress resistance, we suppose that stress resistance of *Impatiens* New Guinea would be strengthened. Several other plants have been reported with strengthened stress tolerance under elevated CO_2 [31–35]. For instance, Sehmer et al. [31] found that under elevated CO_2, the activities of SOD and APX of Norway spruce were enhanced when imposed to O_3 stress to reduce the harm of O_3 on its leaf tissue. Besides, as plant senescence is usually associated with the reduction of antioxidant enzymes activities [33], so we predict that elevated CO_2 would delay the senescence of *Impatiens* New Guinea because of the significant enhancement of antioxidant activities. Rae et al. [34] proved at gene level that senescence was delayed under elevated CO_2. In the investigation of Wang et al. [35] on effects of elevated CO_2 on cut chrysanthemum, which were in great accordance with our findings, also found that elevated CO_2 enhanced activities of POD and SOD, but reduced MDA content, therefore indirectly retarded cell degeneration and finally extended the life of cut flowers.

4.4. Leaf Ultrastructure. Many studies about effects of CO_2 enrichment on plant leaf ultrastructure found that elevated CO_2 increased the number and size of starch grains in chloroplast [6, 36, 37], which were consistent with our findings in biochemical assay of nonstructural carbohydrates (Figure 4) as well as leaf ultrastructure analysis (Figure 5) of *Impatiens* New Guinea. However, in an earlier study on young wheat leaves, more starch accumulation was observed in leaves in ambient CO_2, whereas small starch grains were found to disperse throughout the stroma of chloroplasts in leaves grown under elevated CO_2 [38]. A possible reason provided by the authors for this phenomenon was that during the early developmental stages of plants under elevated CO_2, seedlings had a higher need for energy and carbon skeletons because of fast growth than those under ambient CO_2, thus consuming more starch and leading to relatively less starch accumulated in young wheat leaves grown in elevated CO_2.

In addition, elevated CO_2 increased the ratio of stroma thylakoid to grana thylakoid, which agreed with many previous reports [9, 33, 39]. Griffin et al. [39] conjectured that these changes in the chloroplast structure maybe an approach to maintain leaf-level energy balance. They determined that, as elevated CO_2 lead to a higher photosynthetic rate, increasing the demand for reductant to be used in carbon fixation and stroma thylakoids are enriched in photosystem I centers where NADPH is produced for reduction of CO_2, so increased ratio of stroma thylakoid to grana thylakoid can ensure plant more efficiently fixation of CO_2 into sugar products by increasing reductant production. Meanwhile, we found that grana thylakoids in ambient CO_2 were closely and orderly aligned, whereas grana thylakoids in elevated CO_2 were relatively loosely aligned, and this phenomenon was significantly observed in grana next to starch grains. There were several previous reports on elevated CO_2-induced structural changes of thylakoids which were always accompanied with excess accumulation of starch grains [40, 41]. These findings indicated that structural changes of thylakoids may be caused by excess accumulation of starch grains which press against and separate grana thylakoids, thus changing the original close arrangement of grana thylakoids into loose. Taken together with elevated CO_2-induced structural changes of thylakoids, these findings indicate that elevated CO_2 damage the structure of leaf chloroplast to a certain extent, so we doubt whether the positive effects of elevated CO_2 on growth and development of *Impatiens* New Guinea will continue or not with the increase of CO_2 concentration or application time. Therefore, more work is needed to examine the detailed effects of higher concentration or long-term CO_2 enrichment on growth and development of *Impatiens* New Guinea.

5. Conclusions

According to the above results, we are able to make a conclusion that CO_2 enrichment has positive effects on *Impatiens* New Guinea, which can improve the visual value by increasing flower number and leaf area, promote growth by raising photosynthetic rate, nonstructural carbohydrates content, and ratio of stroma thylakoid to grana thylakoid, and enhance antioxidant capacity by improving activity of antioxidant enzymes. However, whether these positive effects of CO_2 enrichment on *Impatiens* New Guinea will continue with the increase of CO_2 concentration or application time still needs more investigations.

Acknowledgments

This work was supported by the National Science Fund of China (31171983, 30700081, 30870436), the Natural Science Fund of Jiangsu Province (BK2010447), the funding from the International Foundation for Science for Dr. Nian-Jun Teng (Reference no. C/4560-1), and the Student Research Training Project of Nanjing Agricultural University to Fan-Fan Zhang (0903A06).

References

[1] K. Y. Kondratyev and C. Varotsos, "Atmospheric greenhouse effect in the context of global climate change," *Nuovo Cimento della Societa Italiana di Fisica C-Geophysics and Space Physics*, vol. 18, no. 2, pp. 123–151, 1995.

[2] C. Varotsos, J. Ondov, and M. Efstathiou, "Scaling properties of air pollution in Athens, Greece and Baltimore, Maryland," *Atmospheric Environment*, vol. 39, no. 22, pp. 4041–4047, 2005.

[3] C. Varotsos, M. N. Assimakopoulos, and M. Efstathiou, "Technical note: long-term memory effect in the atmospheric CO_2 concentration at Mauna Loa," *Atmospheric Chemistry and Physics*, vol. 7, no. 3, pp. 629–634, 2007.

[4] B. Jin, L. Wang, J. Wang et al., "The effect of experimental warming on leaf functional traits, leaf structure and leaf biochemistry in Arabidopsis thaliana," *BMC Plant Biology*, vol. 11, article 35, 2011.

[5] N. Teng, B. Jin, Q. Wang et al., "No detectable maternal effects of elevated CO_2 on Arabidopsis thaliana over 15 generations," *PLoS One*, vol. 4, no. 6, Article ID e6035, 2009.

[6] H. T. Pan, X. L. Liu, and Q. X. Zhang, "Influences of CO_2 enrichment on ornamental plant production," *Journal of Beijing Forestry University*, vol. 25, no. 1, pp. 93–99, 2003.

[7] F. F. Zhang and N. J. Teng, "A review on effects of CO_2 enrichment on greenhouse flowers," *Journal of Anhui Agricultural Sciences*, vol. 38, no. 31, pp. 17482–17495, 2010.

[8] J. A. Hui, Y. H. Li, Z. Li, and Q. S. Ye, "Effect of elevated CO_2 concentration on photosynthesis, growth and development of Guzmania 'Luna'," *Acta Horticulturae Sinica*, vol. 33, no. 5, pp. 1027–1032, 2006.

[9] N. Teng, J. Wang, T. Chen, X. Wu, Y. Wang, and J. Lin, "Elevated CO_2 induces physiological, biochemical and structural changes in leaves of *Arabidopsis thaliana*," *New Phytologist*, vol. 172, no. 1, pp. 92–103, 2006.

[10] N. J. Teng, T. Chen, and J. X. Lin, "A review on responses of plant sexual reproduction to elevated CO_2," *Journal of Plant Ecology*, vol. 30, no. 6, pp. 1054–1063, 2006.

[11] R. Pandey, P. M. Chacko, M. L. Choudhary, K. V. Prasad, and M. Pal, "Higher than optimum temperature under CO_2 enrichment influences stomata anatomical characters in rose (*Rosa hybrida*)," *Scientia Horticulturae*, vol. 113, no. 1, pp. 74–81, 2007.

[12] S. Croonenborghs, J. Ceusters, E. Londers, and M. P. De Proft, "Effects of elevated CO_2 on growth and morphological characteristics of ornamental bromeliads," *Scientia Horticulturae*, vol. 121, no. 2, pp. 192–198, 2009.

[13] R. H. Li and Z. Fang, "Effects of aluminium on growth and flower color of *Impatiens hawkeri*," *Journal of Agricultural University of Hebei*, vol. 29, no. 5, pp. 32–36, 2006.

[14] C. J. Xu, W. J. Chen, K. S. Chen, and S. L. Zhang, "A simple method for determining the content of starch—iodine colorimety," *Biotechnology*, vol. 8, no. 2, pp. 41–43, 1998.

[15] X. K. Wang, *Principles and Techniques of Plant Physiological Biochemical Experiment*, Higher Education Press, Beijing, China, 2nd edition, 2006.

[16] J. X. Chen and X. F. Wang, *Guidance for Plant Physiological Experiment*, South China University of Technology Press, Guangzhou, China, 2006.

[17] L. M. Mortensen and R. Moe, "Effects of temperature, carbon dioxide concentration, day length and photo flux density on growth, morphology and flowering of miniature roses," *Acta Horticulturae*, vol. 378, pp. 63–70, 1995.

[18] M. C. Van Labeke and P. Dambre, "Effect of supplementary lighting and CO_2 enrichment on yield and flower stem quality of *Alstroemeria* cultivars," *Scientia Horticulturae*, vol. 74, no. 4, pp. 269–278, 1998.

[19] S. L. Wei, Y. H. Liu, H. Y. Qu, S. L. Fu, and Y. L. Fu, "Effects of high CO_2 concentration on physiological and biochemical processes in Lily(*Lilium Dauricum*)," *Acta Phytoecologica Sinica*, vol. 25, no. 4, pp. 410–413, 2001.

[20] L. M. Jablonski, X. Wang, and P. S. Curtis, "Plant reproduction under elevated CO_2 conditions: a meta-analysis of reports on 79 crop and wild species," *New Phytologist*, vol. 156, no. 1, pp. 9–26, 2002.

[21] J. M. Wang, Y. H. Li, S. Q. Huang, L. N. Liu, and Q. S. Ye, "Effects of elevated CO_2 concentration on Photosynthetic rate, growth and development in Anthurium andraeanum Lind. Leaves," *Acta Horticulturae Sinica*, vol. 32, no. 2, pp. 335–338, 2005.

[22] Y. H. Li, R. H. Wu, Q. S. Yang, and Q. S. Ye, "Effects of short-term CO_2 enrichment on photosynthetic rate and growth in *Anthurium andraeanum* L. Seedlings," *Journal of Henan Agricultural University*, vol. 40, no. 6, pp. 607–610, 2006.

[23] E. A. Ainsworth, A. Rogers, L. O. Vodkin, A. Walter, and U. Schurr, "The effects of elevated CO_2 concentration on soybean gene expression. An analysis of growing and mature leaves," *Plant Physiology*, vol. 142, no. 1, pp. 135–147, 2006.

[24] S. P. Long, E. A. Ainsworth, A. Rogers, and D. R. Ort, "Rising atmospheric carbon dioxide: plants face the future," *Annual Review of Plant Biology*, vol. 55, pp. 591–628, 2004.

[25] D. C. Uprety, K. Sunita, D. Neeta, and M. Rajat, "Effect of elevated CO_2 on the growth and yield of rice," *Indian Journal of Plant Physiology*, vol. 5, pp. 105–107, 2006.

[26] Z. J. Mao, G. M. Jia, L. X. Liu, and M. Zhao, "Combined effects of elevated temperature, elevated CO_2 and nitrogen supply on non-structural carbohydrate accumulation and allocation in Quercus mongolica seedlings," *Chinese Journal of Plant Ecology*, vol. 34, no. 10, pp. 1174–1184, 2010.

[27] S. P. Long and B. G. Drake, "Photosynthetic CO_2 assimilation and rising atmospheric CO_2 concentrations," in *Crop Photosynthesis: Spatial and Temporal Determinants*, pp. 69–95, 1992.

[28] J. Sun, K. M. Gibson, O. Kiirats, T. W. Okita, and G. E. Edwards, "Interactions of nitrate and CO_2 enrichment on growth, carbohydrates, and rubisco in arabidopsis starch mutants. Significance of starch and hexose," *Plant Physiology*, vol. 130, no. 3, pp. 1573–1583, 2002.

[29] Y. Gao, Z. Fang, D. F. Chen, W. J. Li, and S. H. Zhang, "Comparison of physiological characteristics of different *Impatiens hawkeri* varieties under solution culture," *Journal of Agricultural University of Hebei*, vol. 5, no. 26, pp. 134–136, 2003.

[30] Y. Zhang, Z. Fang, Y. A. Li, R. H. Li, and H. P. Zhao, "Effect of different aeration hours on the growth and development of New Guinea impatiens under solution culture," *Journal of Agricultural University of Hebei*, vol. 29, no. 4, pp. 23–26, 2006.

[31] L. Sehmer, V. Fontaine, F. Antoni, and P. Dizengremel, "Effects of ozone and elevated atmospheric carbon dioxide on carbohydrates metabolism of spruce needles. Catabolic and detoxification pathways," *Physiologia Plantarum*, vol. 102, no. 4, pp. 605–611, 1998.

[32] V. Velikova, T. Tsonev, C. Barta et al., "BVOC emissions, photosynthetic characteristics and changes in chloroplast ultrastructure of *Platanus orientalis* L. exposed to elevated CO_2 and high temperature," *Environmental Pollution*, vol. 157, no. 10, pp. 2629–2637, 2009.

[33] X. Li, H. Wang, H. Li et al., "Awns play a dominant role in carbohydrate production during the grain-filling stages in wheat (*Triticum aestivum* L.)," *Physiologia Plantarum*, vol. 127, no. 4, pp. 701–709, 2006.

[34] A. M. Rae, R. Ferris, M. J. Tallis, and G. Taylor, "Elucidating genomic regions determining enhanced leaf growth and delayed senescence in elevated CO_2," *Plant, Cell and Environment*, vol. 29, no. 9, pp. 1730–1741, 2006.

[35] Y. L. Wang, Z. Z. Huang, R. Sun, F. D. Chen, and N. J. Teng, "Effects of elevated CO_2 on vase quality, physiological and structural characteristics of cut Chrysanthemum," *Scientia Agricultura Sinica*, vol. 43, no. 21, pp. 4463–4472, 2010.

[36] B. Y. Zuo, Q. Zhang, G. Z. Jiang, K. Z. Bai, and T. Y. Kuang, "Effects of doubled-CO_2 concentration on ultrastructure, supramolecular architecture and spectral characteristics of chloroplasts from wheat," *Acta Botanica Sinica*, vol. 44, no. 8, pp. 908–912, 2002.

[37] X. Wang, O. R. Anderson, and K. L. Griffin, "Chloroplast numbers, mitochondrion numbers and carbon assimilation physiology of *Nicotiana sylvestris* as affected by CO_2 concentration," *Environmental and Experimental Botany*, vol. 51, no. 1, pp. 21–31, 2004.

[38] E. J. Robertson and R. M. Leech, "Significant changes in cell and chloroplast develoment in young wheat leaves (*Triticum aestivum* cv Hereward) grown in elevated CO_2," *Plant Physiology*, vol. 107, no. 1, pp. 63–71, 1995.

[39] K. L. Griffin, O. R. Anderson, M. D. Gastrich et al., "Plant growth in elevated CO_2 alters mitochondrial number and chloroplast fine structure," *Proceedings of the National Academy of Sciences of the United States of America*, vol. 98, no. 5, pp. 2473–2478, 2001.

[40] S. G. Pritchard, C. M. Peterson, S. A. Prior, and H. H. Rogers, "Elevated atmospheric CO_2 differentially affects needle chloroplast ultrastructure and phloem anatomy in *Pinus palustris*: interactions with soil resource availability," *Plant, Cell and Environment*, vol. 20, no. 4, pp. 461–471, 1997.

[41] J. Utriainen, S. Janhunen, H. S. Helmisaari, and T. Holopainen, "Biomass allocation, needle structural characteristics and nutrient composition in scots pine seedlings exposed to elevated CO_2 and O_3 concentrations," *Trees*, vol. 14, no. 8, pp. 475–484, 2000.

Localisation of Abundant and Organ-Specific Genes Expressed in *Rosa hybrida* Leaves and Flower Buds by Direct *In Situ* RT-PCR

Agata Jedrzejuk,[1,2] Heiko Mibus,[1] and Margrethe Serek[1]

[1] *Faculty of Natural Sciences, Institute for Ornamental and Woody Plant Science, University of Hannover, Herrenhauser Street 2, 30419 Hannover, Germany*
[2] *Department of Ornamental Plants, Faculty of Horticulture and Landscape Architecture, Warsaw University of Life Sciences, Nowoursynowska 166, 02-787 Warsaw, Poland*

Correspondence should be addressed to Agata Jedrzejuk, agata.jedrzejuk@wp.pl

Academic Editors: Y. Shoyama, M. Simmonds, and B. Vyskot

In situ PCR is a technique that allows specific nucleic acid sequences to be detected in individual cells and tissues. *In situ* PCR and IS-RT-PCR are elegant techniques that can increase both sensitivity and throughput, but they are, at best, only semiquantitative; therefore, it is desirable first to ascertain the expression pattern by conventional means to establish the suitable conditions for each probe. In plants, *in situ* RT-PCR is widely used in the expression localisation of specific genes, including MADS-box and other function-specific genes or housekeeping genes in floral buds and other organs. This method is especially useful in small organs or during early developmental stages when the separation of particular parts is impossible. In this paper, we compared three different labelling and immunodetection methods by using *in situ* RT-PCR in *Rosa hybrida* flower buds and leaves. As target genes, we used the abundant β-actin and *RhFUL* gene, which is expressed only in the leaves and petals/sepals of flower buds. We used digoxygenin-11-dUTP, biotin-11-dUTP, and fluorescein-12-dUTP-labelled nucleotides and antidig-AP/ streptavidin-fluorescein-labelled antibodies. All of the used methods gave strong, specific signal and all of them may be used in localization of gene expression on tissue level in rose organs.

1. Introduction

Knowledge regarding the cellular localisation of gene transcripts is essential to assess gene function in an integrated context. There are essentially three different experimental procedures in the field of molecular histology, all of which possess inherent advantages and drawbacks. Promoter-reporter gene fusions may be used to analyse the promoter activity of a target gene [1]. Tissue print RNA hybridisation, based on the transfer of the cytoplasmic contents of fresh tissue sections onto a membrane by hand pressure and subsequent hybridisation with a labelled probe, is an extremely rapid and easy procedure with potential for high-throughput applications [2]. The third and perhaps most widely used method is *in situ* hybridisation (ISH) [3], which may be applied to intact plants (whole-mount *in situ*; [4] or, more classically, to tissue sections [5, 6]). Although procedures based on direct signal visualisation (tissue printing and ISH) have produced a plethora of results, they are essentially limited to target genes with relatively high levels of expression. To overcome this limitation, PCR-based localisation procedures have been established [3, 7], principally with animal tissues and cells, and they are often used in medical applications and to a lesser extent in plants. *In situ* PCR (ISPCR) is a technique that allows specific nucleic acid sequences to be detected in individual cells and tissues [7, 8]. The technique is based on PCR performed on fixed, whole cells or sections; ideally, the PCR product is to be detected at the site of synthesis where it aggregates. Thus far, ISPCR has only been used in animal cells and tissues and predominantly to detect viruses, such as HIV [7–9] or hepatitis C [8, 10].

ISPCR and RT-ISPCR are elegant techniques that can increase both sensitivity and throughput, but they are, at best, merely semi-quantitative [6]; therefore, it is desirable first to ascertain the expression pattern by conventional means to establish the suitable conditions for each probe [11]. *In situ* RT-PCR is a technique that allows the *in situ* visualisation of gene expression at much lower levels than by using *in situ* hybridisation [12, 13]. This technique consists of the reverse-transcription of a targeted RNA within a tissue and the subsequent PCR amplification of the resulting cDNA. Therefore, *in situ* RT-PCR defines a powerful tool for the detection of low-abundance transcripts [6] because the revealing threshold can be as low as one or two copies per cell. In comparison, *in situ* hybridisation detects 10 to 20 copies per cell [14, 15]. The first application of *in situ* RT-PCR for plant tissues was reported by Woo et al. [16] and described the expression of the *HIS 3; 2* gene (encoding the H1 histone) in single, detached border cells of pea seedlings. The subsequent reports concerning the application of the *in situ* RT-PCR technique to plant material has included several different plants, tissues, and genes [15].

IS-RT-PCR can be further divided into two types, either direct or indirect, based on whether the label is incorporated into the actual PCR product (direct signal detection) [17] or the PCR product is subsequently detected by hybridisation with a labelled probe (indirect detection) [18]. During the direct *in situ* RT-PCR procedure, digoxygenin (biotin or fluorescein) labelled nucleotides [17, 19] or primers [20] are incorporated into the PCR product, leading to a direct signal detection. In contrast, the indirect signal detection for *in situ* RT-PCR occurs when the PCR product is subsequently visualised by hybridisation with a specifically labelled probe [18]. The direct *in situ* RT-PCR technique can be a rapid alternative to the indirect technique because it avoids the subsequent *in situ* hybridisation step [15].

The combination of these two methods, called *in situ* PCR, which was first described by Haase et al. [21], is a highly sensitive technique that is used to localise a single gene copy at the level of individual cells [6, 22].

Since the first successfully optimised *in situ* RT-PCR method was published in 1995 [16], a number of variations on the traditional *in situ* protocols have been reported, including whole-mount ISH (WISH), in-well *in situ* RT-PCR, and the use of vibratome-sectioned tissues [14]. Furthermore, various steps in the tissue preparation and PCR (including sequential pectinase, roteinase, and DNase digestion) have been optimised for *in situ* RT-PCR [11, 17].

In plants, *in situ* hybridisation and *in situ* RT-PCR are widely used in the expression localisation of specific genes, including MADS box and other function-specific genes in floral buds and other organs. This method is especially useful in small organs or during early developmental stages when the separation of particular parts is impossible.

In this report, we present a simplified protocol for *in situ* RT-PCR in the floral buds and leaves of *Rosa hybrida*.

2. Material and Methods

2.1. Plant Material and Tissue Preparation. The flower buds of *Rosa hybrida* (76/72) are similar to a classic class C-function mutant (flower organs: sepals-petals-petals-sepals) and were selected from an F1 population of the "Lavender Kordana" and "Vanilla Kordana" cultivars (W. Kordes' Rosenschulen Co., Germany) (according to [23]). The plants were propagated from cuttings (four cuttings per pot) under the following greenhouse conditions: temperature at 22°C/18°C (day/night) and a day length extended to 16 h by SON-T lamps (Osram, 400 W, Philips Co.), supplying $600 \, \mu mol \, m^{-2} \, s^{-1}$. For restoration of the fertility, the plants were cultivated under conditions of constant humidity at 24°C without assimilation lighting.

Fertile and sterile buds between 2–5 mm in length and mature/young leaves were fixed in PFA fixative (4% paraformaldehyde, 0.4% DMSO, 0.05 M phosphate-buffered saline [PBS, pH 7.0], and DEPC-treated water) or in 4% FAA (4% formaldehyde, 50% ethanol, and 5% glacial acetic acid) for 2 h under a slight vacuum and, subsequently, for 12–24 h at $+4^\circ$C. Next, the samples were washed twice for 30 min. in PBS, dehydrated in a graded ethanol series (30%, 50%, 70%, 80%, 95%, and 100%) for 1 h in each series at RT (room temperature) under a slight vacuum and twice in histoclear (Histochoice clearing agent, Sigma) for 30 m each time. As a last step before embedding, paraplast pellets (Rotiplast, Roth) were added to the last series of histoclear in a paraffin oven twice a day for 5–7 days at a temperature 56–58°C until the histoclear completely evaporated, and the tissue was embedded in clear paraplast (Rotiplast, Roth).

Semithin sections were prepared in a rotary microtome (Reichert Jung 2040), and the thickness of the preparations ranged between 10–22 μm. All of the preparations were placed on superfrost, RNase-, and DNase-free objective slides (Thermo Scientific MenzelGläser) and dried at 42°C for 2–4 days.

2.2. Hydration, Proteinase K, Pectinase, and DNase Treatment. Before the RT step, the slides were dewaxed in histoclear (Histochoice Clearing Agent, Sigma) twice for 10 min and hydrated in a graded ethanol series and PBS. The DNA was digested either with 10 U of DNaseI (Fermentas) in the supplied buffer with 25 mM $MgCl_2$ or with 25 mM $MnCl_2$ or in a prepared buffer containing 40 mM Tris (pH 7.9), 10 mM NaCl, 6 mM $MgCl_2$, and 10 mM $CaCl_2$ for 30 min to 8 h at 37°C.

In all cases, the DNaseI was removed by thermal heating at 70°C for 10 m and after rinsing in PBS for 2 m. Optionally, the samples were digested with pectinase (Onozuka) for 10 min at RT. Before the pectinase digestion, the slides were incubated for 2 min in pectinase buffer (0.1 M sodium acetate and 5 mM EDTA, pH 4.5) for 2 min at RT. As a last step, the samples were incubated in proteinase K buffer containing 250 mM Tris-HCl (pH 7.5) and 100 mM Na_2EDTA and were digested with proteinase K, dissolved in proteinase K buffer (1 mg/mL), for 10–60 min at RT or at 37°C. After the proteinase K digestion, the samples were rinsed in PBS and PBS plus 0.2% glycine and were postfixed in 4% PFA in PBS for 10 min. Before dehydration, the sections were rinsed in 10 mM triethanolamine (Sigma) and 0.25% acetic anhydride (Sigma) for 10 min to reduce the electrostatic binding of the probe during the PCR step; the samples were subsequently dehydrated.

2.3. Reverse Transcription and PCR Thermal Cycling. For the reverse transcription step, 50 μL containing 400 U of M-MLV revertase (Promega), the buffer supplied by Promega, 0.5 mM of each dNTP (Roche), 1 μM each of forward and reverse primers and DEPC-treated H_2O. For the amplification, the following primer pairs were used: *Rhβactin* (GenBank: AB239794) forward, 5'-TGCTCCCGCTATGTATGTTG-3', and reverse, 5'-GGACTTCTGGGCATCTGAAA-3', and the class A gene *RhFUL* (GenBank:FJ970028) forward, 5'-TCATCCTCCTTTCCCCTTTC-3', and reverse, 5'-GGACCAGTTTCCCTGTGATT-3'.

The sections were first denatured at 70°C for 5 min and were incubated with the RT reaction mix for 1 h at 42°C. Deactivation of revertase was carried out at 70°C for 10 min.

Immediately after the RT step, the PCR step was carried out in 50 μL containing 0.5 U/μL of DNA polymerase (DNA Cloning Service), Williams buffer, 0.3 mM dNTPs (Roche), 25 mM digoxygenin-11-dUTP (Roche), biotin-11-dUTP (Fermentas) or fluorescein-12-dUTP (Fermentas) and 1 μM of each primer as described above.

The PCR amplification was performed in a thermocycler (Hybaid PCR express) with a flat block under the following conditions: 30 s at 94°C followed by 10, 25, 30, and 40 cycles consisting of 30 s at 94°C for, 1 min at 65°C, 1 min at 72°C, and a final step of 72°C for 10 min. As a negative control, some DNaseI-treated sections were not reverse transcribed in the case of the sections treated with digoxygenin-11-dUTP. The PCR was performed without primers in the case of the biotinylated and fluoresceinated samples. The slides treated with Fluorescein-12-dUTP after the PCR reaction were rinsed in PBS for 2 min, dehydrated, air-dried, and enclosed in mounting medium (Sigma).

2.4. Signal Detection. After the PCR step, the samples were denatured in 100% ethanol and stored in PBS overnight at 4°C. The next day, the slides were washed twice in PBS for 30 min at RT, incubated for 1 h in blocking buffer (1% BSA in PBS) and immunoblotted. For the sections treated with Digoxygenin-11-dUTP, immunodetection was carried out with antidigoxygenin-conjugated alkaline phosphatase (Roche) dissolved 1 : 100 in blocking buffer (1% BSA in 1x PBS) for 2 hrs at RT. The samples treated with biotin-11-dUTP were incubated (1 : 20) with FITC-conjugated streptavidin (Sigma, Streptavidin from *Streptomyces avidinii*) primary antibody and, optionally, with antiavidinbiotinylated (Sigma, monoclonal antiavidin—a biotin antibody produced in mice) secondary antibody (1 : 20) for 1 h each at 37°C. After immunodetection with antidigoxygenin, the digoxigenylated samples were washed in blocking buffer and detection buffer (50 mM NaCl, 50 mM Tris-HCl, pH 9.7, and 25 mM $MgCl_2$) for 30 min each and were incubated with NBT/BCIP solution (Sigma) diluted 1 : 50 in the detection buffer (50 mM NaCl, 50 mM Tris-HCl, pH 9.7, and 25 mM $MgCl_2$) for 30 min to overnight in the dark. After immunodetection, the samples were rinsed in PBS, dehydrated, air-dried, and immersed in mounting medium (Sigma).

The results for the slides treated with fluorescein-12-dUTP and the biotin-11-dUTP-streptavidin-fluorescein system were visualised using an epifluorescence microscopy (Axioscop Zeiss) with a mercury lamp at 50 W (HBO 50/AC and a camera Axio Cam Color 412-312) under an excitation filter of 470–490 nm and under bright-field microscopy (Axioscop Zeiss) and the camera Axio Cam Color 412-312 for the digoxygenin-treated slides.

3. Results and Discussion

3.1. Tissue Fixation. The first step of the *in situ* transcript localisation involves the preparation of the samples in a way that ensures the optimal preservation of the tissue and cell structure without any deleterious effects on the stability of the RNA [6]. Among the large number of different fixatives, those that are useful for *in situ* techniques may be divided to two groups, specifically, crosslinking and precipitating fixatives. Crosslinking fixatives, such as formalin or (para) formaldehyde and precipitating fixatives, such as simple alcohols and acetone, can give excellent IS-PCR results. Precipitating fixatives are less damaging to nucleic acids but are not as capable of maintaining cellular integrity. For consistent results, the cross-linking fixative should have a neutral pH and be adequately buffered if it is not prepared fresh, the reagents should be of the highest quality, and the length of fixation should not exceed 24 h. An excellent fixative is 4% formaldehyde in a phosphate buffer at pH 7.0–7.4, prepared within 24 h of use. Prolonged fixation offers no advantages and serves only to introduce unwarranted template damage and to extend the permeabilisation steps [24, 25]. For the rose buds, we used two cross-linking fixatives based on 4% FAA and 4% PFA. The PFA was much more efficient with young flower buds (2-3 mm) and young leaves, and a 12-h fixation was sufficient for a good preservation of the tissue, but a 24-h fixation did not damage the tissue. Young organs were too sensitive for FAA, even during a 12-h fixation, which resulted in tissue damage that was noted as cytolysis and cell wall caving. Mature leaves and flower buds (5 mm) fixed with greater integrity in FAA between 12–24 h whereas PFA fixation in these organs resulted in incomplete fixation and lower signal detection. A longer proteinase K digestion (1 h) was necessary. Similar results have been reported by Johansen [8] during the fixation of sugar cane leaves. Even though clear evidence of DNA damage by many fixatives can occur (DNA fragmentation up to 20 kb during glutaraldehyde use and 8–10 kb has been observed during FAA and PFA use) [26], because of the good tissue structure, a much lower incidence of damage occurs when the tissue is stored at 4°C. These fixatives are commonly used for paraffin- or plastic resin-embedded tissue.

3.2. DNaseI Digestion and Tissue Permeabilisation. To avoid the nonspecific binding of primers to DNA, the samples were treated with DNaseI with different buffers that were supplied by the manufacturer (Fermentas) along with Mg^{2+} or Mn^{2+} ions. DNaseI activity is strictly dependent on Ca^{2+} and is activated by Mg^{2+} or Mn^{2+} ions, which cleave the DNA strand in two different ways:

(i) in the presence of Mg^{2+}, DNaseI cleaves each strand of dsDNA independently in a statistically random fashion;

(ii) in the presence of Mn^{2+}, the enzyme cleaves both DNA strands at approximately the same site, producing DNA fragments with blunt ends or with overhanging termini of only one or two nucleotides (http://www.fermentas.com/en/home). As a third buffer, we prepared a DNaseI buffer that differed in the Tris concentration (40 mM versus 100 mM from the supplier), the pH (7.9 versus 7.5 from the supplier), and the inclusion of NaCl (in our buffer, 10 mM NaCl was added to stabilise the pH, which is convenient during longer storage). In all cases, DNaseI digested the entire genomic DNA during 30 min at 37°C, which was clearly visible in the sections treated as a negative control with an omitted RT step or with the full RT-PCR procedure but lacking primers in the PCR step (Figures 1(a)–1(d)), and further digestion was unnecessary. In our case, we used 10 U of DNaseI (Fermentas) in 50 μL of the solution, which was sufficient for a 30 min digestion, but other authors recommend longer incubation periods of 3 h to overnight with different concentrations of DNaseI ranging from 4–10 U [6, 15, 17, 19].

For more effective probe penetration during classical *in situ* hybridisation or *in situ* RT-PCR, the samples are permeabilised by proteolytic or, optionally, pectolytic enzymes. In *Rosaceae*, the presence of secondary metabolites, such as polyphenols, tannins, and polysaccharides, may significantly inhibit polymerase activity (according to the work in [27]). To avoid enzyme deactivation by abundant polysaccharides, we used the optional pectinase digestion for 10 min at RT according to Urbanczyk et al. and Przybecki et al. [15, 19] and non-digested slides as a control. Our results did not show any increase in signal, whereas the pectinase was adjusted for comparison of the samples where pectinase digestion was avoided. In this case, we suggest that the polysaccharides present in the *Rosa hybrida* flower buds and leaves did not significantly block the polymerase activity during the *in situ* RT-PCR.

The most crucial and important step in successful *in situ* hybridisation or *in situ* RT-PCR is the proteolytic digestion that makes the crosslinked fixed protein matrix permeable to allow the penetration of the probe or polymerase [28]. The most popular enzymes are proteinase K, pepsin, pepsinogen, and trypsin, and each has its own optimal pH. Until the cellular organisation of DNA or RNA is fully understood, the best enzymes are those with broad substrate specificity. Optimal permeabilisation is largely determined by the type of cell or tissue and the conditions of fixation. Consequently, it is notably difficult to extrapolate these conditions for different samples and protocols, and these conditions should always be determined empirically. In our experiment, we chose proteinase K as the most appropriate proteolytic enzyme for plant tissues [6, 14, 16–18, 29] for 10 to 60 min at RT or 37°C. We noticed that successful proteinase digestion depended on the fixative that was used. PFA-fixed mature leaves and large

flower buds (5 mm) required a longer proteinase digestion (up to 60 min) at 37°C. A shorter digestion or a digestion at RT resulted in a weak hybridisation signal. Mature leaves and 5 mm long flower buds fixed in FAA provided the best results with a 30 min digestion at 37°C or 60 min at RT. A 1 h digestion at 37°C resulted in overdigestion and characteristic "bubbles" occurring, especially in the flower buds. The young leaves and small (2-3 mm) flower buds fixed in PFA provided the best digestion results after 30 min at 37°C or RT. The 1-h digestion that was performed under both temperature conditions resulted in tissue damage and the appearance of overdigestion bubbles.

3.3. RT-PCR and Signal Detection. For more than 10 years, *in situ* RT-PCR methods have been carried out as a one-step reaction based on the use of rTtH polymerase (Perkin Elmer) [19] or other one-step polymerases. In our experiment, we used M-MLV revertase in the RT step and DCS polymerase in the PCR step. For the RT step, we used specific primers rather than random oligo primers to increase the specificity of the reaction. The samples were first denatured at 70°C for 5 min. Because the last step of the RT reaction was a revertase deactivation at 70°C, which also resulted in cDNA denaturation, the PCR mix was applied to the tissue immediately after the RT reaction, and the PCR reaction was carried out. According to the literature [17, 18, 25, 28, 30–32], there are two basic strategies for labelling the amplified product. One method is to tag the amplicon during the PCR and is generally known as direct IS-PCR or IS-PCR. The direct labelling of the amplicon during the PCR can be accomplished in two ways. The reporter molecule (typically biotin, digoxygenin, or fluorescein) is either attached to a nucleotide (typically dUTP) and added to the PCR or is incorporated during the synthesis of one or both of the primers, usually at the 5' end. This method of labelling is the easiest way, but it may result in a false-positive signal [33–35]. Strategies for inhibiting this nonspecific incorporation, including hot start, 3' to 5' exonuclease-deficient DNA polymerases, and capping, have proven unsuccessful [14, 34]. Although labelled oligonucleotides may provide problems with false-positive signals as unspecific labelling or unspecific background as cytoplasm staining, most of the *in situ* PCR procedures recommend this type of labelling. The unspecific labelling may be reduced by carefully optimising the annealing temperature of the PCR and confirming the specificity of the signal by performing a parallel indirect IS-PCR. In our experiment, we performed three different labelling and signal detection methods during the direct *in situ* RT-PCR. For the labelled nucleotides, we used digoxygenin-11-dUTP, which was immunolabeled with alkaline phosphatase-bound antidigoxygenin and NBT/BCIP solution, biotin-11-dUTP, which was immunolabeled with streptavidin-fluorescein and, optionally, antiavidin biotinylated to strengthen the signal. As the most direct *in situ* RT-PCR method, we used fluorescein-12-dUTP as one of the labeled nucleotides and the same, omitted long immunolocalisation procedure. As a target gene, we chose the abundant *β-actin* that is highly expressed in all organs, including flower buds and leaves, and an MADS box-specific class A gene,

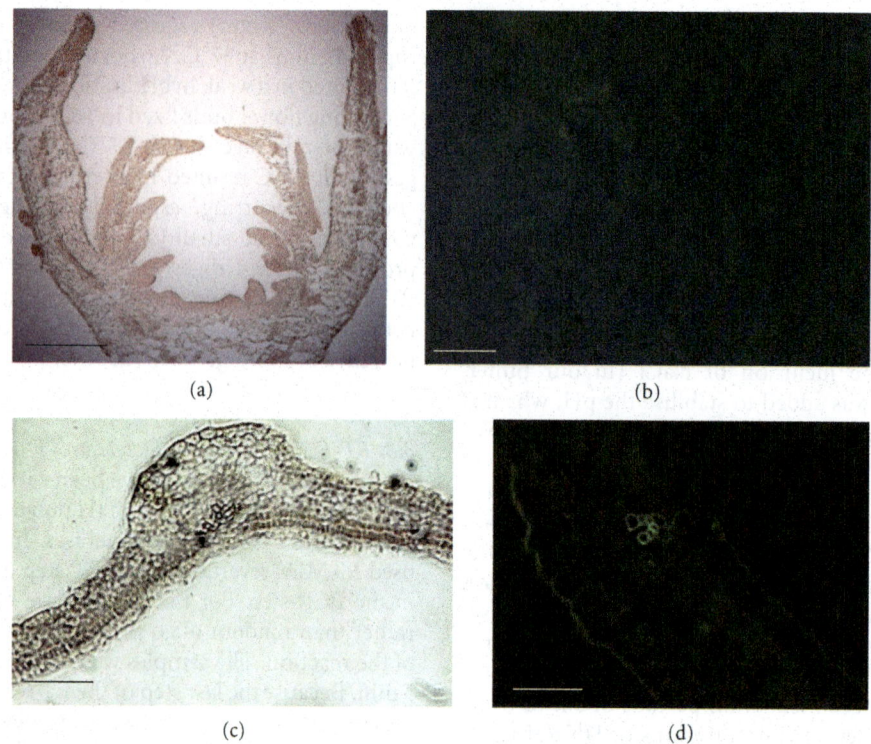

(a)

(b)

(c)

(d)

FIGURE 1: Negative control of *Rosa hybrida* flower buds and leaves (a), (c) Dnase and PCR treated sections. RT step is omitted; (b), (d) PCR-treated sections without primers. Bar = 50 μm.

which is highly expressed in leaves and only in the sepals/petals of flower buds, to confirm the specificity of the signal detection. We also used flower buds and leaves in different developmental stages to determine the intensity of the product amplification during the immunolocalisation. To achieve successful results, the most important variables during the PCR reaction are the concentrations of Mg^{2+}, a thermostable DNA polymerase, the primers, the annealing temperature, and the cycle number. During the PCR step, we strictly maintained the principle of not exceeding 40 cycles. In theory, 10 PCR cycles should generate enough signal for detection if the amplification is close to exponential [36, 37], but, in our case, even during the amplification of such an abundant gene as *β-actin* in young flower buds and leaves, 10 cycles were insufficient; we found that 25 cycles were sufficient for all of the labelling methods (Figure 2(a)–2(n)). When we used an organ-specific gene (*RhFUL* gene), a clear signal was evident after 30 cycles in the labelling with digoxygenin/fluorescein, but nonspecific binding appeared in the biotin-labelled samples even with only 10 PCR cycles. Consequently, the recommendations for the thermal and biochemical parameters for performing PCR on slides can only be described in general, and a brief inspection of the literature reveals little uniformity in the published protocols [14, 35, 38].

It has been reported that streptavidin can bind to biotin-containing proteins in tissue, resulting in nonspecific signals [16, 39], which is a situation that is not observed when digoxygenin or fluorescein are used as labels. Another point may be that during the thermal cycling, some of the proteins

may be denatured, which may also cause unspecific binding of the antibody during immunolocalisation. During signal detection, we tested antidigoxygenin AP Fab fragments (Roche antibody) for digoxygenin, streptavidin-fluorescein, and, optionally, the antiavidin-biotin system for biotin.

The results of our experiment clearly showed that the most specific binding was achieved when we used digoxygenin and fluorescein as labels. The signal was similarly strong in the tissues where *β-actin* was localised and, specifically, in the petals whereas the target (*RhFUL*) gene was localised in the flower buds (Figures 2(a), 2(b), 2(d)–2(j), 2(m) and 2(n)). The results of our investigations showed unspecific binding of streptavidin-fluorescein to all of the organs in the flower bud when the *RhFUL* gene was used (Figure 2(c)). This background was probably caused by the unspecific binding of streptavidin to the endogenous biotin present in the *Rosa hybrida* buds, although there is insufficient data about the natural biotin content in rose organs, especially in the leaves and flower buds.

Special attention should be paid to the Primed *in situ* DNA labelling (PRINS) method, which was first described by Koch [39] as the most indirect method of *in situ* RT-PCR, in which one of the oligonucleotides is fluorochrome-labelled, thereby rendering further immune detection unnecessary.

PRINS is widely used during the localisation of repetitive and telomeric sequences in plant chromosomes [40, 41], but it is not sensitive enough for the localisation of particular genes. Our results showed that direct *in situ* PCR with use of fluorescein-12-dUTP as a labelled nucleotide gives

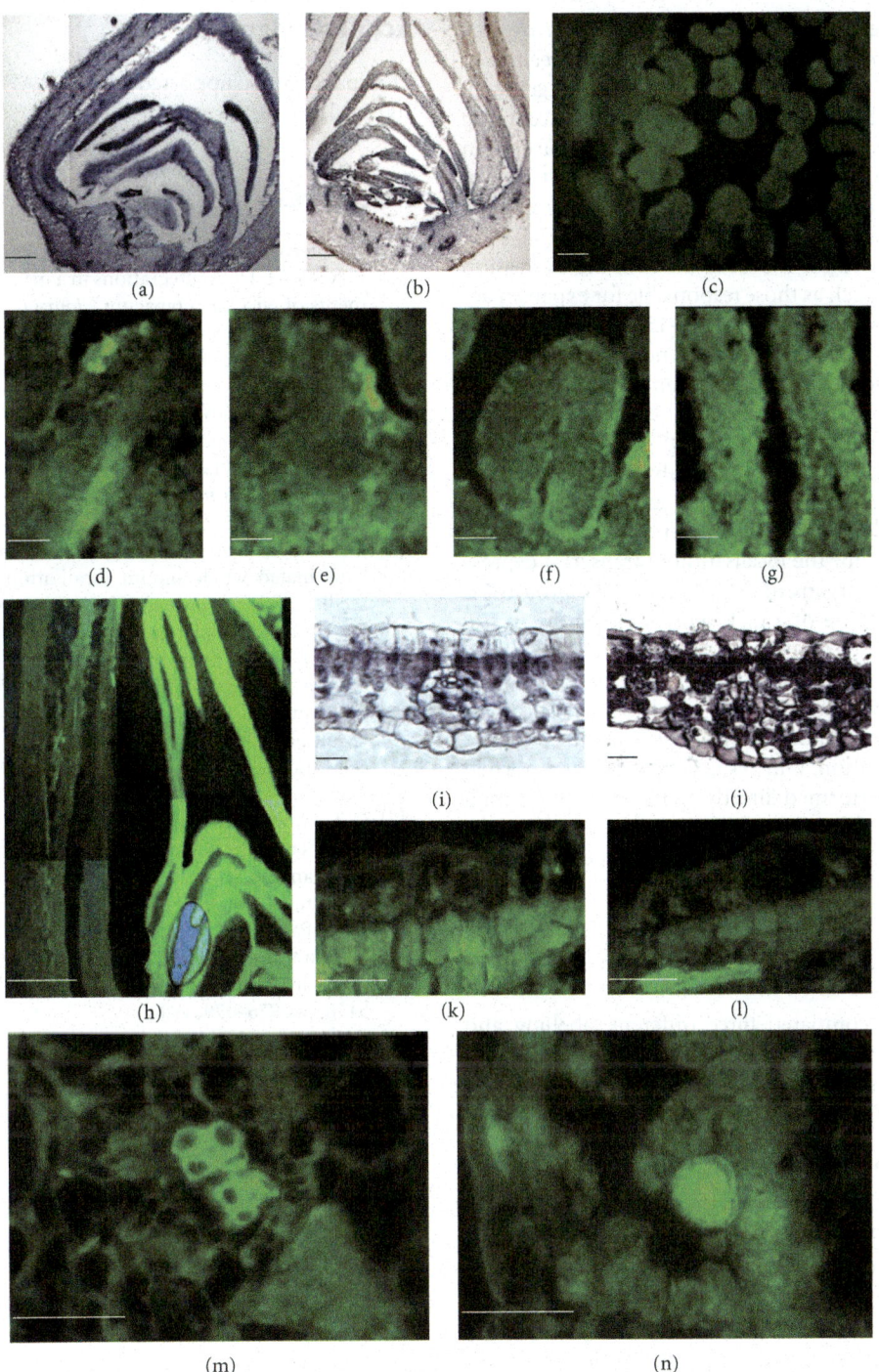

FIGURE 2: *In situ* RT-PCR on *Rosa hybrida* buds and leaves (a) digoxygenilated sections of flower bud, β-actin, (b) digoxygenilated sections of flower bud, Rh FUL gene, (c) biotinylated sections of flower bud, RhFUL gene, (d)–(g) fluoresceinated sections of flower bud, β-actin, (h) fluoresceinated sections of flower bud, RhFUL gene, (i) digoxygeninylated sections of leaf, β-actin, (j) digoxygeninylated sections, RhFUL gene, (k) biotinylated sections, β-actin, (l) biotinylated sections of leaf, RhFUL gene, (m) fluoresceinated sections of leaf, β-actin, (n) fluoresceinated sections, of leaf, RhFUL gene. Bar = 50 μm.

comparable signal strength and specificity as PCR labelled with digoxygenin and may be used to detect abundant and site-specific gene expression. Roses are one of the most economically important groups of ornamental plants, and a number of varieties have been selected based on flower traits, such as petal form, colour, and number [42, 43]. The highest level of interest by researchers and breeders is regarding flower colour and the number of petals, which is connected to MADS-box genes expression, particularly flower organs. For several years, *in situ* RT-PCR has commonly been used in the localisation of the expression of different genes in different tissues in herbaceous and woody plants. According to many literature resources, successful results have been achieved by *in situ* RT-PCR, even in the vascular tissue of such woody plants as *Populus tremula* [43]. According to the literature [14, 16, 18, 25, 28, 29, 35], the *in situ* RT-PCR method is mostly used for the localisation of the expression of abundant genes, such as those responsible for expansin activation or virus-associated genes [43–45]. In much of the available literature, specific-function genes, such as MADS-box genes, in herbaceous and woody plants are mostly localised by traditional *in situ* hybridisation (e.g., an *AGAMOUS* homolog in black spruce [46], an MADS-box family gene in Monterey pine [47], an MADS-box family gene in eucalyptus [48], a *DEFICIENS* homologue [49], and an MADS-box family gene in apple [50, 51]. In this report, we presented a convenient protocol for the localisation of transcript expression in the different organs of *Rosa hybrida*. The protocol is more appealing because of a high sensitivity for the *in situ* RT-PCR reaction and its speed. We demonstrated that a two-step reaction can be completed in two days, and a one-step reaction with fluorescein-12-dUTP used as a label can be completed in one day. Another convenience is the avoidance of the probe preparation. Our results showed that a normal RT-PCR reaction performed directly on tissue showed a high specific expression of the chosen genes, namely, the abundant and widely expressed β-actin and sepal/petal-specific *RhFUL* gene.

4. Conclusions

In this report, we compared three different labelling and immunodetection methods by using *in situ* RT-PCR in *Rosa hybrida* flower buds and leaves. As target genes, we used the abundant β-actin and *RhFUL* gene, which is expressed only in the leaves and petals/sepals of flower buds. We used digoxygenin-11-dUTP, biotin-11-dUTP, and fluorescein-12-dUTP-labelled nucleotides and antidigoxygenin-alkaline phosphatase/streptavidin-fluorescein labeled antibodies.

We conclude that 25 PCR cycles are sufficient for clear evidence of abundant gene expression and 30 cycles are sufficient for site-specific genes.

The fastest method of transcript localisation in leaves and flower buds is direct PCR with fluorescein used as one of the labelled nucleotides. The highest signal sensitivity was achieved using digoxygenin-11-dUTP or fluorescein-12-dUTP as the labelled nucleotides. The biotin-streptavidin labelling system failed because of the unspecific background associated with the localisation of *RhFUL* gene expression in the flower buds. The chosen fixatives (4% PFA and 4% FAA) confirmed the general thesis that PFA preservatives work better in young tissue. We also optimised the digestion conditions and enzyme concentrations for DNaseI and proteinase K. We proved that the optional digestion of pectins is not required to achieve clear and strong signals in rose buds and leaves during *in situ* RT-PCR.

Acknowledgment

This paper was supported by a DAAD grant in the years 2009–2011.

References

[1] C. Suter-Crazzolara, B. Brzobohaty, B. Gazdova, J. Schell, and B. Reiss, "T-DNA integrations in a new family of repetitive elements of Nicotiana tabacum," *Journal of Molecular Evolution*, vol. 41, no. 4, pp. 498–504, 1995.

[2] R. Ziegler, D. L. Engler, and N. T. Davis, "Biotin-containing proteins of the insect nervous system, a potential source of interference with immunocytochemical localization procedures," *Insect Biochemistry and Molecular Biology*, vol. 25, no. 5, pp. 569–574, 1995.

[3] JG Fournier, *Histologie Molé culaire*, Techniques et documentation Lavoisier, Paris, France, 1994.

[4] J. Friml, E. Benková, U. Mayer, K. Palme, and G. Muster, "Automated whole mount localisation techniques for plant seedlings," *Plant Journal*, vol. 34, no. 1, pp. 115–124, 2003.

[5] J. De Almeida Engler, R. De Groodt, M. Van Montagu, and G. Engler, "In situ hybridization to mRNA of Arabidopsis tissue sections," *Methods*, vol. 23, no. 4, pp. 325–334, 2001.

[6] E. Pesquet, O. Barbier, P. Ranocha, A. Jauneau, and D. Goffner, "Multiple gene detection by in situ RT-PCR in isolated plant cells and tissues," *Plant Journal*, vol. 39, no. 6, pp. 947–959, 2004.

[7] G. J. Nuovo, *PCR In Situ Hybridization. Protocols and Applications*, Raven Press, New York, NY, USA, 1992.

[8] B. Johansen, "In situ PCR on plant material with sub-cellular resolution," *Annals of Botany*, vol. 80, no. 5, pp. 697–700, 1997.

[9] B. K. Patterson, M. Till, P. Otto et al., "Detection of HIV-1 DNA and messenger RNA in individual cells by PCR- driven in situ hybridization and flow cytometry," *Science*, vol. 260, no. 5110, pp. 976–979, 1993.

[10] G. J. Nuovo, K. Lidonnici, P. MacConnell, and B. Lane, "Intracellular localization of polymerase chain reaction (PCR)-amplified hepatitis C cDNA," *American Journal of Surgical Pathology*, vol. 17, no. 7, pp. 683–690, 1993.

[11] S. Drea, J. Corsar, B. Crawford, P. Shaw, M. Dolan, and J. H. Doonan, "A streamlined method for systematic, high resolution in situ analysis of mRNA distribution in plants," *Plant Methods*, vol. 1, article 8, 2005.

[12] G. Morel, M. Berger, B. Ronsin et al., "In situ reverse transcription-polymerase chain reaction. Applications for light and electron microscopy," *Biology of the Cell*, vol. 90, no. 2, pp. 137–154, 1998.

[13] I. M. Van Aarle, G. Viennois, L. K. Amenc, M. V. Tatry, D. T. Luu, and C. Plassard, "Fluorescent in situ RT-PCR to visualise the expression of a phosphate transporter gene from an ectomycorrhizal fungus," *Mycorrhiza*, vol. 17, no. 6, pp. 487–494, 2007.

[14] G. J. Nuovo, "Co-labeling using in situ PCR: a review," *Journal of Histochemistry and Cytochemistry*, vol. 49, no. 11, pp. 1329–1339, 2001.

[15] Z. Przybecki, E. Siedlecka, M. Filipecki, and E. Urbanczyk-Wochniak, "In situ reverse transcription PCR on plant tissues," in *PRINS and In Situ PCR Protocols*, F. Pellestor, Ed., Humana Press, Tototwa, NJ, USA, 2006.

[16] H. H. Woo, L. A. Brigham, and M. C. Hawes, "In-cell RT-PCR in a single, detached plant cell," *Plant Molecular Biology Reporter*, vol. 13, no. 4, pp. 355–362, 1995.

[17] H. Koltai and D. McKenzie Bird, "High throughput cellular localization of specific plant mRNAs by liquid-phase in situ reverse transcription-polymerase chain reaction of tissue sections," *Plant Physiology*, vol. 123, no. 4, pp. 1203–1212, 2000.

[18] M. Steinhoff, H. Hesse, B. Göke, A. Steinhoff, R. Eissele, and E. P. Slater, "Indirect RT-PCR in-situ hybridization: a novel non-radioactive method for detecting glucose-dependent insulinotropic peptide," *Regulatory Peptides*, vol. 97, no. 2-3, pp. 187–194, 2001.

[19] E. Urbańczyk-Wochniak, M. Filipecki, and Z. Przybecki, "A useful protocol for in situ RT-PCR on plant tissues," *Cellular and Molecular Biology Letters*, vol. 7, no. 1, pp. 7–18, 2002.

[20] D. Bettinger, C. Mougin, B. Fouqué, B. Kantelip, J. P. Miguet, and M. Lab, "Direct in situ reverse transcriptase-linked polymerase chain reaction with biotinylated primers for the detection of hepatitis C virus RNA in liver biopsies," *Journal of Clinical Virology*, vol. 12, no. 3, pp. 233–241, 1999.

[21] A. T. Haase, E. F. Retzel, and K. A. Staskus, "Amplification and detection of lentiviral DNA inside cells," *Proceedings of the National Academy of Sciences of the United States of America*, vol. 87, no. 13, pp. 4971–4975, 1990.

[22] Y. Hishikawa, S. An, T. Yamamoto-Fukuda, Y. Shibata, and T. Koji, "Improvement of in situ PCR by optimization of PCR cycle number and proteinase K concentration: Localization of X chromosome-linked phosphoglycerate kinase-1 gene in mouse reproductive organs," *Acta Histochemica et Cytochemica*, vol. 42, no. 2, pp. 15–21, 2009.

[23] N. Ahmadi, H. Mibus, and M. Serek, "Characterization of ethylene-induced organ abscission in F1 breeding lines of miniature roses (Rosa hybrida L.)," *Postharvest Biology and Technology*, vol. 52, no. 3, pp. 260–266, 2009.

[24] Y. Tokuda, T. Nakamura, K. Satonaka et al., "Fundamental study on the mechanism of DNA degradation in tissues fixed in formaldehyde," *Journal of Clinical Pathology*, vol. 43, no. 9, pp. 748–751, 1990.

[25] J. R. Hully, "In situ PCR," in *PCR Applications. Protocol for Functional Genomics*, M. A. Innis, D. H. Gelfand, and J. J. Sninsky, Eds., pp. 169–194, Academic Press, London, UK, 1999.

[26] M. P. Douglas and S. O. Rogers, "DNA damage caused by common cytological fixatives," *Mutation Research*, vol. 401, no. 1-2, pp. 77–88, 1998.

[27] S. Porebski, L. G. Bailey, and B. R. Baum, "Modification of a CTAB DNA extraction protocol for plants containing high polysaccharide and polyphenol components," *Plant Molecular Biology Reporter*, vol. 15, no. 1, pp. 8–15, 1997.

[28] IA Darby and TD Hewitson, *In Situ Hybridization Protocols*, Humana Press, New York, NY, USA, 2010.

[29] R. H. Chen and S. V. Fuggle, "In situ cDNA polymerase chain reaction: a novel technique for detecting mRNA expression," *American Journal of Pathology*, vol. 143, no. 6, pp. 1527–1534, 1993.

[30] M. Clark, *In Situ Hybridization: Laboratory Companion*, Chapman and Hall, London, UK, 1996.

[31] D. G. Wilkinson, *In Situ Hybridization: A Practical Approach*, IRL Press, Oxford, UK, 1992.

[32] J. M. Polak and J. McGee, *In Situ Hybridization: Principles and Practice*, Oxford University Press, Oxford, UK, 1999.

[33] P. Komminoth, V. Adams, A. A. Long et al., "Evaluation of methods for hepatitis C virus detection in archival liver biopsies. Comparison of histology, immunohistochemistry, in-situ hybridization, reverse transcriptase polymerase chain reaction (RT-PCR) and in-situ RT-PCR," *Pathology Research and Practice*, vol. 190, no. 11, pp. 1017–1025, 1994.

[34] J. F. Sallstrom, I. Zehbe, M. Alemi, and E. Wilander, "Pitfalls of in situ polymerase chain reaction (PCR) using direct incorporation of labelled nucleotides," *Anticancer Research*, vol. 13, no. 4, pp. 1153–1154, 1993.

[35] I. A. Teo and S. Shaunak, "PCR in situ: aspects which reduce amplification and generate false-positive results," *Histochemical Journal*, vol. 27, no. 9, pp. 660–669, 1995.

[36] B. W. Heniford, A. Shum-Siu, M. Leonberger, and F. J. Hendler, "Variation in cellular EGF receptor mRNA expression demonstrated by in situ reverse transcriptase polymerase chain reaction," *Nucleic Acids Research*, vol. 21, no. 14, pp. 3159–3166, 1993.

[37] S. A. Kovalenko, P. J. Harms, M. Tanaka et al., "Method for in situ investigation of mitochondrial DNA deletions," *Human Mutation*, vol. 10, no. 6, pp. 489–495, 1997.

[38] J. J. O'Leary, R. Chetty, A. K. Graham, and J. O. McGee, "In situ PCR: pathologist's dream or nightmare?" *Journal of Pathology*, vol. 178, no. 1, pp. 11–20, 1996.

[39] J. E. Koch, S. Kolvraa, K. B. Petersen, N. Gregersen, and L. Bolund, "Oligonucleotide-priming methods for the chromosome-specific labelling of alpha satellite DNA in situ," *Chromosoma*, vol. 98, no. 4, pp. 259–265, 1989.

[40] M. Kubaláková, J. Macas, and J. Doležel, "Mapping of repeated DNA sequences in plant chromosomes by PRINS and C-PRINS," *Theoretical and Applied Genetics*, vol. 94, no. 6-7, pp. 758–763, 1997.

[41] M. Menke, J. Fuchs, and I. Schubert, "A comparison of sequence resolution on plant chromosomes: PRINS versus FISH," *Theoretical and Applied Genetics*, vol. 97, no. 8, pp. 1314–1320, 1998.

[42] A. Dubois, O. Raymond, M. Maene et al., "Tinkering with the C-function: a molecular frame for the selection of double flowers in cultivated roses," *PLoS One*, vol. 5, no. 2, Article ID e9288, 2010.

[43] Y. Hibino, K. Kitahara, S. Hirai, and S. Matsumoto, "Structural and functional analysis of rose class B MADS-box genes 'MASAKO BP, euB3, and B3': Paleo-type AP3 homologue 'MASAKO B3' association with petal development," *Plant Science*, vol. 170, no. 4, pp. 778–785, 2006.

[44] M. Gray-Mitsumune, E. J. Mellerowicz, H. Abe et al., "Expansins abundant in secondary xylem belong to subgroup A of the α-expansin gene family," *Plant Physiology*, vol. 135, no. 3, pp. 1552–1564, 2004.

[45] C. Silva, S. Tereso, G. Nolasco, and M. M. Oliveira, "Cellular location of Prune dwarf virus in almond sections by in situ reverse transcription-polymerase chain reaction," *Phytopathology*, vol. 93, no. 3, pp. 278–285, 2003.

[46] R. Rutledge, S. Regan, O. Nicolas et al., "Characterization of an AGAMOUS homologue from the conifer black spruce (Picea mariana) that produces floral homeotic conversions when expressed in Arabidopsis," *Plant Journal*, vol. 15, no. 5, pp. 625–634, 1998.

[47] A. Mouradov, T. V. Glassick, B. A. Hamdorf et al., "Family of MADS-box genes expressed early in male and female reproductive structures of monterey pine," *Plant Physiology*, vol. 117, no. 1, pp. 55–61, 1998.

[48] S. G. Southerton, H. Marshall, A. Mouradov, and R. D. Teasdale, "Eucalypt MADS-box genes expressed in developing flowers," *Plant Physiology*, vol. 118, no. 2, pp. 365–372, 1998.

[49] J. E. Sallström, "Nonspecific amplification in in situ PCR by directincorporation of reporter molecules," *Cell Vision*, vol. 1, pp. 243–251, 1994.

[50] S. K. Sung, G. H. Yu, and G. An, "Characterization of MdMADS2, a member of the SQUAMOSA subfamily of genes, in apple," *Plant Physiology*, vol. 120, no. 4, pp. 969–978, 1999.

[51] S. K. Sung, G. H. Yu, J. Nam, D. H. Jeong, and G. An, "Developmentally regulated expression of two MADS-box genes, MdMADS3 and MdMADS4, in the morphogenesis of flower buds and fruits in apple," *Planta*, vol. 210, no. 4, pp. 519–528, 2000.

mRNA Expression of EgCHI1, EgCHI2, and EgCHI3 in Oil Palm Leaves (*Elaeis guineesis* Jacq.) after Treatment with *Ganoderma boninense* Pat. and *Trichoderma harzianum* Rifai

Laila Naher,[1] Soon Guan Tan,[2] Chai Ling Ho,[2,3] Umi Kalsom Yusuf,[1] Siti Hazar Ahmad,[4] and Faridah Abdullah[1]

[1] *Department of Biology, Faculty of Science, Universiti Putra Malaysia, Selangor, 43400 Serdang, Malaysia*
[2] *Department of Cell and Molecular Biology, Faculty of Biotechnology and Biomolecular Sciences, Universiti Putra Malaysia, Selangor, 43400 Serdang, Malaysia*
[3] *Institute of Tropical Agriculture, Universiti Putra Malaysia, Selangor, 43400 Serdang, Malaysia*
[4] *Department of Crop Science, Faculty of Agriculture, Universiti Putra Malaysia, Selangor, 43400 Serdang, Malaysia*

Correspondence should be addressed to Laila Naher, lailanaherupm@gmail.com

Academic Editors: P. Andrade and D. Neureiter

Background. Basal stem rot (BSR) disease caused by the fungus *Ganoderma boninense* is the most serious disease affecting the oil palm; this is because the disease escapes the early disease detection. The biocontrol agent *Trichoderma harzianum* can protect the disease only at the early stage of the disease. In the present study, the expression levels of three oil palm (*Elaeis guineensis* Jacq.) chitinases encoding EgCHI1, EgCHI2, and EgCHI3 at 2, 5, and 8 weeks inoculation were measured in oil palm leaves from plants treated with *G. boninense* or *T. harzianum* alone or both. *Methods.* The five-month-old oil palm seedlings were treated with Ganowood blocks inoculum and trichomulch. Expression of EgCHI1, EgCHI2, and EgCHI3 in treated leaves tissue was determined by real-time PCR. *Results.* Oil palm chitinases were not strongly expressed in oil palm leaves of plants treated with *G. boninense* alone compared to other treatments. Throughout the 8-week experiment, expression of EgCHI1 increased more than 3-fold in leaves of plants treated with *T. harzianum* and *G. boninense* when compared to those of control and other treated plants. *Conclusion.* The data illustrated that chitinase cDNA expression varied depending on tissue and the type of treatment.

1. Introduction

The oil palm (*Elaeis guineensis* Jacq.) is an important economic crop that produces two types of oils: palm oil from the fibrous mesocarp and kernel oil from the seeds. Currently, the oil palm industry is under threat from a fungal disease called basal stem rot (BSR), which is caused by the fungus *Ganoderma boninense* Pat. [1, 2]. Other fungal diseases, such as vascular wilt (caused by *Fusarium oxysporum* f.sp. *Elaeidis*) and sudden wilt (caused by *Phytomonas staheli* McGhee) [1], also affected the oil palm but BSR is by far the most serious among them; it causes tree loss in palm stands and subsequent loss in yield of palm oil [1, 3]. The disease escapes early detection: by the time fruiting bodies are detected, the disease is too advanced to response to any chemical treatments.

The use of the fungus *Trichoderma* spp. as a biocontrol agent for controlling plant disease was first recognized in the early 1930s [4]. Subsequently, many studies have shown that *Trichoderma* spp. are the most effective biocontrol agents for managing plant disease. *Trichoderma* controls the pathogen via a mycoparasitism process in which it grows towards the pathogenic fungi, coils around them, and secretes cell wall degrading enzymes that limit their growth [5]. An *in vitro* study showed that *Trichoderma* produced trichodermin and antimycotin, which are compounds that inhibited the growth of *Rhizoctonia solani* [6]. Harman et al. [7] proposed a mechanism of disease control that involves the release of cell wall degrading enzymes from *Trichoderma* which activates the expression of genes involved in the plant defence system. *Trichoderma* sp. has been proven to be highly effective for controlling *Ganoderma boninense*/BSR disease in oil palms but only at the early stage of slightly infected palms [8–10]. Therefore, to date, there is no adequate control measure to control BSR improvements of the oil palm defence system against *G. boninense* is the alternative option.

Plants use various defence mechanisms during plant-microbe interactions, including the strengthening of physical barriers (e.g., lignin and cellulose), synthesis of antimicrobial compounds (phytoalexins), and synthesis of pathogenesis-related (PR) proteins. Chitinases are PR proteins that hydrolize the β-1, 4 glycosidic bond in chitin, which is found in most fungal cell walls and is a common constituent of insect cuticles and crustacean shells [11, 12]. Moreover, the breakdown products of chitin may serve as elicitors of the plant defence reaction [13]. Chitinase also expressed at low levels under normal conditions during plant developments [14]. Thus, the physiological expression of chitinases in plants can be both constitutive and induced by biotic or abiotic stresses or induced in pathogen infection [15–18].

The plant chitinases are divided into seven classes (I through VII) based on their structural properties and amino acid sequence similarities [19]. The current view is that not all chitinases are induced in response to pathogen attack: instead, only specific chitinases are stimulated by a particular pathogen. For example, class I chitinase from tobacco showed antifungal activity against *Fusarium solani* germlings, whereas class II chitinases showed only slight antifungal activity when used with β-1,3 glucanase [20]. In Norway spruce (*Picea abies* L. Karst.) plants (clones 409 and 589) when expression of chitinase classes I, II, and IV was monitored after wounding and infection by the fungus *Heterobasidion annosum* Fr., maximum transcript levels for classes II and IV were found in both clones compared to class I [21]. Apart from their role in pathogen defence, chitinases also have a role in symbiotic or biocontrol agent interaction in plant. Our previous study showed that chitinases expression was high in oil palm root tissues when *G. boninense* infection first appeared in root tissues but the expressions declined during the development of the disease while in *T. harzianum* alone or together with *G. boninense* treated oil palm plants; chitinases expression remained upregulated at the end of the experiment in oil palm root tissues [22]. However, some chitinases are developmentally regulated or induced by specific organs. Thus, the purpose of this present study was to investigate the chitinases expression in oil palm leaf tissues treated with the pathogen *G. boninense* Pat. and the biocontrol agent *T. harzianum* Rifai either alone or in combination.

In this study, three oil palm chitinase cDNAs, previously isolated from oil palm encoding EgCHI1 (GenBank accession number ADC55619) which matched with plant chitinase, chitinase class I from *Arabidopsis thaliana* (AAF29391.1), EgCHI2 (HQ831445) which matched with plant chitinase, chitinase class II from *Fragaria* x *Ananassa* (AAF00131.1), and EgCHI3 (HQ831446) which matched with plant chitinase chitinase class III from *Bambusa Oldham* (ABW75909.1) (Naher et al. [22]), were used to investigate the expression levels of chitinases in the leaves of oil palms artificially inoculated with *G. boninense*. Whether the

presence of the BSR biocontrol agent *T. harzianum* affected chitinase expression was also evaluated.

2. Materials and Methods

2.1. Preparation of Plant Treatment Materials. The cultures of seven-day-old *G. boninense* and *T. harzianum* were used for prepared Gano-wood blocks and trichomulch, respectively. Freshly cut rubber wood blocks were used carrier for Gano-wood blocks and palm-pressed mesocarp fibers were used carrier for trichomulch. The preparation of plant treatment materials consisting of Gano-wood blocks and Tricho-mulch have been described previously [22].

2.2. Plant Treatments. The experiments were conducted in a glass greenhouse over 8-week period. The 5-month-old oil palm seedlings used in the experiment were provided by Sime Darby Seeds & Agricultural Services Sdn Bhd (Banting, Selangor, Malaysia). Each of four treatments (control, *G. boninense* Pat., *T. harzianum* Rifai, and *G. boninense* + *T. harzianum*) was replicated three times.

The control treatment consisted of an oil palm plant in a garden pot. The artificial inoculation of oil palms with *G. boninense* Pat. followed by Naher et al. [22]. Briefly, a Gano-wood block was placed in direct contact with the roots of a plant in a garden pot and then covered with soil. For *Trichoderma*-inoculated treatments, 600 g of Trichomulch were placed on the surface of the soil. Plants in the *G. boninense* + *T. harzianum* group were treated with both a Gano-wood block and Trichomulch. The seedlings were watered twice daily using tap water.

In this study, the gene expressions of oil palm chitinases at the early stage of the plant-microbe interaction were investigated. *Ganoderma* is a slow-growing fungus that requires more than 1 week to develop mycelia on the root surface. Thus, the first samples were collected at 2 and the plants then sampled again at 5 and 8. Control and treated leaves were excised using a clean scissors, dried with paper towels, and then weighed. Then, they were wrapped in aluminium foil (1 g/pack) for RNA extraction. The leaf tissues were frozen immediately in liquid nitrogen and stored at $-80°C$.

2.3. RNA Extraction. Total RNA was extracted from treated and untreated leaf tissues using a modified cetyl trimethyl ammonium bromide (CTAB) method [23]. RNA extraction from oil palm was previously described in detail [22]. Briefly, 1 g of tissue was ground in liquid nitrogen into very fine powder which was immediately transferred to a 50 mL polypropylene tube containing 15 mL of CTAB extraction buffer. Next, an equal volume of chloroform : isoamyl alcohol (C : I) (24 : 1) was added to the tube and then centrifuged at 12, 857 g for 15 min at 4°C. The upper layer was carefully transferred to a new 50 mL polypropylene and 15 mL of phenol:chloroform:isoamyl alcohol (P :C: I) (25 : 24 : 1) was added. Centrifugation was performed using the conditions as described above. The final supernatant was adjusted to a final concentration of 2 M LiCl for incubation at 4°C overnight.

mRNA Expression of EgCHI1, EgCHI2, and EgCHI3 in Oil Palm Leaves (Elaeis guineesis Jacq.) after Treatment with
Ganoderma boninense Pat. and Trichoderma harzianum Rifai

65

After overnight incubation, the homogenate was centrifuged at 12,857 g for 30 min at 4°C. The pellet was dissolved in 5 mL of diethylpyrocarbonate-(DEPC) treated water, and then an equal volume of C : I was added and centrifuged at 12,857 g for 15 min at 4°C. The supernatant was transferred to a new tube and the RNA was precipitated by adding 0.1 volume of 3 M sodium acetate pH 5.2 and 2.5 volumes of 100% ethanol, followed by incubation at −80°C overnight. After centrifugation, the resulting pellet was washed with 70% (v/v) ethanol. The pellet was air dried and resuspended in DEPC-treated water. The RNA purity was examined using a spectrophotometer at 230, 260, and 280 nm and integrity of RNA was examined using 1% denaturing formaldehyde agarose gel electrophoresis [24]. Then RNA was treated with DNase I (Qiagen, USA) according to manufacturer's instructions.

2.4. Designing of cDNA Primers. To measure the expression of chitinases in oil palm, the primers were designed using Primer 3 software version 0.4.0 based on the 3′ untranslated region (UTR) of oil palm chitinase cDNAs as already isolated in our previous study [22]. The following primers were used for real-time RT : PCR : EgCHI1-F, 5′-GCT GTC CAT CAA TTG GAT CCT C-3′ and EgCHI1-R, 5′-CTT TAC TGG CGT GGT TCG AGT-3′; EgCHI2-F, 5′-TCG GAA TTT TTG GTC CTT TTT-3′ and EgCHI2-R, 5′-GTT TAG GGC TTG ATC AGC- 3′; and EgCHI3-F, 5′-TGTCATATCATCTCCAGT-TCCAG-3′ and EgCHI3-R, 5′–GAG TTT GTA CGG TTG CCC CTG-3′; actin-F, 5′-CCC ACC TGA ACG GAA ATA CA-3′ and actin-R, 5′-CGG ATG GCA CCT CAG TCT TA-3′. The actin gene (Genbank accession number EL691466) was used as an endogenous control.

2.5. cDNA Translation and Reverse Transcriptase (RT-)PCR. To conduct the chitinase expression analysis, total RNA was translated into cDNA. Equal amounts of DNase-treated RNA (1 μg) of control and treated samples were converted into cDNA using the quantitative reverse transcript cDNA synthesis kit following the manufacturer's instructions (Qiagen, USA). Briefly, 1 μg of total RNA and 2 μL 7X gDNA wipe buffer (provided in the kit) were transferred into a clean PCR tube, followed by the addition of DEPC-treated water to a total volume of 14 μL. The mixture was then incubated at 42°C for 2 min and chilled on ice quickly. The remaining components of the kit were added to the reaction mixture, followed by 4 μL of 5X quantiscript RT buffer, 1 μL of RT primer mix, and 1 μL of reverse transcriptase enzyme; the mixture was incubated at 42°C for 30 min. Finally, the reaction was heated at 95°C for 3 min to terminate the cDNA synthesis reaction, and the cDNA was stored at −20°C.

2.6. Real-Time RT-PCR. Real-time RT-PCR was performed using the Bio-Rad iQ5 real-time PCR system (Bio-Rad, USA). Equal amounts of RNA (1 μg) extracted from control and treated oil palm leaves samples were converted into cDNA by using the quantitative reverse transcript cDNA synthesis kit (Qiagen, USA) following the manufacturer's instructions. Real-time RT-PCR was performed on EgCHI1,

EgCHI2, and EgCHI3 together with actin in three replicates in one 96-well plate and PCR conditions were as follows: 1 cycle of 95°C for 10 min followed by 40 cycles of 95°C for 30 s, 60°C for 1 min, and 72°C for 1 min. The annealing temperature for all targets and the endogenous control was 60°C.

2.7. Real-Time PCR Analysis. Real-time PCR was used to analyze the mRNA expression level of each transcript encoding EgCHI1, EgCHI2, and EgCHI3 in oil palms leaves in interaction with *G. boninense* Pat. and *T. harzianum* Rifai. The relative expression of each transcript was calculated by the $\Delta\Delta C_T$ method [25] using iQ5 software (Bio-Rad); the expression levels of EgCHI1, EgCHI2, and EgCHI3 were estimated after being normalized to the endogenous control gene and the significant expression levels were considered if the standard error ≤0.5.

3. Results and Discussion

BSR which is caused by the fungus *G. boninense* Pat. is a serious disease that affects the oil palm and is a major threat to the oil palm industry. To date, there is no adequate measure to control this disease, and researchers are looking for ways to improve the oil palm's defence system against *G. boninense* Pat. Plant chitinases are PR proteins that belong to the repertoire of plant defence mechanisms that are believed to constitute the early defence response in plants. Generally, chitinase induction is considered to be part of the nonspecific defence reaction initiated in a plant after pathogen attack or exposure to physical, chemical, or environmental stresses [26]. Thus, the plant chitinases may be involved in the oil palm's reaction to infection by *G. boninense* Pat.

The goal of this study was to investigate the potential role of chitinase mRNA expression in oil palms infected by *G. boninense* Pat. as well as in samples treated by *T. harzianum*, which a biocontrol agent is used to combat BSR disease. Prior to running real-time RT-PCR for the expression study, primers of EgCHI1, EgCHI2, and EgCHI3, and an endogenous control (actin) were optimized for annealing temperature. The annealing temperature of all of the primers optimized at 60°C. Afterwards, real-time PCR was performed for the target transcripts. The PCR efficiencies of all targets and the endogenous control were approximately equal (91-92%).

Figures 1(a), 1(b), and 1(c) show the relative expression levels of EgCHI1, EgCHI2, and EgCHI3, respectively, in leaves in response to inoculation with *G. boninense* Pat. and *T. harzianum* Rifai alone or in combination at different time points compared with that of the control plants. In *G. boninense* alone treated plants, no significant upregulation (SE > 0.5) in expression of any of the transcripts was observed at any time points. It was reported from previous study that the pathogenesis-related (PR) protein chitinase was elicited in plants during early response to the pathogen attack [27, 28]. However, none of the oil palm chitinases studied was strongly induced against *G. boninense*. It could

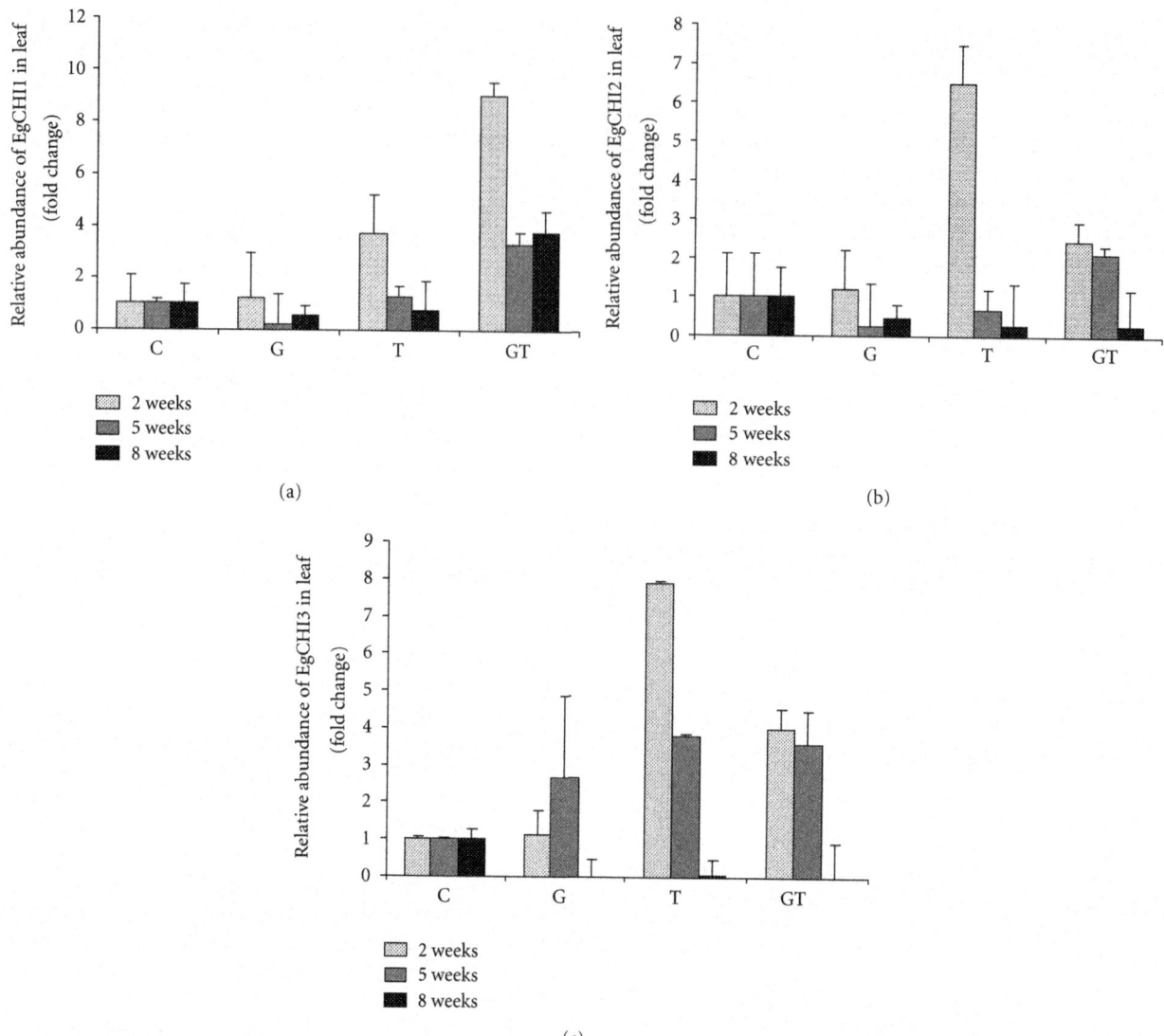

FIGURE 1: Relatives abundances of EgCHI1, EgCHI2, and EgCHI3 at various time points in oil palm leaf tissues, inoculated with *G. boninense* and *T. harzianum* either alone or together. To compare the levels of the transcripts among the treatments, the value of the control plant transcripts was set at 1 and the data for the treatments were then normalized to this value. Error bars indicate standard errors. The expression level was considered significant if the standard errors ≤ 0.5. C = Control, G = *Ganoderma*, T = *Trichoderma*, GT = *Ganoderma* + *Trichoderma*.

be that the earliest time point of this study was too late to reflect gene expression against the pathogen or that oil palm chitinases might serve as a local or as an organ-specific defence mechanism, as the chitinase expression results from oil palm leaves found in this study differed from those detected in root tissues of plants treated in the same way [22]. This result was in contrast to that of a previous study [22] in which at 5 weeks after the disease was first observed only in root tissue, all the transcripts were upregulated.

In the plants treated with *T. harzianum* alone, significant upregulation was detected only for expression of EgCHI3: a 7.9-fold and 3.8-fold increased at 2 and 5 wpi (Figure 1(c)), respectively. In the *T. harzianum* + *G. boninense* treatment, the expression of EgCHI1 (Figure 1(a)) was dramatically increased (9.03-fold, 3.3-fold and 3.8-fold at 2, 5, and 8 wpi, resp.) and that of EgCHI2 (Figure 1(b)) was also

up-regulated (2.44-fold and 2.1-fold at 5 and 8 wpi, resp.). In the same treatment, the EgCHI3 expression (Figure 1(c)) was up-regulated (4-fold and 3.57-fold) at 2 wpi and 5 wpi, respectively, and at week 8 the EgCHI3 expression was very low in the presence of all fungi-treated oil palm plants. A similar result was observed in grapevine (*Vitis vinifera* L.) infected with *Plasmopara viticola*: chitinase class III was expressed only twice in the early stages (days 2 and 6) of a 10-day experiment, but in healthy leaves, the expression was low at day 2 and increased at the late stage of 6 to 8 days [29]. In another study, chitinase class III was not expressed in the leaves of *Vitis vinifera* infected by the pathogen *Botrytis cinerea*, but class I was expressed; moreover, in the control plant both classes were constitutively expressed [30]. These authors suggested that the timing of sporulation of the fungi and the nature of the infection were reflected in the plant's

mRNA Expression of EgCHI1, EgCHI2, and EgCHI3 in Oil Palm Leaves (Elaeis guineesis Jacq.) after Treatment with
Ganoderma boninense Pat. and Trichoderma harzianum Rifai

67

gene expression. Overall, the expression data showed that oil palm chitinases expression was higher when plants were treated with *Trichoderma* and it might be that *Trichoderma* helped to induce chitinases in plants. Shores and Harman [31] found that chitinase activity was higher in plant from *Trichoderma*-treated seeds than from untreated seeds. Hence, they suggested that *Trichoderma*-treated plants expressed chitinase as defence mechanism to be more resistance to disease.

In conclusion, chitinase expression was not significantly increased in *Ganoderma*-alone-treated plants. In terms of susceptibility, plant chitinases that are expressed at low levels under normal conditions may not be strong enough to ward off fungal pathogen growth [32, 33]. Since not all classes of chitinases were investigated in this study, the roles of other chitinase classes in oil palm during pathogen attack deserved further attention. Our data also illustrated that the presence of *Trichoderma* might be involved in inducing chitinase expression especially EgCHI1 in oil palms to enhance defence mechanism.

Acknowledgment

The authors would like to thank the Ministry of Higher Education, Malaysia, for funding the paper via FRGS Grant number 01-01-07-138FR.

References

[1] D. Ariffin, "Major diseases of oil palm," in *Advances in Oil Palm Research*, B. Yusof, S. Jalani, and K.W. Chan, Eds., vol. 1, pp. 596–622, Malaysia Palm Oil Board, Bangi, Malaysia, 2000.

[2] J. R. Al-Obaidi, Y. Mohd-Yusuf, T. Chin-Chong, N. Mhd-Noh, and R. Y. Othman, "Identification of a partial oil palm polygalacturonase-inhibiting protein (EgPGIP) gene and its expression during basal stem rot infection caused by *Ganoderma boninense*," *African Journal of Biotechnology*, vol. 9, no. 46, pp. 7788–7797, 2010.

[3] M. S. Mazliham, L. Pierre, and A.S. Idris, "Towards automatic recognition and grading of *Ganoderma* infection pattern using fuzzy systems," *World Academy of Science Engineering and Technology*, vol. 25, pp. 51–56, 2007.

[4] R. Weindling, "Studies on lethal principle effective in the parasitic action of *Trichoderma lignorum* on *Rhizoctinia solani* and other soil fungi," *Phytopathology*, vol. 24, pp. 1153–1179, 1934.

[5] F. Vinale, K. Sivasithamparam, E. L. Ghisalberti et al., "A novel role for Trichoderma secondary metabolites in the interactions with plants," *Physiological and Molecular Plant Pathology*, vol. 72, no. 1–3, pp. 80–86, 2008.

[6] B. L. Bertagnolli, S. Daly, and J. B. Sinclair, "Antimycotic compounds from the plant pathogen *Rhizoctonia solani* and its antagonist *Trichoderma harzianum*," *Journal of Phytopathology*, vol. 146, no. 2-3, pp. 131–135, 1998.

[7] G. E. Harman, C. R. Howell, A. Viterbo, I. Chet, and M. Lorito, "Trichoderma species—opportunistic, avirulent plant symbionts," *Nature Reviews Microbiology*, vol. 2, no. 1, pp. 43–56, 2004.

[8] M. Sariah, C. W. Choo, H. Zakaria, and M. S. Norihan, "Quantification and characterisation of *Trichoderma* spp. from different ecosystems," *Mycopathologia*, vol. 159, no. 1, pp. 113–117, 2005.

[9] A. Susanto, P. S. Sudharto, and R. Y. Purba, "Enhancing biological control of basal stem rot disease (*Ganoderma boninense*) in oil palm plantations," *Mycopathologia*, vol. 159, no. 1, pp. 153–157, 2005.

[10] G. N. M. Ilias, *Trichoderma and its efficacy as a bio-control agent of basal stem rot of oil Oil palm (Elaeis guineensis Jacq.)* [Doctoral dissertation], Universiti Putra Malaysia, Selangor, Malaysia, 2000.

[11] H. Li and L. H. Greene, "Sequence and structural analysis of the chitinase insertion domain reveals two conserved motifs involved in chitin-binding," *PloS one*, vol. 5, no. 1, article e8654, 2010.

[12] F. Mauch, L.A. Hadwiger, and T. Boller, "Ethylene: symptom, not signal for induction of chitinase and ß-1, 3-glucanase in pea pods by pathogens and elicitors," *Plant Physiology*, vol. 76, pp. 607–611, 1984.

[13] C. R. Voisey and A. J. Slusarenko, "Chitinase mRNA and enzyme activity in *Phaseolus vulgaris* (L.) increase more rapidly in response to avirulent than to virulent cells of *Pseudomonas syringae* pv. *phaseolicola*," *Physiological and Molecular Plant Pathology*, vol. 35, no. 5, pp. 403–412, 1989.

[14] J. J. Liu, A. K. M. Ekramoddoullah, and A. Zamani, "A Class IV chitinase is up-regulated by fungal infection and abiotic stresses and associated with slow-canker-growth resistance to *Cronartium ribicola* in western white pine (*Pinus monticola*)," *Phytopathology*, vol. 95, no. 3, pp. 284–291, 2005.

[15] I. Blilou, J. A. Ocampo, and J. M. García-Garrido, "Induction of LTP (lipid transfer protein) and Pal (phenylalanine ammonia-lyase) gene expression in rice roots colonized by the arbuscular mycorrhizal fungus *Glomus mosseae*," *Journal of Experimental Botany*, vol. 51, no. 353, pp. 1969–1977, 2000.

[16] H. S. Coventry and I. A. Dubery, "Lipopolysaccharides from *Burkholderia cepacia* contribute to an enhanced defensive capacity and the induction of pathogenesis-related proteins in *Nicotianae tabacum*," *Physiological and Molecular Plant Pathology*, vol. 58, no. 4, pp. 149–158, 2001.

[17] I. Yedidia, N. Benhamou, Y. Kapulnik, and I. Chet, "Induction and accumulation of PR proteins activity during early stages of root colonization by the mycoparasite *Trichoderma harzianum* strain T-203," *Plant Physiology and Biochemistry*, vol. 38, no. 11, pp. 863–873, 2000.

[18] G. W. Zehnder, J. F. Murphy, E. J. Sikora, and J. W. Kloepper, "Application of rhizobacteria for induced resistance," *European Journal of Plant Pathology*, vol. 107, no. 1, pp. 39–50, 2001.

[19] F. Brunner, A. Stintzi, B. Fritig, and M. Legrand, "Substrate specificities of tobacco chitinases," *Plant Journal*, vol. 14, no. 2, pp. 225–234, 1998.

[20] M. B. Sela-Buurlage, A. S. Ponstein, S. A. Bres-Vloemans, L. S. Melchers, P. J. M. Van Den Elzen, and B. J. C. Cornelissen, "Only specific tobacco (*Nicotiana tabacum*) chitinases and β-1,3-glucanases exhibit antifungal activity," *Plant Physiology*, vol. 101, no. 3, pp. 857–863, 1993.

[21] A. M. Hietala, H. Kvaalen, A. Schmidt, N. Jøhnk, H. Solheim, and C. G. Fossdal, "Temporal and spatial profiles of chitinase expression by Norway spruce in response to bark colonization by *Heterobasidion annosum*," *Applied and Environmental Microbiology*, vol. 70, no. 7, pp. 3948–3953, 2004.

[22] L. Naher, C. Ho, S.G. Tan, U.K. Yusuf, and F. Abdullah, "Cloning of transcripts encoding chitinases from *Elaeis guineensis* Jacq. and their expression profiles in response to fungal infections," *Physiology and Molecular Plant Pathology*, vol. 76, pp. 96–103, 2011.

[23] S. Chang, J. Puryear, and J. Cairney, "A simple and efficient method for isolating RNA from pine trees," *Plant Molecular Biology Reporter*, vol. 11, no. 2, pp. 113–116, 1993.

[24] J. Sambrook and D. W Russell, *Molecular Cloning A Laboratory Manual*, Cold Spring Harbor Laboratory Press, New York, NY, USA, 3rd edition, 2001.

[25] K. J. Livak and T. D. Schmittgen, "Analysis of relative gene expression data using real-time quantitative PCR and the $2^{-\Delta\Delta C}$T method," *Methods*, vol. 25, no. 4, pp. 402–408, 2001.

[26] M. J. Pozo, C. Azcón-Aguilar, E. Dumas-Gaudot, and J. M. Barea, "Chitosanase and chitinase activities in tomato roots during interactions with arbuscular mycorrhizal fungi or *Phytophthora parasitica*," *Journal of Experimental Botany*, vol. 49, no. 327, pp. 1729–1739, 1998.

[27] C. Rinaldi, A. Kohler, P. Frey et al., "Transcript profiling of poplar leaves upon infection with compatible and incompatible strains of the foliar rust *Melampsora larici-populina*," *Plant Physiology*, vol. 144, no. 1, pp. 347–366, 2007.

[28] L. C. Van Loon and E. A. Van Strien, "The families of pathogenesis-related proteins, their activities, and comparative analysis of PR-1 type proteins," *Physiological and Molecular Plant Pathology*, vol. 55, no. 2, pp. 85–97, 1999.

[29] G. Busam, H. H. Kassemeyer, and U. Matern, "Differential expression of chitinases in *Vitis vinifera* L. Responding to systemic acquired resistance activators or fungal challenge," *Plant Physiology*, vol. 115, no. 3, pp. 1029–1038, 1997.

[30] N. Robert, K. Roche, Y. Lebeau et al., "Expression of grapevine chitinase genes in berries and leaves infected by fungal or bacterial pathogens," *Plant Science*, vol. 162, no. 3, pp. 389–400, 2002.

[31] M. Shoresh and G. E. Harman, "Differential expression of maize chitinases in the presence or absence of *Trichoderma harzianum* strain T22 and indications of a novel exo- endo-heterodimeric chitinase activity," *BMC Plant Biology*, vol. 10, article 136, 2010.

[32] J. A. Baldé, R. Francisco, Á. Queiroz, A. P. Regalado, C. P. Ricardo, and M. M. Veloso, "Immunolocalization of a class III chitinase in two muskmelon cultivars reacting differently to *Fusarium oxysporum* f. sp. melonis," *Journal of Plant Physiology*, vol. 163, no. 1, pp. 19–25, 2006.

[33] D. A. Samac and D. M. Shah, "Effect of chitinase antisense RNA expression on disease susceptibility of Arabidopsis plants," *Plant Molecular Biology*, vol. 25, no. 4, pp. 587–596, 1994.

The Effect of High Concentrations of Glufosinate Ammonium on the Yield Components of Transgenic Spring Wheat (*Triticum aestivum* L.) Constitutively Expressing the *bar* Gene

Zoltán Áy,[1] **Róbert Mihály,**[1] **Mátyás Cserháti,**[2] **Éva Kótai,**[1] **and János Pauk**[1]

[1] *Department of Biotechnology, Cereal Research Non-Profit Ltd. Co., Alsó kikötő sor 9, 6726 Szeged, Hungary*
[2] *Biological Research Centre, Institute of Plant Biology, Hungarian Academy of Sciences, Temesvári körút 62, 6726 Szeged, Hungary*

Correspondence should be addressed to János Pauk, janos.pauk@gabonakutato.hu

Academic Editor: Victor Fedorenko

We present an experiment done on a *bar*⁺ wheat line treated with 14 different concentrations of glufosinate ammonium—an effective component of nonselective herbicides—during seed germination in a closed experimental system. Yield components as number of spikes per plant, number of grains per spike, thousand kernel weight, and yield per plant were thoroughly analysed and statistically evaluated after harvesting. We found that a concentration of glufosinate ammonium 5000 times the lethal dose was not enough to inhibit the germination of transgenic plants expressing the *bar* gene. Extremely high concentrations of glufosinate ammonium caused a bushy phenotype, significantly lower numbers of grains per spike, and thousand kernel weights. Concerning the productivity, we observed that concentrations of glufosinate ammonium 64 times the lethal dose did not lead to yield depression. Our results draw attention to the possibilities implied in the transgenic approaches.

1. Introduction

Effective weed control has become one of the most significant procedures in cropping operations to ensure good quality harvests. Due to the high costs of energy required, mechanical weed control practices are now viewed as unsatisfactory and have been largely replaced by chemical weed control using herbicides. Herbicides generally function by disrupting unique and essential processes in plants, for example, photosynthesis, pigment biosynthesis, mitosis, or essential amino acid biosynthesis [1].

Amino acid biosynthesis is one of the pathways targeted most by herbicides. The discovery of a peptide antibiotic produced by the actinomycetes *Streptomyces viridochromogenes* and *S. hygroscopicus* was reported several decades ago [2, 3]. The antibiotic, named PTT (phosphinothricin-tripeptide = phosphinothricyl-alanyl-alanine = bialaphos), consists of two molecules of L-alanine and one molecule of the rare amino acid PT (L-phosphinothricin). According to the

postulated biosynthetic pathway, PT is generated from two molecules of phosphoenolpyruvate, one molecule of acetyl coenzyme A and one methyl group of methylcobalamin in thirteen biosynthetic steps [4, 5]. The bioactive component of the PTT molecule is the PT which, as a structural analogue of glutamic acid, interferes with amino acid synthesis through the competitive, irreversible inhibition of GS (glutamine synthetase), the key enzyme of nitrogen metabolism [6, 7]. The inhibition of GS reduces glutamine acid levels and triggers ammonium ion accumulation to levels up to 100-fold higher than in control cells [8, 9]. Due to this, PT has bactericidal, fungicidal, and herbicidal properties. In the case of plants, two to four hours after application of PT, photosynthesis slows down and plants yellow and die in two to five days [10].

Since many herbicides are nonselective, both crops and weeds share the processes mentioned above. For instance, over 40 monocotyledonous and more than 150 dicotyledonous species are sensitive to PT [11]. Consequently, selectivity must be based on the different ways herbicides act upon weeds and crops. The most effective approach to achieve this goal is the development of crop cultivars with tolerance to the so-called broad-spectrum herbicides by using plant biotechnology techniques such as *in vitro* cell culture, mutagenesis, or genetic transformation followed by selection under herbicide pressure. Tolerance via genetic transformation can be conferred by modification of the herbicide target enzyme in such a way that the herbicide molecule does not bind to the target enzyme or introduction of a gene coding for a herbicide detoxifying enzyme [1, 12].

Usually, genes coding for proteins useful in herbicide resistance in crops can be isolated from herbicide degrading soil microorganisms. The strategy to develop PT resistant crops is based on the mechanism used by PTT-producing actinomycetes, which can protect themselves against the autotoxic effect. This pathway is mediated by the enzyme PAT (phosphinothricin-N-acetyltransferase) which acetylates the free amino group of PT, thereby causing its detoxification. The PAT-encoding *bar* (bialaphos resistance) and *pat* genes were isolated from *Streptomyces hygroscopicus* [13–15] and *S. viridochromogenes* Tü494 [16], respectively. Both genes code for PAT proteins of 183 amino acids, which show 85% homology to each other, variations of the genes being confined to their noncoding regions [17].

Glufosinate ammonium is a proherbicide which is converted by plant cells into PT. Originally it was engineered by Hoechst in the 1970s for preharvest desiccation in potato, legumes, and oilseed rape. Since the discovery of the *bar/pat* gene system, glufosinate ammonium has found its applications in weed control and in selection of transgenic plants expressing resistance genes. It is marketed under a number of trade names including Basta, Challenge, Finale, and Radicale. Engineering tolerance to glufosinate ammonium in crops including wheat by genetic modification has been studied by many research groups [18–22].

The present study is the first which describes an experiment with a transgenic line of spring wheat constitutively expressing the gene *bar* in order to determine the extent of herbicide resistance and the complex effect of extremely high concentrated glufosinate ammonium on different yield parameters.

2. Materials and Methods

2.1. Genetic Transformation and Selection of Transgenic Plants. Spring wheat plants (*Triticum aestivum*, L., cv. CY-45) were grown in the greenhouse. Donor spikes were harvested 12–14 days after flowering. Embryos were excised from surface-sterilized immature seeds and plated onto callus induction medium. Gene transfer via particle bombardment was carried out according to Altpeter et al. [23]. The vector pAHC25 [24] containing the gene *bar* regulated by a constitutive maize ubiquitin promoter was used for genetic transformation. Putative transgenic plantlets were transferred to the soil in the greenhouse after a 4–6-week period of *in vitro* regeneration. After molecular studies, plants were sprayed with the wide-range herbicide Finale 14 SL (IUPAC name: *methyl(E)-methoximino-{(E)-a-[1-(a,a,a-trifluoro-m-*

tolyl)ethylide-neaminooxy]-o-tolyl}-acetate; active ingredient: 150 g · L^{-1} glufosinate ammonium) at 1.0% v/v, as recommended by the manufacturer. Survivor plants were grown and harvested. Progenies were also grown in the greenhouse alike and self-pollinated through six generations in order to acquire homozygous wheat lines, thereby eliminating the possibility of the segregation of the *bar* gene. Nontransgenic individuals were selected according to the results of molecular genetic methods and were eliminated by being sprayed with Finale 14 SL solution in every generation.

2.2. Test for Herbicide Resistance. As a benchmark, the lethal dose of glufosinate ammonium was defined in a preliminary experiment. Mature embryos were excised from surface-sterilized seeds of the nontransgenic spring wheat variety CY-45 and were *in vitro* germinated in tubes, containing 5 mL of half-strength MS$_0$ medium [25] supplemented with 0, 1, 2, and 4 mg · L^{-1} of glufosinate ammonium ($C_5H_{15}N_2O_4P$; 198.16 g/mol; Sigma), respectively. Incubation was carried out in a growing chamber (24°C, 16 h light/8 h dark photoperiod) and results were evaluated on the tenth day of culture.

The resistance test was carried out with the transgenic spring wheat line "T-124" in the seventh self-pollinated generation (T_7). The gene *bar* had one integration site in this wheat line. Culture conditions during germination of the mature embryos were the same as in the pilot experiment. Media representing fourteen treatments with different concentrations of glufosinate ammonium added to them were as follows: 2, 4, 8, 16, 32, 64, 128, 200, 400, 600, 800, 1000 and 5000 mg · L^{-1}. Medium of the control treatment contained no herbicide. One embryo was put into every tube and every treatment was repeated eight times. After three weeks of culture, plantlets were transferred to pots filled with soil, acclimatized and grown to maturity in the greenhouse. Plants were sprayed with insecticides and fungicides twice during the growing period. Exclusively mechanical weed control was also applied. Spikes were harvested individually and sorted into two groups termed well filled and low filled according to visual qualification. Yield components as number of spikes per plant, number of grains per spike, and yield per plant were measured while thousand kernel weight was calculated after harvesting.

2.3. Molecular Assays. Plantlets were analyzed by molecular methods in every transgenic generation. At the seedling stage, 30 mg of leaf samples were collected and immediately frozen in liquid nitrogen. For the purification of total RNA, the "SV Total RNA Isolation System" kit (Promega) was applied; the protocol also contained the DNase treatment. To prove not only the presence but also the expression of the *bar* gene, a fragment 375 bp in length derived from its RNA transcript was amplified by RT-PCR (one step reverse transcriptase polymerase chain reaction) with the aid of the specific primers bar5F and bar6R (5′-CAGGAACCG-CAGGAGTGGA-3′ and 5′-CCAGAAACCCACGTCATG-3′, resp.). RT-PCR products were detected by electrophoresis on 1% TAE-agarose gel. Only the *bar*$^+$ plants were grown to maturity and harvested in every generation. Concerning the resistance test population, one out of the eight individuals was randomly chosen in each herbicide treatment and analyzed as described above.

2.4. Experimental Conditions of Transgenic Research. Transgenic experiments were carried out in closed experimental conditions (*in vitro* growing chamber and closed greenhouse cabin). After the observations destruction of experimental plant material was documented in an official report for the Hungarian authorities.

2.5. Statistical Evaluation. Results of well-filled and low-filled groups were evaluated separately. In every treatment, main rates were calculated by averaging of the results of the eight repeats. Data of partially and totally sterile spikes were also included in the statistical analysis using Microsoft Excel 2003 software (Microsoft Inc., USA). The effect of glufosinate ammonium on the agronomical parameters was evaluated by one-way analysis of variance (one-way ANOVA).

3. Results

In a preliminary experiment, we defined the lethal dose of glufosinate ammonium. Embryos excised from the nontransgenic spring wheat variety CY-45 were germinated *in vitro*. During each repeat experiment, only those embryos germinated which were placed onto medium without any glufosinate ammonium while 1–4 mg $\cdot$ L^{-1} effective medium concentration resulted in neither shoots nor roots (Figure 1). This information revealed that, during germination, the lethal dose of glufosinate ammonium must be less than 1 mg $\cdot$ L^{-1} in this experiment.

In the course of the test for herbicide resistance of the transgenic wheat line "T-124," as it was expected, genetic segregation of the *bar* gene was not observed in the experimental plant population. This fact was confirmed by RT-PCR as well (Figure 2). Every embryo germinated under herbicide pressure; consequently, the resistance test was done with 112 transgenic wheat plants. Embryos germinated with the same intensity but, noticeably, the presence of 5000 mg $\cdot$ L^{-1} glufosinate ammonium in the medium led to slower germination. Plantlets had shoots only 1 cm in length on the seventh day of culture while those growing on the other media had shoots 11–12 cm in length at the same timepoint (Figure 3). Those treated with 5000 mg $\cdot$ L^{-1} glufosinate ammonium during germination stayed behind in development and growth compared to the others throughout the entire growing period. They only began to flower when the others had already been ready for harvesting (Figure 4), and finally, their growing period was prolonged by three weeks. In spite of these observations, every plantlet grew to maturity and developed 773 spikes in total (100%). According to visual qualification of the seeds, 311 spikes (40.2%) were considered as well filled while 462 others (59.8%) proved to be low filled (Figure 5). Obviously, partial and total sterility occurred only among the low-filled ones (19 spikes (2.4%) and 7 spikes (0.9%), resp.).

The number of spikes per plant varied between 2.375 and 3.125 in the well-filled group. These data represent

Figure 1: Germination of mature embryos of the nontransgenic spring wheat variety CY-45 on media containing 0, 1, 2, and 4 mg $\cdot$ L^{-1} glufosinate ammonium (from left to right) on the tenth day of culture. *bar* * 1.0 cm.

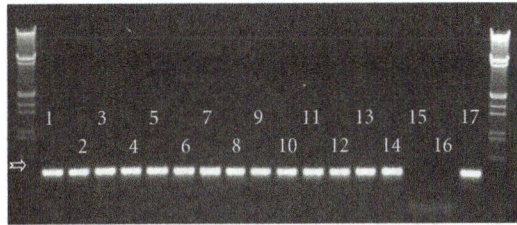

Figure 2: Detection of RNA transcripts derived from the herbicide resistance gene *bar* by electrophoresis of RT-PCR products. The white arrow shows the expected 375 bp fragment. Markers: λ-DNA digested with EcoRI and HindIII restriction enzymes. Samples from left to right: 1–14: according to increasing herbicide concentrations (1 refers to 0 while 14 refers to 5000 mg $\cdot$ L^{-1} of glufosinate ammonium), 15: nontransgenic CY-45 plant, 16: distilled water, 17: pAHC25 plasmid DNA.

the same level of significance (Table 1). By contrast, this parameter was similar in the case of low-filled spikes but strongly increased at the three highest concentrations of glufosinate ammonium. Plants treated with 5000 mg $\cdot$ L^{-1} herbicide showed the most intensive shoot development (Figure 6(a)) causing a bushy phenotype. Data in this group corresponded to three levels of significance (Table 1).

The highest value of the number of grains per well-filled spikes was 21.1 while the lowest was 17.4. The latter one was a result of application of 5000 mg $\cdot$ L^{-1} glufosinate ammonium and it is significantly lower than the other values (Table 1). Compared to this, the number of grains per spike was lower in the low-filled group and varied between 21.4 and 12.6 (Figure 6(b)). These data correspond to three levels of significance. Interestingly, 16 and 200–800 mg $\cdot$ L^{-1} of glufosinate ammonium resulted in the same level of significance (Table 1).

The thousand kernel weight was calculated after the yield of the spikes was harvested. Obviously, drastic differences were found between the two main groups. Representing three

TABLE 1: Significance levels of the averages of the eight repeat experiments at LSD$_{005}$ in the well-filled (i) and in the low-filled (ii) groups and in case of the total yield per plants (iii) according to one-way analysis of variance (one-way ANOVA).

(i)	LSD$_{5\%}$	cont.	2	4	8	16	32	64	128	200	400	600	800	1000	5000
Spikes per plant (pc)	1,051	a	a	a	a	a	a	a	a	a	a	a	a	a	a
Grains per spikes (pc)	1,893	a	a	a	a	a	a	a	a	a	a	a	a	a	b
Thousand kernel weight (g)	1,855	a	a	a	a	b	b	b	b	b	c	c	c	c	c
Sum. yield of spikes (g)	0,836	a	a	a	a	a	a	a	a	a	a	a	a	a	a
(ii)	LSD$_{5\%}$	cont.	2	4	8	16	32	64	128	200	400	600	800	1000	5000
Spikes per plant (pc)	1,111	a	a	a	a	a	a	a	a	a	a	a	B	C	C
Grains per spikes (pc)	2,274	a	a	a	a	b	a	a	a	b	b	b	b	c	c
Thousand kernel weight (g)	1,947	a	a	a	a	a	b	b	b	c	c	c	c	c	d
Sum. yield of spikes (g)	0,734	a	a	a	a	a	a	a	b	b	b	b	a	a	a
(iii)	LSD$_{5\%}$	cont.	2	4	8	16	32	64	128	200	400	600	800	1000	5000
Total yield of plants (g)	0,846	a	a	a	a	a	a	a	b	b	b	b	b	a	b

FIGURE 3: Germination of mature embryos of the transgenic spring wheat line "T-124" on media containing 0, 200, 400, 600, 800, 1000, and 5000 mg · L^{-1} glufosinate ammonium (from left to right) on the seventh day of culture. *bar* * 1.0 cm.

FIGURE 4: Plants treated with 5000 mg · L^{-1} glufosinate ammonium during germination (on the right) had growing period three weeks longer than the untreated control ones (on the left).

FIGURE 5: Grains of the well-filled (on the top) and the low-filled spikes (on the bottom). Control (on the left) and 5000 mg · L^{-1} (on the right) treatments resulted in different size and exterior of grains. *bar* * 0.5 cm.

levels of significance (Table 1), weight values of the well-filled spikes varied from 37.1 g to 28.6 g. Contrary to this, data of the low-filled spikes indicated four levels (Table 1) where the weight value changed between 29.8 g and 16.9 g (Figure 6(c)).

Yield per spikes was summarized before evaluation both in well-filled and low-filled groups in order to receive the yield per plant. This parameter showed similarity between the two groups since values in the well-filled group varied from 2.15 g to 1.42 g and in the other case from 2.18 g to 1.03 g (Figure 6(d)). There were no significant differences between the well-filled spike groups (Table 1) but, noticeably, a significantly lower yield in the low-filled group was due not to treatments with the highest concentration of glufosinate

The Effect of High Concentrations of Glufosinate Ammonium on the Yield Components of Transgenic Spring Wheat
(Triticum aestivum L.) Constitutively Expressing the bar Gene

73

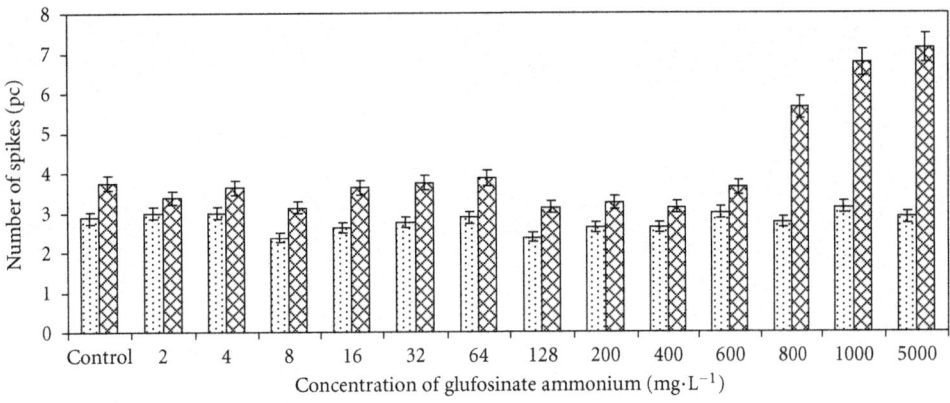

(a)

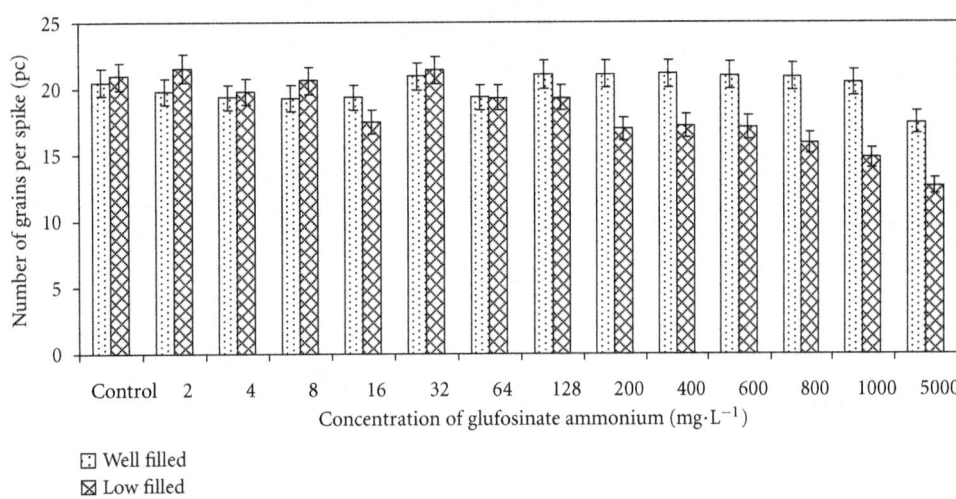

(b)

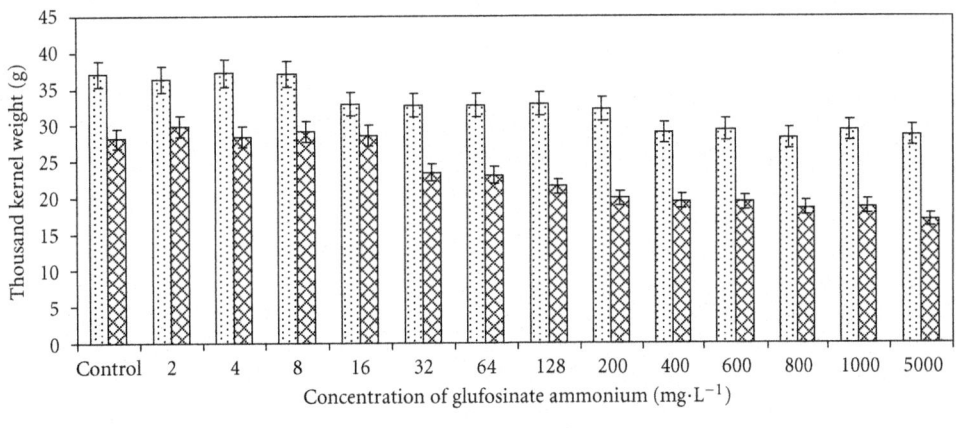

(c)

FIGURE 6: Continued.

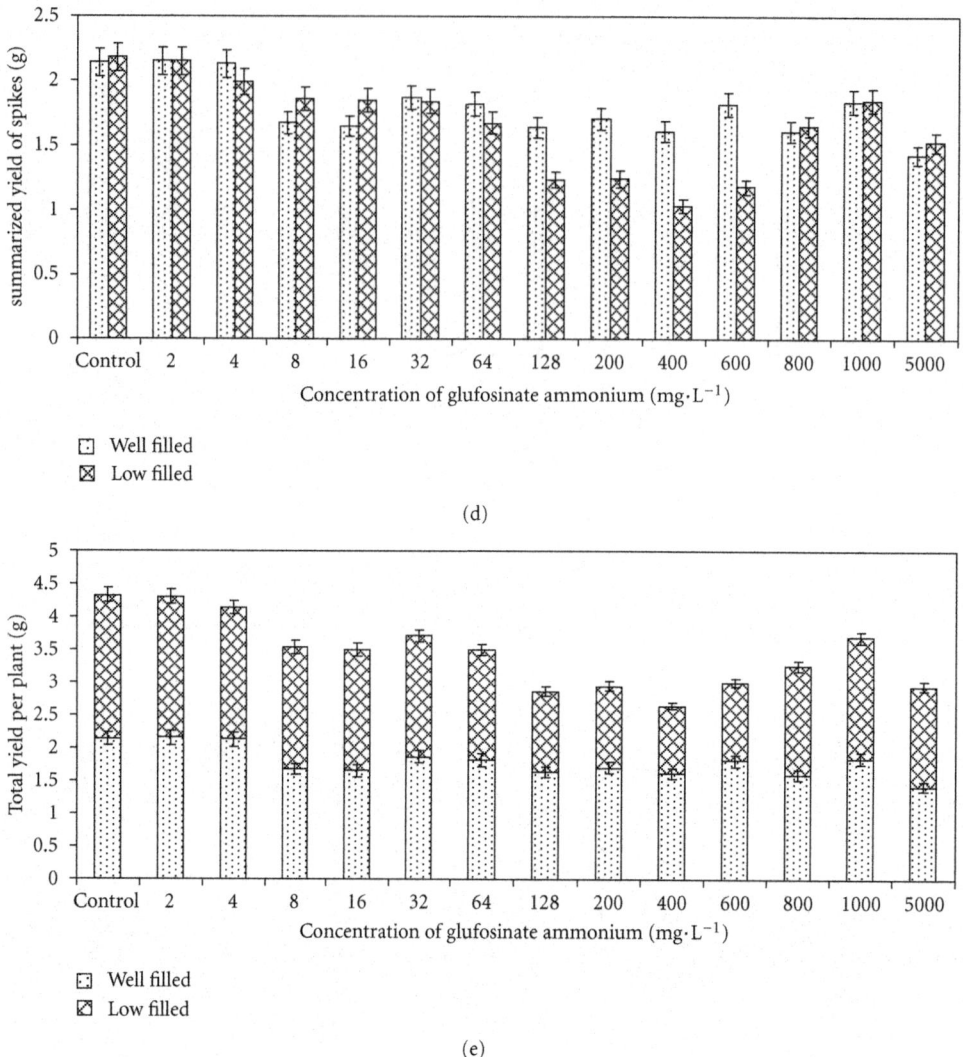

(d)

(e)

FIGURE 6: Effect of different concentrations of the herbicide glufosinate ammonium on the number of spikes (a), number of grains per spike (b), thousand kernel weight (c), summarized yield of spikes (d), and total yield per plant (e) of the transgenic wheat line "T-124." Values are equal to the average of the eight repeat experiments.

ammonium but rather to treatments with a concentration of only 128–600 mg · L^{-1} (Table 1).

Since yield is the most important agronomical parameter, we also represent the total yield per plant by summarizing the results of the well-filled and the low-filled groups. In this case, data varied between 4.32 g and 2.64 g (Figure 6(e)). Compared to the control plants, total yield of those treated with 128–5000 mg · L^{-1} glufosinate ammonium—except the 1000 mg · L^{-1} one—significantly decreased below 3 grams (Table 1).

To form a more detailed picture of the complex effect of glufosinate ammonium on the yield components, we represent the results also in cycle diagrams (Figure 7). The most conspicuous divergence between the well-filled and low-filled groups was the increase in the number of spikes up to 190% under extremely high concentration of the herbicide. Other parameters showed similar changes but not similar

significance levels, showing that the yield parameters changed the same way in both well-filled and low-filled groups.

4. Discussion

Initial growth conditions play a key role in the life cycle of a plant and they determine the vigour during the seedling stage. According to our former observations, wheat was the most sensitive to PT-like herbicides exactly during seed germination (data not shown). Therefore, we exposed transgenic wheat embryos to different concentrations of the herbicide glufosinate ammonium which is converted by plant cells into PT.

In the preliminary experiment, we found that less than 1 mg · L^{-1} of glufosinate ammonium in the culture medium is enough to inhibit CY-45 (wild-type) embryo germination.

The Effect of High Concentrations of Glufosinate Ammonium on the Yield Components of Transgenic Spring Wheat (Triticum aestivum L.) Constitutively Expressing the bar Gene

75

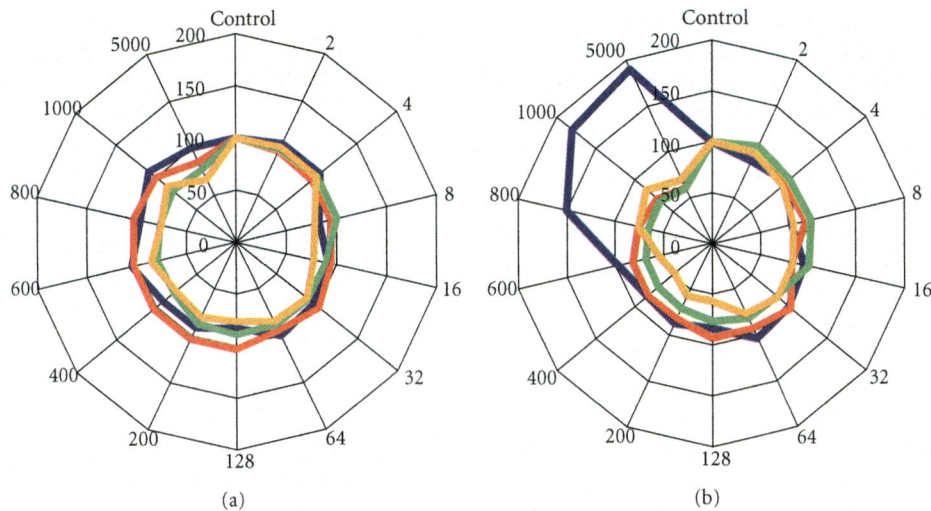

FIGURE 7: Complex effect of glufosinate ammonium on the well-filled (a) and the low-filled spikes (b). Colour key: blue—number of spikes; red—number of grains per spikes; green—thousand kernel weight; yellow—summarized yield of spikes. Control treatment represents 100 per cent. Values are equal to the average of the eight repeat experiments.

Similarly low concentrations of PT-like herbicides made possible the successful selection of transgenic tissues according to pioneer wheat transformation studies [26–28].

Throughout the first three weeks of their life cycle, wheat plantlets derived from the transgenic plant line "T-124" constitutively expressing the bar gene were challenged by 14 different concentrations of glufosinate ammonium. By transferring the plants into the soil, herbicide pressure was stopped and every plantlet was grown to maturity under the same conditions. Nevertheless, those treated with higher concentrations of glufosinate ammonium showed significant differences in the four examined parameters compared to the controls, thus, these divergences were clearly the aftermath of the herbicide treatment and confirm the importance of growth conditions during seed germination.

Our purpose was to determine the extent of herbicide resistance and the complex effect of extremely high concentrations of glufosinate ammonium. Therefore, we evaluated the application of the herbicide not with the well-known scale method but with exact and repeatable measurement of yield parameters such as the number of spikes per plant, grains per spike, thousand kernel weight, and yield per plant as an objective standard. Besides these important traits, we also recorded the length of the growing period of the plants which had been prolonged strikingly by the influence of the highest concentration ($5000\,\mathrm{mg} \cdot \mathrm{L}^{-1}$) of the herbicide. It is probable that so many glufosinate ammonium molecules were converted into PT molecules in the cells that in spite of the constitutive production of the PAT enzyme, plants could detoxify the herbicide only at the expense of slowed down metabolism which led to the absorption of fewer nutrients from the culture medium. Plants tried to compensate for this lag after the transfer into soil made manifest not in the strengthening of the main shoot but in developing several lateral shoots. Those individual plants treated with 800 and $1000\,\mathrm{mg} \cdot \mathrm{L}^{-1}$ of glufosinate ammonium showed a similar stool phenotype at harvest time which suggests that all the three highest concentrations of herbicide targeted the plants seriously. Certain studies reported that PT applied in levels lower than the lethal dose stimulates *in vitro* shoot regeneration in the case of grape [29], snapdragon [30], and rice [31, 32]. Our results reveal that increased ammonium ion level within the plant cell might act as a source of abiotic stress. Therefore, according to the apical dominance theory, inhibition of the apical tissues can lead to more intensive lateral shoot growth. However, this kind of escape was coupled with a weaker condition, which developed low-filled spikes without exception.

The reason why we sorted the spikes into well-filled and low-filled groups was to get a more detailed picture of the complex effect of glufosinate ammonium on the yield components. Table 1 shows the differences between these groups very well. The decrease in the number of grains per spike was caused mainly by the shortening of spikes but in the case of 16, 200 and $5000\,\mathrm{mg} \cdot \mathrm{L}^{-1}$ treatments, this was supplemented with partial or total sterility of low-filled spikes (data not shown). Thousand kernel weight decreased with almost the same intensity in both groups. However, changes in the values of this index did not manifest themselves in the summarized yield of spikes in the well-filled group because the stable number of spikes and number of grains per spike offset them. Quite different phenomena were observed in the low-filled group. Lower thousand kernel weights began to cause a decrease in the summarized yield of spikes at $128\,\mathrm{mg} \cdot \mathrm{L}^{-1}$ of glufosinate ammonium, but this tendency was reversed at $800\,\mathrm{mg} \cdot \mathrm{L}^{-1}$ and higher concentrations. This decrease can be traced back unambiguously to the negative changes in thousand kernel weight and number of grains per spike while the increase was caused by the higher number of spikes. Total yield per plants fluctuated similarly but only the $1000\,\mathrm{mg} \cdot \mathrm{L}^{-1}$ treatment could reverse the reduction. We did not check the quality of the grains in this experiment but we must make it absolutely clear that in the case of the three highest concentrations of glufosinate ammonium, the yield was restored definitely by the increased number of low-filled spikes representing a visibly poor quality. Briefly, plants could compensate the effect of extremely high concentrations of herbicide only at the expense of tissue deterioration, which is a kind of yield depression.

To make the above results comparable with other studies, we consider writing a study on the importance of the extent of resistance in plants, all the more so, since similar publications describe the sensitivity of plants to herbicides in different ways. If we take plant death as a basis we cannot say by how much more resistant our transgenic plants are compared to the control ones since all of them survived the $5000\,mg\cdot L^{-1}$ treatment, thus the lethal dose remained unknown. If we take the slightest significant change in the examined parameters we can see that the $8\,mg\cdot L^{-1}$ treatment was the highest which caused no significant difference. We think that both of these approaches are misleading; therefore we chose the total yield, the most important trait of cereal crops, as a benchmark. We found that $64\,mg\cdot L^{-1}$ was the highest concentration of herbicide which caused no significant loss in the yield. Consequently, a threshold value of resistance to glufosinate ammonium must be between 64 and $128\,mg\cdot L^{-1}$ according to this experiment. Since the lethal dose of the herbicide proved to be less than $1\,mg\cdot L^{-1}$, transgenic plants therefore achieved at least 64-fold resistance. This value is undoubtedly higher than those published in articles not only about PT [11, 31, 33–35] but also about other types of herbicides like imidazolinones [36] and glyphosate [37–40].

We suggest that this kind of high herbicide resistance should not be utilized in practice because it can lead to the rapid natural development of resistant weed populations [41]. We set a rather high theoretical value on our results as researchers will need to analyze the impacts of many transgenes with similar rigour in the near future.

Acknowledgments

This work was supported by a grant from the National Research and Technology Office (Budapest) as a part of a joint German-Hungarian "NAP_BIO 2006 ALAP3-01435/ 2006" Project.

References

[1] R. M. S. Mulwa and L. M. Mwanza, "Biotechnology approaches to developing herbicide tolerance/selectivity in crops," *African Journal of Biotechnology*, vol. 5, no. 5, pp. 396–404, 2006.

[2] E. Bayer, K. H. Gugel, K. Hägele et al., "Metabolic products of microorganisms. 98. Phosphinothricin and phosphinothricyl-alanyl-alanine," *Helvetica Chimica Acta*, vol. 55, no. 1, pp. 224–239, 1972.

[3] P. J. Lea, K. W. Joy, J. L. Ramos, and M. G. Guerrero, "The action of 2-amino-4-(methylphosphinyl)-butanoic acid (phosphinothricin) and its 2-oxo-derivative on the metabolism of cyanobacteria and higher plants," *Phytochemistry*, vol. 23, no. 1, pp. 1–6, 1984.

[4] E. Schinko, K. Schad, S. Eys, U. Keller, and W. Wohlleben, "Phosphinothricin-tripeptide biosynthesis: an original version of bacterial secondary metabolism?" *Phytochemistry*, vol. 70, no. 15-16, pp. 1787–1800, 2009.

[5] D. Schwartz, S. Berger, E. Heinzelmann, K. Muschko, K. Welzel, and W. Wohlleben, "Biosynthetic gene cluster of the herbicide phosphinothricin tripeptide from *Streptomyces viridochromogenes* Tü494," *Applied and Environmental Microbiology*, vol. 70, no. 12, pp. 7093–7102, 2004.

[6] H. Diddens, H. Zaehner, and E. Kraas, "On the transport of tripeptide antibiotics in bacteria," *European Journal of Biochemistry*, vol. 66, no. 1, pp. 11–23, 1976.

[7] G. M. Kishore and D. M. Shah, "Amino acid biosynthesis inhibitors as herbicides," *Annual Review of Biochemistry*, vol. 57, pp. 627–663, 1988.

[8] K. Tachibana, T. Watanabe, Y. Sekizawa, and T. Takematsu, "Action mechanism of bialaphos 2. Accumulation of ammonia in plants treated with bialaphos," *Journal of Pesticide Science*, vol. 1, pp. 33–37, 1986.

[9] A. Wild and R. Manderscheid, "The effect of phosphinothricin on the assimilation of ammonia in plants," *Zeitschrift für Naturforschung*, vol. 5, pp. 500–504, 1984.

[10] C. Wendler, A. Putzer, and A. Wild, "Effect of glufosinate (phosphinothricin) and inhibitors of photorespiration on photosynthesis and ribulose-1,5-bisphosphate carboxylase activity," *Journal of Plant Physiology*, vol. 6, pp. 666–671, 1992.

[11] P. L. J. Metz, W. J. Stiekema, and J. P. Nap, "A transgene-centered approach to the biosafety of transgenic phosphinothricin-tolerant plants," *Molecular Breeding*, vol. 4, no. 4, pp. 335–341, 1998.

[12] J. Dekker and S. O. Duke, "Herbicide-resistant field crops," *Advances in Agronomy*, vol. 54, pp. 69–116, 1995.

[13] Y. Kumada, H. Anzai, E. Takano et al., "The bialaphos resistance gene (*bar*) plays a role in both self-defense and bialaphos biosynthesis in *Streptomyces hygroscopicus*," *Journal of Antibiotics*, vol. 41, no. 12, pp. 1838–1845, 1988.

[14] T. Murakami, H. Anzai, S. Imai, A. Satoh, K. Nagaoka, and C. J. Thompson, "The bialaphos biosynthetic genes of *Streptomyces hygroscopicus*: molecular cloning and characterization of the gene cluster," *Molecular & General Genetics*, vol. 205, no. 1, pp. 42–53, 1986.

[15] C. J. Thompson, N. R. Movva, R. Tizard et al., "Characterization of the herbicide-resistance gene bar from *Streptomyces hygroscopicus*," *The EMBO Journal*, vol. 9, pp. 2519–2523, 1987.

[16] E. Strauch, W. Wohlleben, and A. Puhler, "Cloning of a phosphinothricin N-acetyltransferase gene from *Streptomyces viridochromogenes* Tü494 and its expression in *Streptomyces lividans* and *Escherichia coli*," *Gene*, vol. 63, no. 1, pp. 65–74, 1988.

[17] W. Wohlleben, W. Arnold, I. Broer, D. Hillemann, E. Strauch, and A. Puhler, "Nucleotide sequence of the phosphinothricin N-acetyltransferase gene from *Streptomyces viridochromogenes* Tü494 and its expression in *Nicotiana tabacum*," *Gene*, vol. 70, no. 1, pp. 25–37, 1988.

[18] P. Christou, T. L. Ford, and M. Kofron, "Production of transgenic rice (*Oryza sativa* L.) plants from agronomically important indica and japonica varieties via electric discharge particle acceleration of exogenous DNA into immature zygotic embryos," *Bio/Technology*, vol. 9, no. 10, pp. 957–962, 1991.

[19] W. J. Gordon-Kamm, T. M. Spencer, M. L. Mangano et al., "Transformation of maize cells and regeneration of fertile transgenic plants," *Plant Cell*, vol. 2, no. 7, pp. 603–618, 1990.

[20] V. Janakiraman, M. Steinau, S. B. McCoy, and H. N. Trick, "Recent advances in wheat transformation," *In Vitro Cellular and Developmental Biology - Plant*, vol. 38, no. 5, pp. 404–414, 2002.

[21] S. Tan, R. Evans, and B. Singh, "Herbicidal inhibitors of amino acid biosynthesis and herbicide-tolerant crops," *Amino Acids*, vol. 30, no. 2, pp. 195–204, 2006.

[22] V. Vasil, A. M. Castillo, M. E. Fromm, and I. K. Vasil, "Herbicide resistant fertile transgenic wheat plants obtained by microprojectile bombardment of regenerable embryogenic callus," *Bio-Technology*, vol. 10, no. 6, pp. 667–674, 1992.

[23] F. Altpeter, V. Vasil, V. Srivastava, E. Stöger, and I. K. Vasil, "Accelerated production of transgenic wheat (*Triticum aestivum* L.) plants," *Plant Cell Reports*, vol. 16, no. 1-2, pp. 12–17, 1996.

[24] A. H. Christensen and P. H. Quail, "Ubiquitin promoter-based vectors for high-level expression of selectable and/or screenable marker genes in monocotyledonous plants," *Transgenic Research*, vol. 5, no. 3, pp. 213–218, 1996.

[25] T. Murashige and F. Skoog, "A revised medium for rapid growth and bioassays with tobacco cultures," *Physiologia Plantarum*, vol. 15, pp. 473–497, 1962.

[26] D. Becker, R. Brettschneider, and H. Lorz, "Fertile transgenic wheat from microprojectile bombardment of suctellar tissue," *Plant Journal*, vol. 5, no. 2, pp. 299–307, 1994.

[27] N. S. Nehra, R. N. Chibbar, N. Leung et al., "Self-fertile transgenic wheat plants regenerated from isolated scuteller tissues following microprojectile bombardment with two distinct gene constructs," *Plant Journal*, vol. 5, no. 2, pp. 285–297, 1994.

[28] J. T. Weeks, O. D. Anderson, and A. E. Blechl, "Rapid production of multiple independent lines of fertile transgenic wheat (*Triticum aestivum*)," *Plant Physiology*, vol. 102, no. 4, pp. 1077–1084, 1993.

[29] D. Hebert-Soule, J. R. Kikkert, and B. I. Reisch, "Phosphinothricin stimulates somatic embryogenesis in grape (Vitis sp. L.)," *Plant Cell Reports*, vol. 14, no. 6, pp. 380–384, 1995.

[30] Y. Hoshino and M. Mii, "Bialaphos stimulates shoot regeneration from hairy roots of snapdragon (*Antirrhinum majus* L.) transformed by *Agrobacterium rhizogenes*," *Plant Cell Reports*, vol. 17, no. 4, pp. 256–261, 1998.

[31] Y. W. Liu, W. Y. Chou, and C. Y. Wang, "*In vitro* induction of phosphinothricin tolerance in rice (*Oryza sativa*)," *Plant Protection Bulletin (Taipei)*, vol. 1, pp. 47–58, 2005.

[32] O. Toldi, S. Tóth, A. S. Oreifig, E. Kiss, and B. Jenes, "Production of phosphinothricin-tolerant rice (*Oryza sativa* L.) through the application of phosphinothricin as growth regulator," *Plant Cell Reports*, vol. 19, no. 12, pp. 1226–1231, 2000.

[33] S. Gopalakrishnan, G. K. Garg, D. T. Singh, and N. K. Singh, "Herbicide-tolerant transgenic plants in high yielding commercial wheat cultivars obtained by microprojectile bombardment and selection on Basta," *Current Science*, vol. 79, no. 8, pp. 1094–1100, 2000.

[34] G. Keller, L. Spatola, D. McCabe, B. Martinell, W. Swain, and M. E. John, "Transgenic cotton resistant to herbicide bialaphos," *Transgenic Research*, vol. 6, no. 6, pp. 385–392, 1997.

[35] M. Manickavasagam, A. Ganapathi, V. R. Anbazhagan et al., "Agrobacterium-mediated genetic transformation and development of herbicide-resistant sugarcane (*Saccharum* species hybrids) using axillary buds," *Plant Cell Reports*, vol. 23, no. 3, pp. 134–143, 2004.

[36] K. E. Newhouse, W. A. Smith, M. A. Starrett, T. J. Schaefer, and B. K. Singh, "Tolerance to imidazolinone herbicides in wheat," *Plant Physiology*, vol. 100, no. 2, pp. 882–886, 1992.

[37] G. R. Heck, C. L. Armstrong, J. D. Astwood et al., "Development and characterization of a CP4 EPSPS-based, glyphosate-tolerant corn event," *Crop Science*, vol. 45, no. 1, pp. 329–339, 2005.

[38] A. R. Howe, C. S. Gasser, S. M. Brown et al., "Glyphosate as a selective agent for the production of fertile transgenic maize (*Zea mays* L.) plants," *Molecular Breeding*, vol. 10, no. 3, pp. 153–164, 2002.

[39] S. R. Padgette, K. H. Kolacz, X. Delannay et al., "Development, identification, and characterization of a glyphosate-tolerant soybean line," *Crop Science*, vol. 35, no. 5, pp. 1451–1461, 1995.

[40] H. Zhou, J. D. Berg, S. E. Blank et al., "Field efficacy assessment of transgenic roundup ready wheat," *Crop Science*, vol. 43, no. 3, pp. 1072–1075, 2003.

[41] C. R. Rainbolt, D. C. Thill, J. P. Yenish, and D. A. Ball, "Herbicide-resistant grass weed development in imidazolinone-resistant wheat: weed biology and herbicide rotation," *Weed Technology*, vol. 18, no. 3, pp. 860–868, 2004.

Pollen Viability, Pistil Receptivity, and Embryo Development in Hybridization of *Nelumbo nucifera* Gaertn

Yan-Li Wang, Zhi-Yong Guan, Fa-Di Chen, Wei-Min Fang, and Nian-Jun Teng

College of Horticulture, Nanjing Agricultural University, Nanjing 210095, China

Correspondence should be addressed to Nian-Jun Teng, njteng@njau.edu.cn

Academic Editors: S. Shimeld, M. J. Suso, T. Takamizo, and E. Tyystjarvi

Seed set is usually low and differs for different crosses of flower lotus (*Nelumbo nucifera* Gaertn.). The reasons remain unknown, and this has a negative impact on lotus breeding. To determine the causes, we carried out two crosses of flower lotus, that is, "Jinsenianhua" × "Qinhuaihuadeng" and "Qinhuaihuadeng" × "Jinsenianhua" and pollen viability, pistil receptivity, and embryo development were investigated. The pollen grains collected at 05:00-06:00 hrs had the highest viability, and the viabilities of "Jinsenianhua" and "Qinhuaihuadeng" were 20.6 and 15.7%, respectively. At 4 h after artificial pollination, the number of pollen grains germinating on each stigma reached a peak: 63.0 and 17.2 per stigma in "Jinsenianhua" × "Qinhuaihuadeng" and "Qinhuaihuadeng" × "Jinsenianhua", respectively. At 1 d after artificial pollination, the percentages of normal embryos in the two crosses were 55.0 and 21.9%, respectively; however, at 11 d after pollination, the corresponding percentages were 20.8 and 11.2%. Seed sets of the two crosses were 17.9 and 8.0%, respectively. The results suggested that low pistil receptivity and embryo abortion caused low seed set in "Qinhuaihuadeng" × "Jinsenianhua", whereas low fecundity of "Jinsenianhua" × "Qinhuaihuadeng" was mainly attributable to embryo abortion.

1. Introduction

The lotus (*Nelumbo nucifera* Gaertn.) is an economically important aquatic plant in the family of Nelumbonaceae. In general, lotus is grouped into three categories based on utilization and morphological features: rhizome lotus, seed lotus, and flower lotus. Lotus has been cultivated for more than 2000 years in China and its current cultivation area covers most parts of China [1–3]. The flower lotus is one of the ten most famous traditional flowers in China and is widely cultivated in gardens and scenic spots for environmental beautification and water purification [4–6]. As the Chinese economy, people's living standards and the tourism industry have rapidly improved in recent years, the flower lotus has played an increasingly important role in cleaning and beautifying the environment. Various breeding methods including artificial hybridization, radiation techniques, and multiploid approach have been applied to develop new lotus cultivars [6–8]. Among these methods, artificial hybridization is the most widely used and most effective way to produce new lotus cultivars. Although there have been some new lotus cultivars developed by traditional hybridization breeding in the past two decades in China, reproductive barriers often exist in artificial hybridization and seriously reduce the breeding efficiency of lotus [6]. To date, few studies have examined reproductive barriers in lotus hybridization, thus the factors affecting the breeding efficiency of flower lotus remain unknown.

Because the features of parental reproductive systems and the interaction of their systems are usually related to breeding efficiency in plant cross-breeding, thus the parental reproductive systems and reproductive behaviors after pollination have been extensively investigated in many plants, for example, *Dendranthema grandiflorum*, *Fragaria ananassa*, and *Phaseolus vulgaris* [9–11]. Many of these studies have successfully revealed the reasons leading to reduction in seed production and breeding efficiency. Enlightened by these previous studies, we carried out a systematic investigation of factors that may affect breeding efficiency of flower lotus in the present study. These factors included pollen viability of male parents just before artificial pollination, germination behavior of pollen grains on stigmas after artificial pollination, and embryo development after fertilization. The purpose of this study was to unravel the main factors causing low fecundity in the cross-breeding of water lotus. The expected output will provide valuable information for efficient measures to overcome reproductive barriers and improve breeding efficiency of water lotus and other crops, in the future.

2. Materials and Methods

2.1. Plant Materials. Two crosses of flower lotus (*N. nucifera*) were performed: "Jinsenianhua" (female plant) × "Qinhuaihuadeng" (male) and "Qinhuaihuadeng" (female) × "Jinsenianhua" (male). All plants were grown in Nanjing Yileen, Zhujiang Town, Pukou District, Nanjing, China (32°07′ N, 118°62′ E). In the past several years, we have carried out many lotus crosses and found that seed set was very low in some crosses. Therefore, two representative lotus cultivars were used for reciprocal crosses in the present study, aiming to reveal the factors affecting breeding efficiency of flower lotus.

2.2. Pollen Viability. Lotus pollen is usually viable for a few hours after shedding, and the viability is difficult to determine. We previously systematically investigated lotus pollen viability using different methods: fluorescein diacetate (FDA), triphenyltetrazolium chloride (TTC), germination in vitro, and the peroxidase method. The methods of FDA, TTC, and germination in vitro were not suitable for checking pollen viability, and so we used the peroxidase method. On sunny days, fresh pollen grains were, respectively, collected at 05:00-06:00 hrs, 06:00-07:00 hrs, and 07:00-08:00 hrs. The liquid for determining pollen viability comprised two reagents: reagent I (1 : 1 : 1 of 0.5% benzidine, 0.5%α-naphthol and 0.25% sodium carbonate) and reagent II (0.3% hydrogen peroxide solution). The pollen grains under test were distributed on a glass slide coated in the culture medium comprising one drop of each of reagents I and II and then incubated at 30°C for 30 min. The pollen grains stained red if they were viable, and viable grains were counted in ten optical fields (at least 50 pollen grains per field) under an Olympus BX41 microscope. Each experiment was repeated three times and statistical analysis was performed by Tukey's test.

2.3. Sexual Hybridization. Lotus is an insect-pollinated species and its pistil matures 1-2 d before the stamens, thus we did not emasculate the flowers of female plants and only bagged them at 2 d before their pistils matured. When the female flowers opened and the stigmas were full of bright yellow mucus, this indicated the most suitable period for pollination. We carried out artificial pollination at 06:00-08:00 hrs on sunny days using fresh pollen collected from male plants at 05:00-06:00 hrs. The crosses were performed from late July to early August 2009, when average temperature was approximately 31°C, with range 26–38°C.

2.4. Pollen Germination on Stigmas after Artificial Pollination. Germination of pollen grains on stigmas was examined using the method of Sun et al. [11] with minor modifications. Thirty pistils were, respectively, sampled at 0.5, 1, 2, 4, 6, 8, 10, and 12 h after pollination and fixed and stored in FAA solution (5 : 5 : 90 of formalin, acetic acid and 70% ethanol) at 4°C until use. The samples were softened overnight in $1 \, mol \, L^{-1}$ NaOH, rinsed in water and mounted on a microscope slide with a drop of 0.1% aniline blue ($0.1 \, mol \, L^{-1}$ K_3PO_4 supplemented with 18% glycerol), and then observed under a fluorescence microscope (Zeiss Axioskop 40; Carl Zeiss Shanghai Company Ltd, Shanghai, China) with excitation filter BP 395–440, chromatic beam splitter FT 460, and barrier filter LP 470. Digital images were captured using an Axiocam MRC camera [12]. In addition, some pistils were fixed in 2.5% glutaraldehyde (0.1 M phosphate buffer, pH 7.2). After rinsing in buffer, the samples were postfixed overnight in 1% (wv) buffered osmium tetroxide, washed in buffer, dehydrated in an ethanol series (40, 70, 90 and 100%, 15 min each time), and then critical-point-dried using liquid CO_2. The pistils were then mounted on aluminum specimen stubs with adhesive tabs. After coating with gold, they were examined using a Philips XL-30 environment scanning electron microscope (SEM) (Hitachi S-3000N) [13].

2.5. Embryo Development after Pollination and Seed Set. About 80 ovaries at 1 and 2 d after pollination were, respectively, collected and then immersed in FAA at 4°C until use for examination of embryo development. The ovules were dehydrated through a graded series of alcohol solutions (70, 85, 95, and 100%, 5 min each time), infiltrated with xylene, and embedded in paraffin wax [11]. Sections were cut to a thickness of 8–10 μm using a Leica RM2016 microtome (Shanghai Leica Instruments Co Ltd, China), stained in Heidenhain's haematoxylin, and were observed and photographed under the Zeiss Axioskop 40 microscope. In addition, we collected about 80 ovaries at different periods after pollination in order to examine the percentage of normal embryos, and the samples were observed under a stereomicroscope equipped with a digital camera.

One month after artificial pollination, about 30 lotus seed pods were randomly chosen and plump seeds were collected from them. Then seed set in each cross was calculated with the following formula: seed set = (number of plump seeds/total number of pollinated stigmas) × 100% [11].

2.6. Statistical Analysis. The data were analysed by a one-way analysis of variance using the SPSS software 16.0 (SPSS Inc, Chicago, IL, USA). Tukey's Honestly Significant Difference (HSD) ($P \leq 0.05$) was used to discriminate the values.

3. Results

3.1. Viability of Pollen Grains Collected at Different Times. The collection time of lotus pollen grains significantly affected pollen viability (Figure 1): viabilities of "Jinsenianhua" pollen grains collected at 05:00-06:00 hrs, 06:00-07:00 hrs, and 07:00-08:00 hrs were 20.6, 12.1, and 6.8%, respectively;

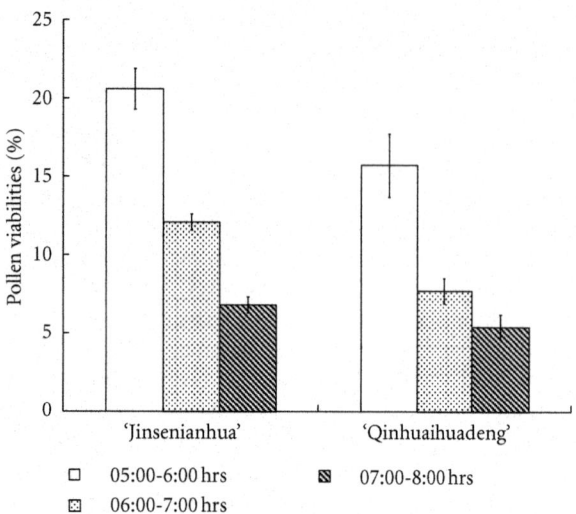

FIGURE 1: Viabilities of pollen grains collected at different times.

TABLE 1: Number of pollen grains germinating on stigma after pollination.

Time after pollination (h)	Number of pollen grains germinating per stigma in the two crosses	
	"Jinsenianhua" × "Qinhuaihuadeng"	"Qinhuaihuadeng" × "Jinsenianhua"
0.5	16.4 ± 1.1	11.2 ± 2.6
1	20.2 ± 2.6	13.6 ± 2.1
2	34.4 ± 3.4	15.0 ± 2.2
4	63.0 ± 6.8	17.2 ± 1.9
6	41.6 ± 3.9	14.2 ± 1.6
8	24.8 ± 2.6	10.0 ± 1.6
10	20.4 ± 3.2	8.2 ± 1.5
12	13.8 ± 2.9	7.4 ± 1.7

the corresponding values for "Qinhuaihuadeng" were 15.7, 7.7 and 5.4%. In addition, there was no significant difference in viabilities of pollen grains collected at 05:00-06:00 hrs between "Jinsenianhua" and "Qinhuaihuadeng". Since pollen grains collected at 5:00-6:00 hrs had the highest viability for both cultivars, we therefore carried out artificial pollination experiments using pollen grains collected in this time span.

3.2. Pollen Germination on Stigmas after Artificial Pollination. In both crosses, some pollen grains germinated within 0.5 h of artificial pollination. After that, more pollen grains germinated on stigmas and the number reached a peak at 4 h after pollination (Table 1; Figures 2(A), 2(B)): there were on average 63.0 and 17.2 germinated pollen grains per stigma at 4 h after pollination in the crosses "Jinsenianhua" × "Qinhuaihuadeng" and "Qinhuaihuadeng" × "Jinsenianhua", respectively. Then, at 12 h, the corresponding numbers of germinated pollen grains on each stigma gradually decreased to 13.8 and 7.4. Although the change pattern of the number of germinated pollen grains on stigmas was very similar for the two crosses, the number in the cross "Jinsenianhua" × "Qinhuaihuadeng" was significantly higher than that of "Qinhuaihuadeng" × "Jinsenianhua" at each time point after pollination (Table 1). Moreover, there were many pollen tubes with abnormalities such as branching, splitting, coiling, and convolution on stigmas in the "Jinsenianhua" × "Qinhuaihuadeng" cross, which resulted in the failure of these pollen tubes penetrating the stigma surface (Figures 2(C)–2(E)). In contrast, most pollen tubes grew normally and entered the stigma surface for "Jinsenianhua" × "Qinhuaihuadeng" (Figure 2(F)).

3.3. Percentage of Normal Embryos. Zygotes divided very quickly in the two crosses. For example, at 1 d after artificial pollination, globular embryos were observed in many ovules, and heart embryos were observed in some ovules at 2 d after artificial pollination (Figures 2(G), 2(H)). At 4 d after pollination, most embryos had reached the cotyledon embryo stage. Normal globular embryos were observed in 55.0% of ovaries at 1 d after pollination, and normal heart embryos were observed in 37.5% of ovaries at 2 d after pollination in the "Jinsenianhua" × "Qinhuaihuadeng" cross (Table 2). However, the corresponding values for the "Qinhuaihuadeng" × "Jinsenianhua" cross were 21.9 and 14.7%, respectively. As the embryos continued to develop, increasing numbers of embryos degenerated in the two crosses (Table 2; Figures 3(A)–3(F)). For instance, there were only 20.8 and 11.2% normal embryos at 11 d after pollination in "Jinsenianhua" × "Qinhuaihuadeng" and "Qinhuaihuadeng" × "Jinsenianhua", respectively (Table 2; Figures 3(G), 3(H)). The percentage of normal embryos at each developmental stage in the cross of "Jinsenianhua" × "Qinhuaihuadeng" was much higher than that at the corresponding time in the other cross.

3.4. Seed Set of the Two Lotus Crosses. Seed set in the "Jinsenianhua" × "Qinhuaihuadeng" cross was 17.9%, which was more than twice that in the "Qinhuaihuadeng" × "Jinsenianhua" cross (8.0%; Table 3). At 25 d after artificial pollination, most normal seeds were nearly mature (Figure 3(E)) and normal seeds became dark brown and were completely mature at approximately 30 d after pollination (Figure 3(F)).

4. Discussion

Seed production of crops is usually closely related to pollen viability [14–18]. If pollen grains with low viability are pollinated onto stigmas of female parents in crop crosses, the probabilities of pollination failure will increase. As a consequence, seed set will be low and breeding efficiency will be reduced [19, 20]. In crosses between rice subspecies, the low pollen viability of male parents was the main factor leading to low fecundity of the crosses [20]. Similar phenomena have been observed in chrysanthemum and soybean [21, 22]. In the present study, the highest pollen viabilities of the two cultivars were only 20.6% and 15.7%, respectively (Figure 1). It initially seemed that poor pollen viability may have some

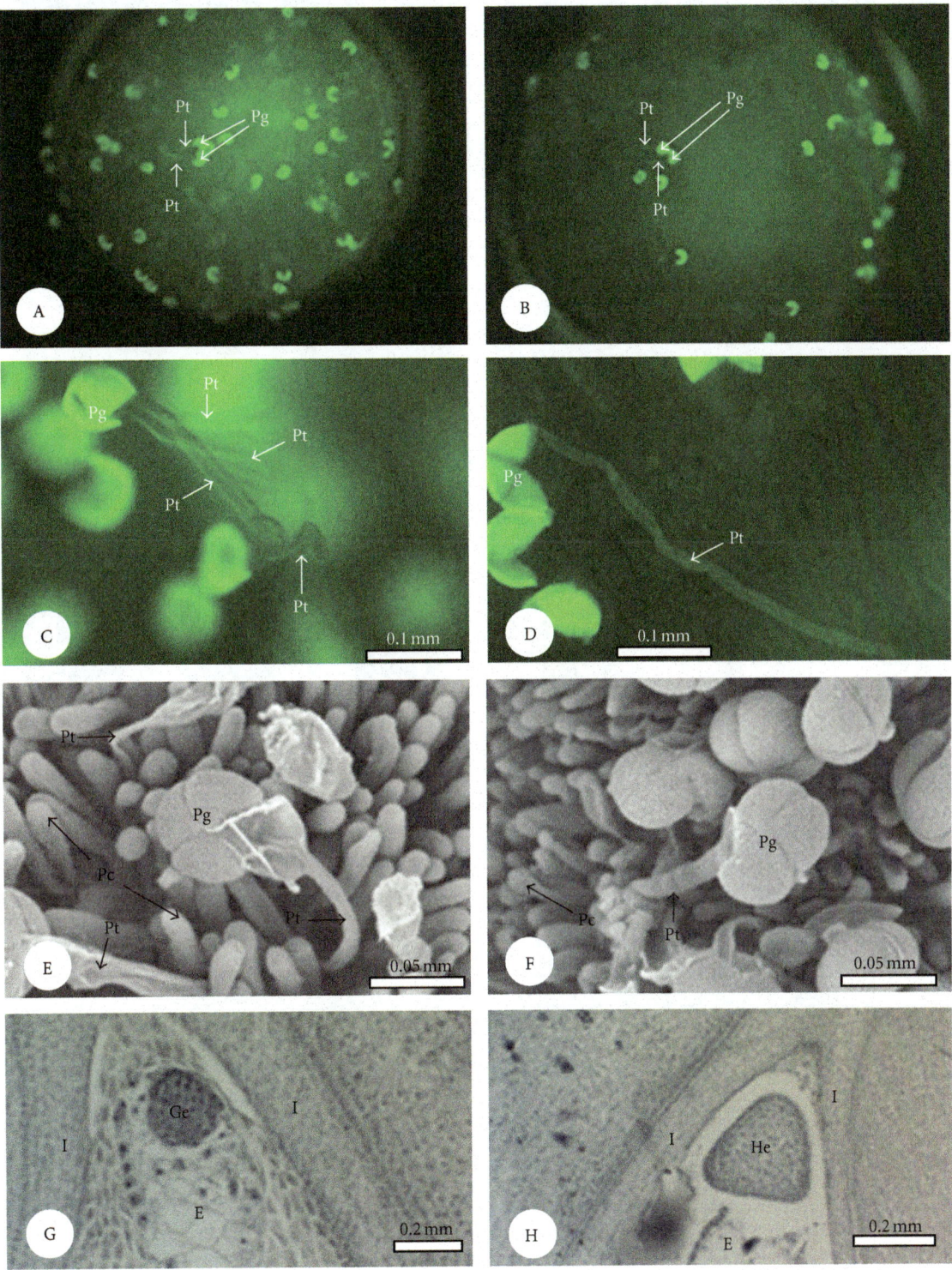

FIGURE 2: Pollen grains and pollen tubes on stigmas at 4 h after pollination and anatomical structure of young embryos. (A): Many pollen grains germinated on stigmas in "Jinsenianhua" × "Qinhuaihuadeng". (B): A few pollen grains germinated on stigmas in "Qinhuaihuadeng" × "Jinsenianhua". (C and D): Some abnormal pollen tubes in "Qinhuaihuadeng" × "Jinsenianhua". (E): SEM micrograph of an abnormal pollen tube in "Qinhuaihuadeng" × "Jinsenianhua". (F): SEM micrograph of a normal pollen tube in "Jinsenianhua" × "Qinhuaihuadeng". (G): A normal globular embryo at 1 d after artificial pollination. (H): A normal heart embryo at 2 d after artificial pollination. Abbreviations: E (endosperm), Ge (globular embryo), He (heart embryo), I (integument), Pc (papilla cell), Pg (pollen grain), and Pt (pollen tube).

FIGURE 3: The morphology of lotus ovary, seed, and seed pod at different times after pollination. (A): Normal ovaries at 5 d after pollination. (B): Abnormal ovaries at 5 d after pollination. (C): Normal ovaries at 11 d after pollination. (D): Abnormal ovaries at 11 d after pollination. (E): Seeds at 25 d after pollination. (F): Seeds at 30 d after pollination. (G): Lotus seed pod at 11 d after pollination in "Jinsenianhua" × "Qinhuaihuadeng" (asterisks indicate normal seeds). (H): Lotus seed pod at 11 d after pollination in "Qinhuaihuadeng" × "Jinsenianhua" (asterisks indicate normal seeds). Abbreviations: Ao (abnormal ovary); No (normal ovary).

negative effects on seed set of the two lotus crosses; however, this is not true for the following reasons. The percentages of normal embryos at 1 d after artificial pollination were 55.0 and 21.9% for "Jinsenianhua" × "Qinhuaihuadeng" and "Qinhuaihuadeng" × "Jinsenianhua", respectively (Table 2).

In addition, the corresponding highest pollen viabilities of male plants were 15.7 and 20.6% in the two crosses (Figure 1). Furthermore, the fertilization rate of ovules not only has a close relationship with pollen viability, but also is more closely related to the absolute quantity of pollen grains

TABLE 2: Number of normal and abnormal embryos at different days after pollination.

Crosses	Days after pollination	Developmental stages	Percentage of normal embryo (%)
"Jinsenianhua" × "Qinhuaihuadeng"	1	Globular embryo	55.0 ± 6.3
	2	Heart embryo	37.5 ± 4.5
	4		29.8 ± 4.2
	6	Cotyledon embryo	25.5 ± 3.7
	8		21.6 ± 3.1
	11		20.8 ± 3.3
"Qinhuaihuadeng" × "Jinsenianhua"	1	Globular embryo	21.9 ± 2.9
	2	Heart embryo	14.7 ± 1.8
	4		12.9 ± 1.6
	6	Cotyledon embryo	12.0 ± 1.5
	8		12.5 ± 1.2
	11		11.2 ± 1.4

TABLE 3: Seed set of the two lotus crosses.

Crosses	Seed set (%)
"Jinsenianhua" × "Qinhuaihuadeng"	17.9 ± 2.5
"Qinhuaihuadeng" × "Jinsenianhua"	8.0 ± 1.4

germinating on stigmas [13, 22]. These suggest that the large difference in percentages of normal embryos at 1 d after artificial pollination in the two crosses may be attributed to other factors, not to pollen viability. We therefore speculated that low pollen viability may not be the main factor influencing seed production of the two lotus crosses.

Whether pollen and stigma can identify each other is also an important factor affecting seed set of plant crosses [23–28]. If pollen and pistil fail to recognize each other normally, most pollen grains may fail to germinate or germinate abnormally even they have very high viability. As a consequence, low pistil receptivity will result in failure of fertilization and low fecundity in the crosses. In this study, the numbers of pollen grains germinated per lotus stigma reached a peak at 4 h after artificial pollination of 63.0 and 17.9 in "Jinsenianhua" × "Qinhuaihuadeng" and "Qinhuaihuadeng" × "Jinsenianhua", respectively. The 17.9 germinated pollen grains per stigma in "Qinhuaihuadeng" × "Jinsenianhua" may not have been sufficient to fertilize one pistil. The possible reasons may be that most pollen tubes had abnormalities and few pollen tubes could grow toward the embryo sac along the long style. Finally, most pollen tubes failed to reach the embryo sac and only a few pollen tubes entered the embryo sac. The low percentage of normal embryos (21.9%) at 1 d after artificial pollination further supports this explanation. However, 63.0 germinated pollen grains in "Jinsenianhua" × "Qinhuaihuadeng" may be sufficient to fertilize a pistil, as confirmed by the high percentage of normal embryos (55.0%) at 1 d after artificial pollination. Therefore, low pistil receptivity may be partly responsible for low seed set in "Qinhuaihuadeng" × "Jinsenianhua", while seed set in "Jinsenianhua" × "Qinhuaihuadeng" was less negatively influenced by the interaction between pollen and stigma.

Embryo development is a further factor influencing seed set, and embryo abortion usually leads to low seed set [10, 19, 29, 30]. For instance, Sun et al. [11] reported that embryo abortion was a critical factor resulting in the failure of the interspecies cross between D. grandiflorum "Yuhuaxingchen" and C. nankingense and a main factor leading to low seed set of D. grandiflorum "Yuhuaxingchen" and D. zawadskii. Deng et al. [31] found that a postfertilization barrier, that is, abortion of many embryos at various developmental stages before maturation significantly reduced fecundity between Chrysanthemum and Ajania. Similarly, in the interspecies cross between P. vulgaris and P. coccineus, Ndoutoumou et al. [10] also found that embryo abortion after fertilization was a key factor resulting in the failure of this cross. In the present study, many embryos aborted during their developmental processes in the two crosses. There were 55.0 and 21.9% normal embryos at 1 d after pollination in "Jinsenianhua" × "Qinhuaihuadeng" and "Qinhuaihuadeng" × "Jinsenianhua", respectively; however, corresponding values at 20 d after pollination decreased to 20.8 and 11.2%. Such results clearly indicate that embryo abortion was a main factor dramatically reducing seed set of both lotus crosses.

Seed set in "Jinsenianhua" × "Qinhuaihuadeng" was 17.9%, about 2.2 times that in "Qinhuaihuadeng" × "Jinsenianhua" with 8.0%. Such a large difference in seed set of the two crosses may be largely due to difference in pistil receptivity. For example, there was no significant difference in the highest pollen viability of the two lotus cultivars. In addition, there were 63.0 germinated pollen grains in "Jinsenianhua" × "Qinhuaihuadeng", about 3.7 times of that in "Qinhuaihuadeng" × "Jinsenianhua" with 17.2. Moreover, the percentages of normal embryos at 1 d after pollination in the two crosses were 55.0 and 21.9%, respectively, A 2.35-fold discrepancy. Therefore, the difference in percentages of normal embryos at 1 d after pollination in the two crosses may be mainly attributed to the difference in the number of germinated pollen grains. Taken together, these results suggest that a large difference in pistil receptivity may be mainly responsible for the large difference in seed set of the two crosses. However, the reasons for the differences in pistil receptivity in the two crosses remain unclear. In addition, female cytoplasm may also have some effects on seed set of the reciprocal crosses.

In conclusion, we systematically investigated the possible factors influencing fecundity of two lotus crosses, mainly including pollen viability, pistil receptivity, and embryo development. There were three findings of note. Firstly, low pollen viability had no significant effects on seed set of the lotus crosses. Secondly, low pistil receptivity and embryo abortion were two main factors causing low seed set in "Qinhuaihuadeng" × "Jinsenianhua", and the low fecundity of "Jinsenianhua" × "Qinhuaihuadeng" was mainly attributable to embryo abortion. Thirdly, the large difference in seed set in the two crosses was largely due to the large difference in pistil receptivity of the two crosses. These results suggest that special pollination methods and embryo rescue techniques may be effective in overcoming reproductive barriers and enhancing breeding efficiency in lotus crosses in future.

Acknowledgments

The authors are very grateful to Prof. Xi-Jin Mu and Prof. Yu-Xi Hu for valuable discussions during the early experimental stages. This study was supported by the Program for New Century Excellent Talents in Universities, Ministry of Education, China (NCET-11-0669), the Natural Science Foundation of China (31171983), the Natural Science Foundation of Jiangsu Province (BK2010447), and the Science & Technology Pillar Program of Jiangsu Province (BE2011325).

References

[1] Y. C. Han, C. Z. Teng, F. H. Chang et al., "Analyses of genetic relationships in *Nelumbo nucifera* using nuclear ribosomal ITS sequence data, ISSR and RAPD markers," *Aquatic Botany*, vol. 87, no. 2, pp. 141–146, 2007.

[2] Y. C. Han, C. Z. Teng, G. R. Wahiti, M. Q. Zhou, Z. L. Hu, and Y. C. Song, "Mating system and genetic diversity in natural populations of *Nelumbo nucifera* (Nelumbonaceae) detected by ISSR markers," *Plant Systematics and Evolution*, vol. 277, no. 1-2, pp. 13–20, 2009.

[3] H. B. Guo, "Cultivation of lotus (*Nelumbo nucifera* Gaertn. ssp. *nucifera*) and its utilization in China," *Genetic Resources and Crop Evolution*, vol. 56, no. 3, pp. 323–330, 2009.

[4] G. Z. Huang, "Studies on the biology of anthesis and the technique of artificial pollination of the lotus (*Nelumbo nucifera* Gaertn.)," *Acta Horticulturae Sinica*, vol. 9, no. 1, pp. 51–56, 1982.

[5] Q. C. Wang and X. Y. Zhang, *Chinese Lotus*, China Forestry Press, Beijing, China, 2005.

[6] L. Jiang, F. D. Chen, N. X. Cui, and J. J. Gu, "Study on seed setting of hybridization, selfing and open pollination of six cultivars of *Nelumbo nucifera*," *Acta Agricuturae Shanghai*, vol. 24, no. 3, pp. 61–64, 2008.

[7] X. L. Chen, J. Z. Bao, C. G. Liu, H. Cao, and J. Q. Zhai, "Preliminary study on irradiation breeding of ornamental lotus," *Acta Agricuturae Nucleatae Sinica*, vol. 18, no. 3, pp. 201–203, 2004.

[8] Y. H. Li, Y. Z. Pan, and Y. Q. Chen, "Advance in researches on the genetic diversity of *Nelumbo nucifera*," *Journal Sichuan Forestry Science and Technology*, vol. 29, no. 4, pp. 66–70, 2008.

[9] A. E. Marta, E. L. Camadro, J. C. Díaz-Ricci, and A. P. Castagnaro, "Breeding barriers between the cultivated strawberry, *Fragaria* × *ananassa*, and related wild germplasm," *Euphytica*, vol. 136, no. 2, pp. 139–150, 2004.

[10] P. N. Ndoutoumou, A. Toussaint, and J. P. Baudoin, "Embryo abortion and histological features in the interspecific cross between *Phaseolus vulgaris* L. and *P. coccineus* L," *Plant Cell, Tissue and Organ Culture*, vol. 88, no. 3, pp. 329–332, 2007.

[11] C. Q. Sun, F. D. Chen, N. J. Teng, Z. L. Liu, W. M. Fang, and X. L. Hou, "Factors affecting seed set in the crosses between *Dendranthema grandiflorum* (Ramat.) Kitamura and its wild species," *Euphytica*, vol. 171, no. 2, pp. 181–192, 2010.

[12] N. Teng, T. Chen, B. Jin et al., "Abnormalities in pistil development result in low seed set in *Leymus chinensis* (Poaceae)," *Flora*, vol. 201, no. 8, pp. 658–667, 2006.

[13] B. Jin, L. Wang, J. Wang et al., "The structure and roles of sterile flowers in *Viburnum macrocephalum* f. *keteleeri* (Adoxaceae)," *Plant Biology*, vol. 12, no. 6, pp. 853–862, 2010.

[14] A. Dafni and D. Firmage, "Pollen viability and longevity: practical, ecological and evolutionary implications," *Plant Systematics and Evolution*, vol. 222, no. 1–4, pp. 113–132, 2000.

[15] C. Wilcock and R. Neiland, "Pollination failure in plants: why it happens and when it matters," *Trends in Plant Science*, vol. 7, no. 6, pp. 270–277, 2002.

[16] S. Y. Hu, *Reproductive Biology of Angiosperms*, High Education Press, Beijing, China, 2005.

[17] H. Li, S. An, Y. Zhi et al., "Protogynous, pollen limitation and low seed production reasoned for the dieback of Spartina anglica in coastal China," *Plant Science*, vol. 174, no. 3, pp. 299–309, 2008.

[18] N. I. Park, E. C. Yeung, and D. G. Muench, "Mago Nashi is involved in meristem organization, pollen formation, and seed development in Arabidopsis," *Plant Science*, vol. 176, no. 4, pp. 461–469, 2009.

[19] J. L. Meng, *Genetics of Plant Reproduction*, Science Press, Beijing, China, 1977.

[20] F. L. Fan and X. R. Tang, "Recent progress on the study of the fecundity of cross rice between subspecies," *Crop Research*, vol. 5, no. 2, pp. 220–225, 2002.

[21] L. M. Zhao, H. Sun, M. Huang, S. M. Wang, and Y. Q. Wang, "The relationship between seed setting rate and pollen sterility rate for soybean," *Soybean Science*, vol. 23, no. 4, pp. 250–252, 2004.

[22] C. Q. Sun, Z. Z. Huang, Y. L. Wang et al., "Overcoming pre-fertilization barriers in the wide cross between *Chrysanthemumgrandiflorum* (Ramat.) Kitamura and *C. nankingense* (Nakai) Tzvel. by using special pollination techniques," *Euphytica*, vol. 178, no. 2, pp. 195–202, 2011.

[23] A. Mazzucato, I. Olimpieri, F. Ciampolini, M. Cresti, and G. P. Soressi, "A defective pollen-pistil interaction contributes to hamper seed set in the *parthenocarpic* fruit tomato mutant," *Sexual Plant Reproduction*, vol. 16, no. 4, pp. 157–164, 2003.

[24] Z. Huang, J. Zhu, X. Mu, and J. Lin, "Pollen dispersion, pollen viability and pistil receptivity in *Leymus chinensis*," *Annals of Botany*, vol. 93, no. 3, pp. 295–301, 2004.

[25] O. M. Aliyu, "Pollen-style compatibility in cashew (*Anacardium occidentale* L.)," *Euphytica*, vol. 158, no. 1-2, pp. 249–260, 2007.

[26] C. B. Lee, L. E. Page, B. A. McClure, and T. P. Holtsford, "Postpollination hybridization barriers in *Nicotiana* section *Alatae*," *Sexual Plant Reproduction*, vol. 21, no. 3, pp. 183–195, 2008.

[27] S. Ganesh Ram, S. Hari Ramakrishnan, V. Thiruvengadam, and J. R. Kannan Bapu, "Prefertilization barriers to interspecific hybridization involving *Gossypium hirsutum* and four diploid wild species," *Plant Breeding*, vol. 127, no. 3, pp. 295–300, 2008.

[28] M. J. Wheeler, B. H. J. De Graaf, N. Hadjiosif et al., "Identification of the pollen self-incompatibility determinant in *Papaver rhoeas*," *Nature*, vol. 459, no. 7249, pp. 992–995, 2009.

[29] N. Mallikarjuna and K. B. Saxena, "Production of hybrids between *Cajanus acutifolius* and *C. cajan*," *Euphytica*, vol. 124, no. 1, pp. 107–110, 2002.

[30] P. M. Datson, B. G. Murray, and K. R. W. Hammett, "Pollination systems, hybridization barriers and meiotic chromosome behaviour in Nemesia hybrids," *Euphytica*, vol. 151, no. 2, pp. 173–185, 2006.

[31] Y. Deng, S. Chen, N. Teng et al., "Flower morphologic anatomy and embryological characteristics in *Chrysanthemum multicaule* (Asteraceae)," *Scientia Horticulturae*, vol. 124, no. 4, pp. 500–505, 2010.

11

Erratic Male Meiosis Resulting in $2n$ Pollen Grain Formation in a 4x Cytotype ($2n = 28$) of *Ranunculus laetus* Wall. ex Royle

Puneet Kumar and Vijay Kumar Singhal

Department of Botany, Punjabi University, Patiala 147 002, Punjab, India

Correspondence should be addressed to Vijay Kumar Singhal, vksinghal53@gmail.com

Academic Editors: A. Kulharya and B. Vyskot

Two accessions were studied for male meiosis in *Ranunculus laetus* from the cold regions of Northwest Himalayas. One accession showed the presence of 14 bivalents at diakinesis and regular segregation of bivalents at anaphase I which lead to normal tetrad formation with four n microspores and consequently n pollen grains and 100% pollen fertility. Second accession from the same locality revealed the erratic meiosis characterized by the presence of all the 28 chromosomes as univalents in meiocytes at metaphase I. Univalent chromosomes failed to segregate during anaphases and produced restitution nuclei at meiosis I and II. These restitution nuclei resulted into dyads and triads which subsequently produced two types of apparently fertile pollen grains. On the basis of size, the two types of pollen grains were categorized as n (normal reduced) and $2n$ (unreduced, 1.5-times larger than the n pollen grains). The estimated frequency of $2n$ pollen grains from dyads and triads (61.59%) was almost the same as that of the observed one (59.90%), which indicated that $2n$ pollen grains in *R. laetus* were the result of dyads and triads. The present paper herein may provide an insight into the mechanisms of the formation of various intraspecific polyploids through sexual polyploidization in *R. laetus*.

1. Introduction

Ranunculus laetus Wall. ex Royle (family: Ranunculaceae), a highly polymorphic species [1, 2], has been studied chromosomally quite extensively from various regions of the Himalayas in India and outside of India from the hills of Nepal, Russia, China, and Pakistan (Figures 1(a) and 1(b)). The species exhibited a great amount of heterogeneity in chromosome number and level of ploidy with 2x (($2n = 14$) [3], Cangshan Mountains, Yunnan, China; ($2n = 16$) [4], Nepal), 4x (($2n = 28$), [1, 5, 6], Kashmir Himalayas; [7, 8], Eastern Himalayas; [9], Northwest Himalayas; [10, 11], Russia; [12, 13], Garhwal Himalayas; [14], Kinnaur in Himachal Pradesh; [15, 16], Chamba, Lahaul-Spiti, Kinnaur and Dalhousie hills in Himachal Pradesh; ($2n = 32$) [1], Shimla hills in Himachal Pradesh; [17], Indian Himalayas; [18], Kashmir Himalayas; [19, 20], Western Pakistan; [21, 22] Pakistan), 6x (($2n = 42$) [8], Eastern Himalayas), 8x (($2n = 56$) [20], Pakistan) based on two different basic chromosome numbers (2x, 4x, 6x, 8x on x = 7, 2x, 4x on x = 8).

Despite these intraspecific chromosomal variations and levels of ploidy (2x, 4x, 6x, and 8x) nothing is known about the origin of various polyploids in the species. All the previous studies carried out in the species were restricted either to count the chromosome number or to study the karyotype or DNA content. Previous communications [15, 16] from this laboratory have addressed in detail the cytological behaviour in 12 different accessions from the cold deserts of India, focusing on male meiosis. These accessions which uniformly existed at 4x level ($2n = 28$) showed normal bivalent formation and equal segregation of chromosomes at anaphases. However, these accessions depicted some irregularities during male meiosis such as cytomixis, chromosome stickiness, pycnotic chromatin material, out of plate bivalents at metaphase I, nonsynchronous disjunction of bivalents, and laggards at anaphases/telophases which resulted into PMCs (pollen mother cells) with abnormal microsporogenesis and 9–31% pollen sterility [16]. While studying the male meiosis in the species from the cold regions of Northwest Himalayas in Chamba district we have noticed

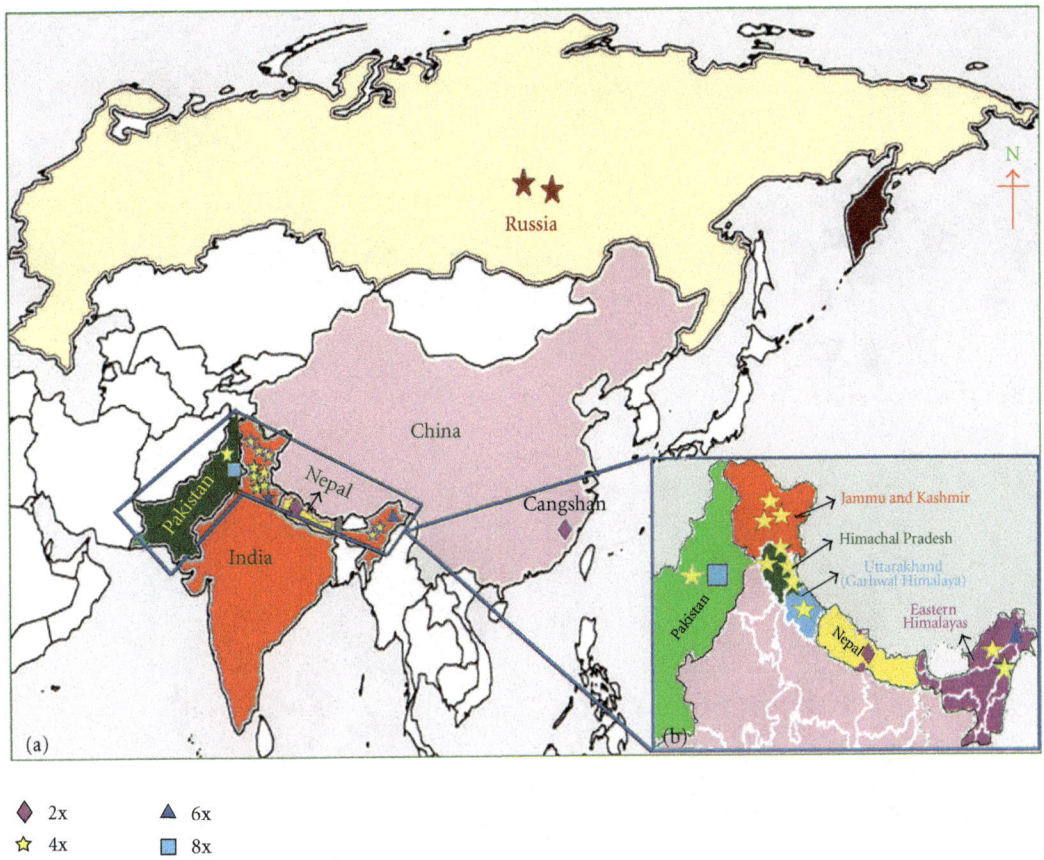

♦ 2x   ▲ 6x
★ 4x   ■ 8x

FIGURE 1: (a) Map showing distribution of 2x, 4x, 6x, and 8x (indicated with symbols) cytotypes in India, China, Nepal, Pakistan, and Russia. (b) Distribution of different cytotypes in Himalayan regions of India (2x, 4x, 6x), Nepal (2x), and Pakistan (4x, 8x).

in one accession that all the 28 chromosomes remained as univalents in the PMCs at metaphase I (M-I) during meiosis I. The products of such PMCs produced restitution nuclei and consequently yielded 2n (unreduced) pollen grains.

The present study herein aims to analyze the detailed meiotic course, microsporogenesis and to elucidate the cytological mechanism that lead to the formation of 2n pollen grains. The study may also provide an insight into the mechanisms of the formation of various intraspecific polyploids in R. laetus.

2. Material and Methods

2.1. Plant Material.
Material for male meiotic studies was collected from the wild plants growing on open moist slopes around the apple orchards in Bharmour in Chamba district (32°26′24″N, 76°33′31, altitude, 2,350 m) of Himachal Pradesh in June-July of 2009. The cytologically worked-out plants were identified using regional floras and compared with the specimens deposited at the Herbarium of Botanical Survey of India, Northern Circle, Dehra Dun. The voucher specimens (PUN, 51345, 51346) were deposited in the Herbarium, Department of Botany, Punjabi University, Patiala (PUN).

2.2. Meiotic Studies.
For meiotic chromosome counts, unopened floral buds of suitable sizes were fixed in a freshly prepared Carnoy's fixative (mixture of alcohol, chloroform, and glacial acetic acid in a volume ratio 6:3:1) for 24 h. These were subsequently transferred to 70% alcohol and stored in refrigerator at 4°C until used for meiotic analysis. Meiocytes were prepared by squashing the developing anthers, and stained with acetocarmine (1%). Chromosome number was determined at M-I from freshly prepared slides with light microscope Olympus. 500–600 pollen mother cells were analyzed for meiotic behaviour at different stages, M-I/II, anaphase I/II (A-I/II), telophase I/II (T-I/II).

2.3. Pollen Grain Analysis.
Pollen fertility was estimated through stainability tests using glycerol-acetocarmine (1:1) mixture and aniline blue (1%). Up to 450–800 pollen grains were examined for pollen fertility and size frequencies. Well-filled pollen grains with stained nuclei were taken as apparently fertile while shriveled and unstained pollen were counted as sterile. In each case, the size of 200 pollen grains was measured using an occulomicrometre. As per Xue et al. [23] pollen grain which measures 1.5-times larger than the n (normal reduced) pollen in diameter was taken as 2n (unreduced) pollen. Estimation of the theoretical frequency

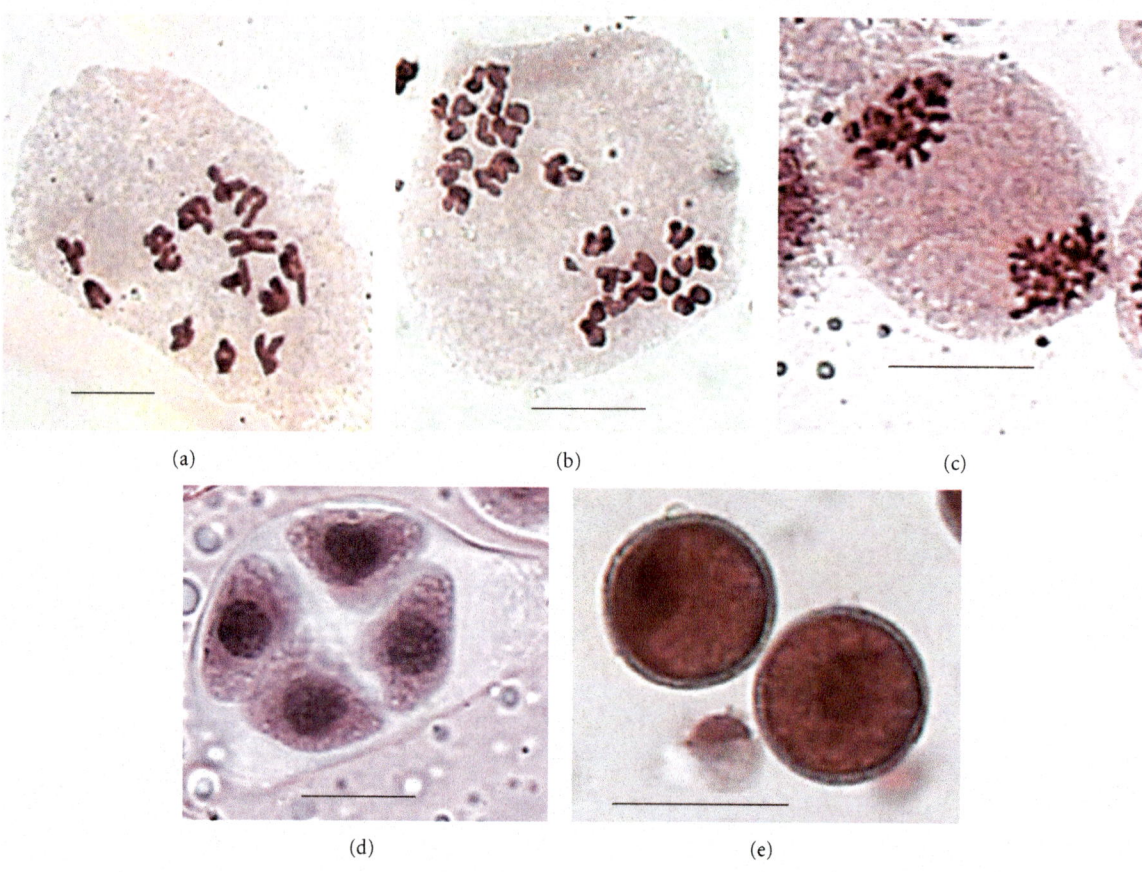

(a)   (b)   (c)

(d)   (e)

FIGURE 2: (a–e) Meiocytes with normal meiotic behaviour in *R. laetus*. (a) A PMC with 14 bivalents at diakinesis. (b) A PMC showing 14 : 14 chromosomes distributions at A-I. (c) A PMC showing two poles at A-I. (d) A tetrad with four *n* (reduced) microspores. (e) Apparently stained fertile *n* (reduced) pollen grains. Scale bar = 10 μm (a–d); 20 μm (e).

of 2*n* pollen grains was made from the number of observed dyads, triads, and tetrads during microsporogenesis. Generally a dyad resulted into two unreduced pollen grains, a triad produced one unreduced pollen grain and two reduced pollen grains, and each tetrad gave rise to four reduced pollen grains. The frequency of 2*n* pollen grains was calculated following Xue et al. [23]:

$$\text{Frequency of 2}n\text{ pollen grains} = \frac{2 \times \text{dy} + \text{tri}}{2 \times \text{dy} + 3 \times \text{tri} + 4 \times \text{tet}},$$

$$\times 100\%$$

(1)

dy = total number of dyads observed; tri = total number of triads observed; tet = total number of tetrads observed.

2.4. Photomicrographs. Photomicrographs from the freshly prepared desirable slides having clear chromosome counts, dyads, triads, tetrads, and pollen grains were taken with a digital imaging system of *Leica QWin*.

3. Results

Meiosis in one of the accession collected from Bharmour, 2,300 m was totally normal with the presence of 14 bivalents

at diakinesis (Figure 2(a)) and regular 14 : 14 segregation of chromosomes at opposite poles (Figures 2(a) and 2(c)) leading to normal tetrads with four *n* microspores (Figure 2(d)) and consequently *n* pollen grains (21.98–26.35 μm × 20.82–24.01 μm, Figure 2(e)) and 100% pollen fertility. However, the second accession also collected from Bharmour, 2,300 m showed highly abnormal meiosis characterized by the erratic behaviour of chromosomes at different stages of meiosis I and II.

3.1. Chromosomal Behaviour during Meiosis I. Analysis of PMCs at M-I of meiosis I revealed that all the PMCs showed the presence of 28 chromosomes as univalents which either remained randomly dispersed in the cytoplasm or shifted towards the periphery or in the centre or in 2–5 groups in the PMCs (Figures 3(a)–3(c)). Furthermore, the movements of chromosomes at A-I is very irregular, and in most of the PMCs they lagged behind (57.83%, Figure 3(d)). In the majority of the cases these lagging chromosomes did not get included into the telophase nuclei and formed micronuclei at T-I (Figure 3(e)). In some of the PMCs, segregation of chromosomes at A-I was irregular and the most common distribution was observed to be 11 : 17 (Figure 3(f)). In many PMCs it was also noticed that chromosome failed to move towards the A-I poles and remained in the centre of the

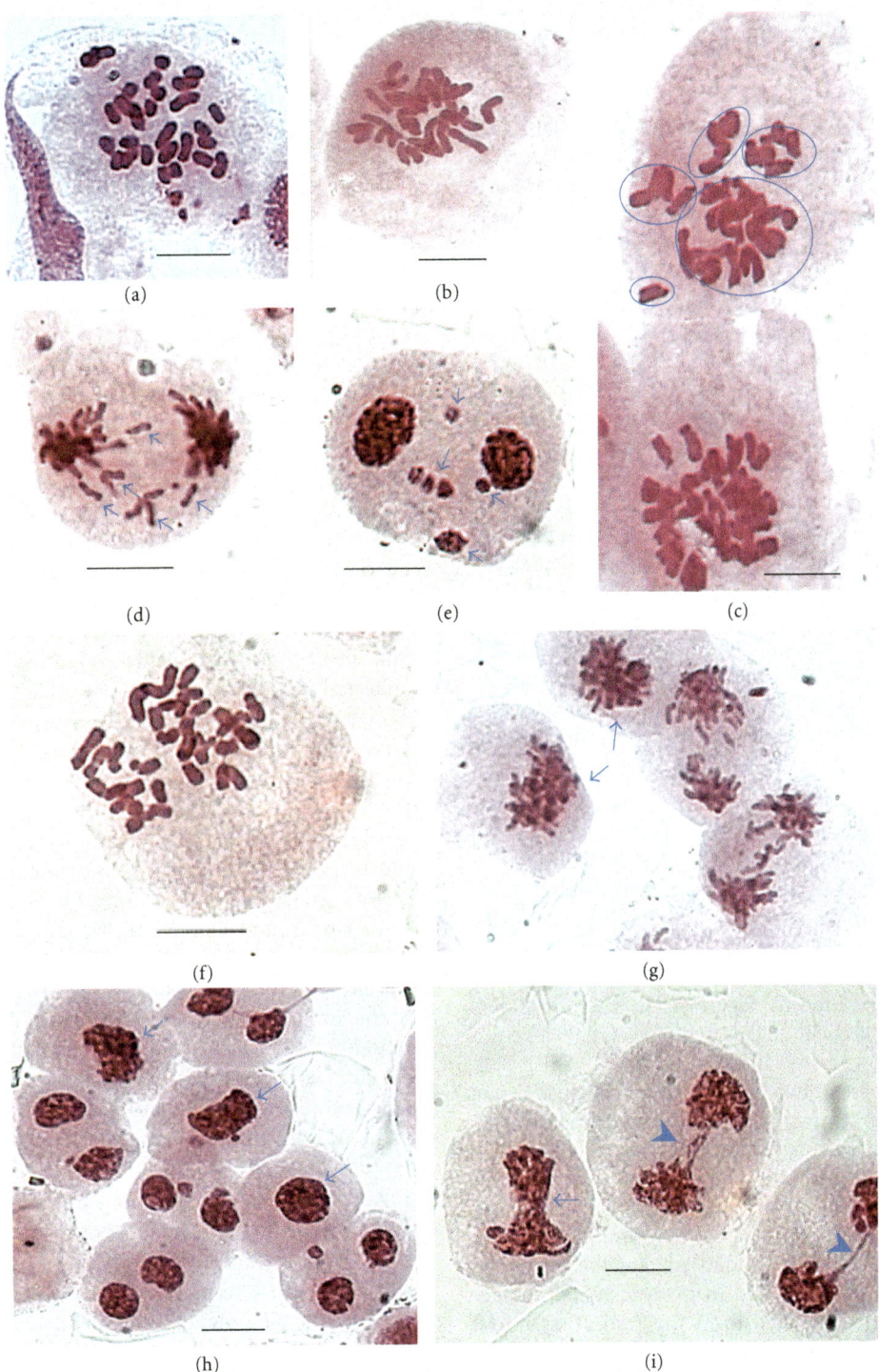

FIGURE 3: (a–i) Meiocytes with erratic male meiosis at first meiotic division in *R. laetus*. (a) A PMC with 28 randomly dispersed univalent chromosomes at M-I. (b) 28 univalent chromosomes positioned towards the periphery of PMC at M-I. (c) In one of the PMC univalent chromosomes lying in groups (encircled) and in an adjacent PMC in a single group in the centre. (d) A PMC at A-I with lagging chromosomes (arrowed). (e) Micronuclei at T-I (arrowed). (f) Unequal distribution of chromosomes (11 : 17) at A-I. (g) Restitution nuclei at A-I (arrowed). (h) Restitution nuclei at T-I (arrowed). (i) Thick (arrowed) and thin (arrowhead) chromatin bridge at T-I. Scale bar = 10 μm.

PMC to form restitution nuclei (Figures 3(g) and 3(h)). Even some of the PMCs showed thick chromatin bridges at anaphases and telophases which did not allow the separation of chromatin material and thus formed restitution nuclei (Figure 3(i)).

3.2. Chromosomal Behaviour during Meiosis II.

3.2. Chromosomal Behaviour during Meiosis II. Behaviour of chromosomes during different stages of meiosis II was also erratic and was characterized by the irregular segregation of chromosome at two poles of M-II. The most common distribution of chromosomes in the PMCs at M-II was 10 : 18 (Figure 4(a)). In some of the PMCs the chromosome remained scattered and unoriented during M-II and anaphase II (Figures 4(b) and 4(c)). Some of the chromosomes (1–6) lagged behind at A-II (62.69%, Figure 4(d)) and did not get included in four haploid nuclei at T-II and constituted micronuclei during sporad stage (Figure 4(e)). Formation of restitution nuclei at second meiotic division can be seen in Figure 4(f) where one PMC showed four haploid nuclei at T-II and the adjacent PMC with only two restitution (unreduced) nuclei which probably resulted into dyad formation as evidenced from the presence of the dyads during microsporogenesis. Analysis of 1445 sporads during microsporogenesis revealed the presence of dyads in 73.77% (1066/1445) of the observed sporads (without micronuclei 299/1445, 20.70%, Figure 4(g), or with micronuclei 767/1445, 53.07%, Figure 4(g)) or, triads (71/1445, 4.91%, Figure 4(g)) and tetrads with micronuclei (308/1445, 21.32%, Figure 4(g)).

3.3. 2n (Unreduced) Pollen Grain Formation. Although pollen fertility was not affected significantly (92%, Figure 4(h)), the erratic meiotic behaviour resulted in two sizes of pollen grains. Depending on the size, these apparently fertile pollen grains were categorized as n (22.48–27.98 μm × 19.27–24.77 μm, normal reduced) and $2n$ (29.82–33.49 μm × 25.23–32.56 μm, unreduced; Figure 4(i)). The $2n$ pollen grains were noticed to be in higher frequency (59.90%) compared to the n pollen grains (40.10%).

Each dyad give rise to two $2n$ microspores whereas a triad produced only one $2n$ microspore and two n microspores. The frequency of apparently fertile $2n$ pollen grains which was estimated from different types of sporads found to be 61.59%. The frequency of $2n$ pollen grains estimated collectively from dyads and triads was almost near the observed one, 59.90% which indicated that the $2n$ pollen grains in *R. laetus* were resulted from dyads and triads at sporad stage which originated from the restitution nuclei observed during meiosis I and II.

4. Discussion

The chromosome number in sexually reproducing eukaryotes does not get doubled at each generation which is ensured through a precise, systematic, and specialized process of meiosis [24]. Vital events of this dynamic process are the recognition of homologues chromosomes and their subsequent pairing and synapsis, which are the prerequisites for genetic recombination and balanced gamete formation. Successful completion of meiosis relies on the above mentioned events during the cell cycle. Interactions between homologous chromosomes during recognition, pairing, and synapsis are highly coordinated and controlled by a large number of genes [25–29]. Dysfunctioning of any one of these events generally resulted in serious consequences like failure of chromosome pairing which may have resulted in unbalanced gamete formation. Synaptic mutants represent one such event in the cell cycle where homologous chromosomes either lack pairing during late prophase I [30] or they are not able to generate or retain chiasmata [26, 31, 32]. To describe the condition where homologous chromosomes failed to pair, the term asynapsis is employed. On the other hand, in cases where chromosomes paired at zygotene and pachytene but failed to remain paired during subsequent stages of meiosis refers to desynapsis. In the present investigation all the analysed PMCs did not reveal the expected chromosome associations of 14II, instead they exhibited completely random dispersion of univalents in the cytoplasm at M-I suggesting asynaptic mutation. Peirson et al. [33] were of the opinion that in most of the asynaptic mutants univalents show random distribution in the cytoplasm at M-I and never align at the equatorial plate while in desynapsis bivalents and univalents orient at the equatorial plate during M-I.

Synaptic variation resulting in complete and partial failure of chromosome pairing of homo/homeologous chromosomes has been studied in a large number of species [15, 26, 34–38]. Physical and chemical mutagens are widely reported to induce synaptic mutations [39–41] but only a few reports are available on the spontaneous origin of synaptic variants in natural populations [36–38, 42, 43]. A large number of factors such as drastic temperature fluctuation, ageing, water content and humidity, soil conditions, and gene mutations [26, 43–45] are reported to be responsible for the spontaneous origin of synaptic mutants in natural populations. The accession with completely normal meiotic behaviour and 100 percent pollen fertility was growing along with the individual which showed synaptic mutation. So the genetic factors seem plausible behind the synaptic irregularities in the species.

Another interesting phenomenon in the presently investigated species is the formation of restitution nuclei. Restitution nuclei were formed because univalent chromosomes/daughter chromatids failed to distribute themselves uniformly at the poles during anaphases. These restitution nuclei resulted into the formation of dyads and triads which subsequently produced two types of pollen grains. Different methods had been used to detect production of $2n$ pollen grains in plants. Owing to the relatively close correlation between larger pollen grains and $2n$ status, the presence of large-sized pollen grains had been frequently used as a criteria for the indication of $2n$ pollen [23, 46–52]. Presently, on the basis of size, two types of pollen grains were categorized as n (normal reduced) and $2n$ (unreduced). The pollen grains which were 1.5-times larger than the normal

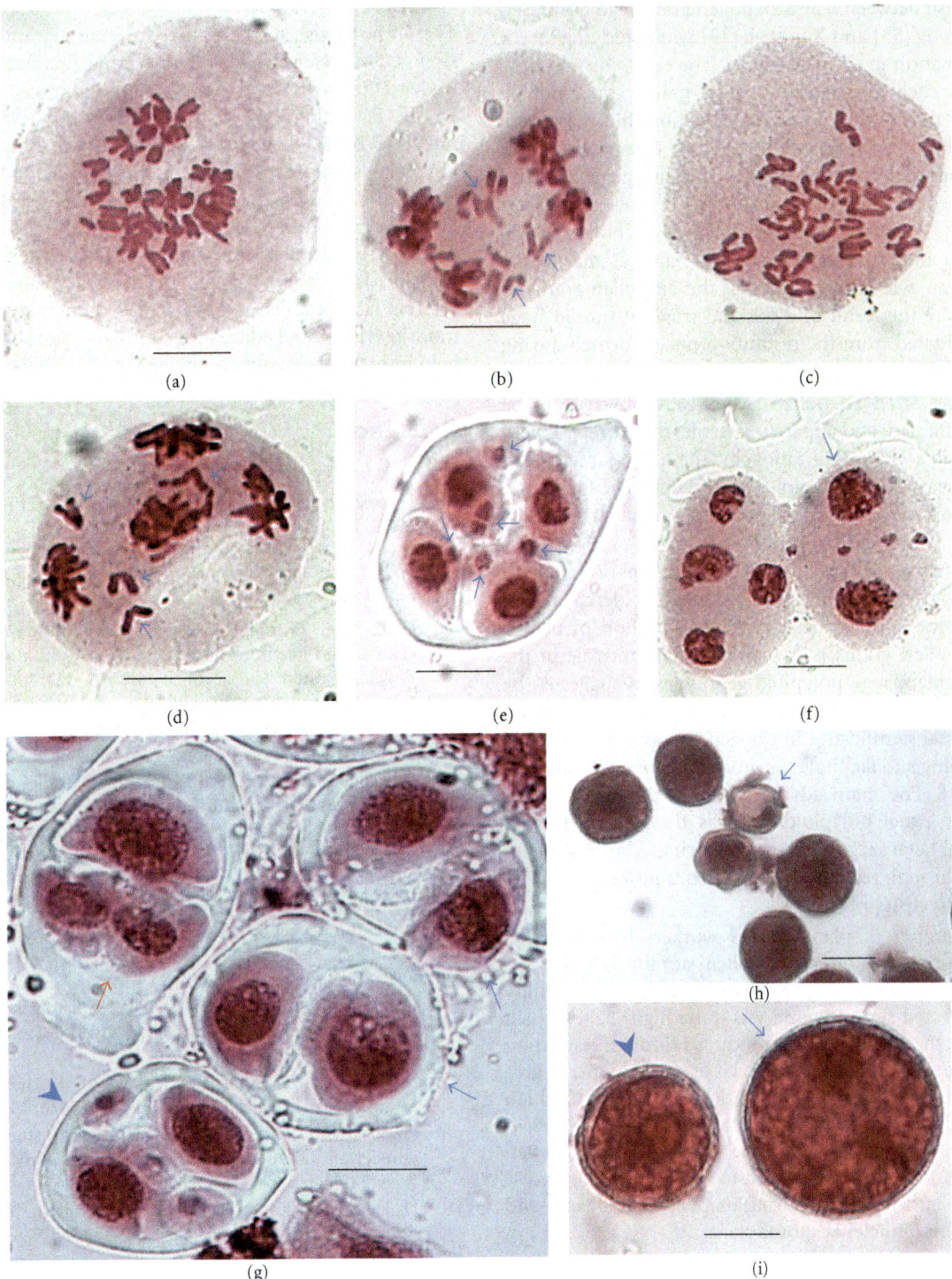

FIGURE 4: (a–i) Meiocytes with erratic male meiosis at second meiotic division in *R. laetus*. (a) Unequal distribution of chromosomes (10 : 18) at M-II. (b) Scattered chromosomes at anaphase-II (arrowed). (c) Random and unoriented distribution of chromosomes at M-II. (d) A PMC showing lagging of chromosomes (arrowed). (e) Tetrad with micronuclei (arrowed). (f) Out of the two PMCs, one showing restitution (unreduced) nuclei (arrowed) and the other with four haploid nuclei at T-II. (g) Sporads in group; dyads with micronuclei (arrowhead), dyads without micronuclei (blue arrows), and triads (red arrow). (h) Apparently stained fertile and unstained or lightly stained, and shriveled sterile (arrowed) pollen grains. (i) Apparently stained fertile *n* (reduced, arrowhead) and 2*n* (unreduced, arrowed) pollen grains. Scale bar = 10 μm (a–d); 20 μm (h).

pollen were regarded here as $2n$ pollen grains. Similar criteria to distinguish between n and $2n$ pollen grains had been used earlier by Peng [53] and Xue et al. [23] while studying the $2n$ pollen formation in Chinese jujube. The exact chromosome number of such double-sized pollen grains could not be ascertained during the present investigations but these were surely of the unreduced in their genetic constitution as is clearly depicted from their size, as increasing nucleus and cytoplasm content may in turn influence pollen diameter [50, 54–58]. The estimated frequency of $2n$ pollen grains from dyads and triads was almost the same as that of the observed one, which indicated that the $2n$ pollen grains in *R. laetus* were the result of dyads and triads at sporad stage which originated from the restitution nuclei formed during meiosis I and II.

The large-sized $2n$ pollen grains were observed to be well filled, stained, and apparently fertile; therefore, it is very much possible that fertilization by these $2n$ gametes can produce intraspecific polyploids [15, 59–63]. The formation of $2n$ gametes is a common phenomenon in the plants which may result from a variety of different meiotic irregularities [63–65]. Unreduced gametes ($2n$ pollen grains) or gametes with somatic chromosome number are considered one of the main processes for natural polyploidization of plants. These $2n$ pollen grains may play an important role in the establishment of new polyploid genotypes as suggested by Dewitte et al. [66] and Silva et al. [67]. Unreduced gametes are of colossal significance in cytogenetics as well as applied plant breeding and facilitate the production of new polyploid species [23]. The main advantage which $2n$ pollen grains offer over asexual polyploidization is the transmission of the parental heterozygosity to the offspring. The $2n$ gametes produced through restitution nuclei can transfer at least 75–80% heterozygosity [47, 68].

In a number of plants earlier workers have reported that synaptic mutation causes pollen sterility [15, 31, 36–38, 69–72]. However, in the present study pollen fertility was not affected seriously and was quite high (92%) which may be due the fact that dyads produced through restitution nuclei are genetically balanced which lead to a higher degree of pollen fertility. Similar observations regarding the high pollen fertility in an asynaptic mutant of *Allium amplectens* had been made by Levan [73]. The presence of some pollen sterility (8%) could be attributed to the presence of unoriented univalents which lag during anaphases/telophases, and constitute micronuclei at sporad stage.

5. Conclusions

Presently studied accession with erratic male meiosis is a spontaneous asynaptic mutant in which univalent chromosomes behaved in a highly irregular manner resulting into restitution nuclei and consequently $2n$ pollen grains. Furthermore, authors safely conclude that $2n$ pollen grains may have played a role in the evolution of species by forming the polyploid genotypes through sexual polyploidization as has been suggested by others [59, 62, 74, 75].

Acknowledgments

The authors wish to thank the University Grants Commission (UGC), New Delhi, for providing financial assistance under DRS SAP I and II, ASIST programme, and UGC-Dr. D. S. Kothari Postdoctoral Fellowship (Award Letter no. F.4-2/2006(BSR)/13-427/2011(BSR)) to Puneet Kumar. Financial support provided by the Council of Scientific and Industrial Research (CSIR) under the Senior Research Fellowship to Puneet Kumar is also greatly acknowledged. Thanks are also due to the head of the Department of Botany, Punjabi, University, Patiala, for necessary laboratory and internet facilities. The authors are also thankful to the anonymous reviewers and editors for comments and suggestions to help improve the quality of the paper.

References

[1] P. N. Mehra and P. Remanandan, "Cytology of some W. Himalayan Ranunculaceae," *Cytologia*, vol. 37, pp. 281–296, 1972.

[2] B. S. Aswal and B. N. Mehrotra, *Flora of Lahaul-Spiti (A Cold Desert in North West Himalaya)*, Bishen Singh Mahendra Pal Singh, Dehra Dun, India, 1994.

[3] Q.-E. Yang, "Cytology of eleven species in the genus *Ranunculus* L. and five in its four related genera from China," *Acta Phytotaxonomica Sinica*, vol. 39, pp. 405–422, 2001.

[4] B. L. Vaidya and K. K. Joshi, "Cytogenetical studies of some species of Himalayan *Anemone* and *Ranunculus* (Ranunculaceae)," *Cytologia*, vol. 68, no. 1, pp. 61–66, 2003.

[5] R. Kabu, B. A. Wafai, and P. Kachroo, "Studies on the genus *Ranunculus* Linn. I. Natural diploidy in *R. laetus* Wall. ex *Hook. et Thoms.* and impact of intraspecific chromosome variability on the phenotype of the species," *Phytomorphology*, vol. 38, pp. 321–325, 1988.

[6] B. K. Bhat, S. K. Bakshi, and M. K. Kaul, "In IOPB chromosome number reports XXXVIII," *Taxon*, vol. 21, pp. 679–684, 1972.

[7] A. K. Sharma and A. K. Sarkar, "Chromosome number reports of plants. In annual report 1967–1968, Cytogenetics Laboratory, Department of Botany, University of Calcutta," *Research Bulletin*, vol. 2, pp. 38–48, 1970.

[8] S. C. Roy and A. K. Sharma, "Cytotaxonomic studies in Indian Ranunculaceae," *Nucleus*, vol. 14, pp. 132–143, 1971.

[9] P. N. Mehra and B. Kaur, "Cytological study of some Himalayan Ranunculaceae," in *Proceedings of the 50th Indian Science Congress*, part 3, pp. 453–454, 1963.

[10] D. Goepfert, "Karyotypes and DNA content in species of *Ranunculus* L. and related genera," *Botaniska Notiser*, pp. 464–489, 1974.

[11] N. D. Agapova and E. A. Zemskova, "Chromosome numbers in some species of the genus *Ranunculus* (Ranunculaceae)," *Botaničeskij Žurnal*, vol. 70, pp. 855–856, 1985.

[12] S. S. Bir and H. Thakur, "SOCGI plant chromosome number reports-II," *The journal Cytology and Genetics*, vol. 19, pp. 114–115, 1984.

[13] S. S. Bir, H. Thakur, and G. S. Chatha, "Chromosomal studies in certain members of Ranunculaceae and Menispermaceae," *Nucleus*, vol. 29, pp. 183–186, 1986.

[14] M. Baltisberger and A. Kocyan, "IAPT/IOPB chromosome data 9," *Taxon*, vol. 59, no. 4, pp. 1298–1302, 2010.

[15] P. Kumar and V. K. Singhal, "Chromosome number, male meiosis and pollen fertility in selected angiosperms of the cold deserts of Lahaul-Spiti and adjoining areas (Himachal Pradesh, India)," *Plant Systematics and Evolution*, vol. 297, no. 3-4, pp. 271–297, 2011.

[16] P. Kumar, V. K. Singhal, P. K. Rana, S. Kaur, and D. Kaur, "Cytology of *Ranunculus laetus* Wall. ex Royle from cold desert regions and adjoining hills of North-west Himalayas (India)," *Caryologia*, vol. 64, no. 1, pp. 25–32, 2011.

[17] C. M. Arora, "New chromosome report II," *Bulletin of the Botanical Survey of India*, vol. 2, p. 305, 1961.

[18] S. N. Sobti and S. D. Singh, "A chromosome survey of Indian medicinal plants-part I," *Proceedings of the Indian Academy of Sciences—Section A*, vol. 54, no. 3, pp. 138–144, 1961.

[19] S. R. Baquar and S. H. Abid Askari, "Chromosome numbers in some flowering plants of West Pakistan," *Génét Ibérica*, vol. 22, pp. 1–11, 1970.

[20] S. R. Baquar and S. H. Abid Askari, "Chromosome numbers in some flowering plants of West Pakistan," *Génét Ibérica*, vol. 22, pp. 41–51, 1970.

[21] S. Khatoon, *Polyploidy in the flora of Pakistan—an analytical study*, Ph.D. thesis, University of Karachi, Karachi, Pakistan, 1991.

[22] S. Khatoon and S. I. Ali, *Chromosome Atlas of the Angiosperms of Pakistan*, University of Karachi, Karachi, Pakistan, 1993.

[23] Z. Xue, P. Liu, and M. Liu, "Cytological mechanism of *2n* pollen formation in Chinese jujube (*Ziziphus jujuba* Mill. 'Linglingzao')," *Euphytica*, vol. 182, no. 2, pp. 231–238, 2011.

[24] E. I. Mikhailova, S. P. Sosnikhina, G. A. Kirillova et al., "Nuclear dispositions of subtelomeric and pericentromeric chromosomal domains during meiosis in asynaptic mutants of rye (*Secale cereale* L.)," *Journal of Cell Science*, vol. 114, no. 10, pp. 1875–1882, 2001.

[25] I. N. Golubovskaya, "Genetic control of meiosis," in *Tsitologiya i genetika meioza (Cytology and Genetics of Meiosis)*, p. 312, Nauka, Moscow, Russia, 1975.

[26] P. R. K. Koduru and M. K. Rao, "Cytogenetics of synaptic mutants in higher plants," *Theoretical and Applied Genetics*, vol. 59, no. 4, pp. 197–214, 1981.

[27] G. S. Roeder, "Meiotic chromosomes: it takes two to tango," *Genes and Development*, vol. 11, no. 20, pp. 2600–2621, 1997.

[28] R. K. Dawe, "Meiotic chromosome organization and segregation in plants," *Annual Review of Plant Biology*, vol. 49, pp. 371–395, 1998.

[29] S. P. Sosnikhina, G. A. Kirillova, E. I. Mikhailova et al., "Genetic control of chromosome synapsis at meiosis in Rye *Secale cereale* L.: the *sy19* gene controlling heterologous synapsis," *Russian Journal of Genetics*, vol. 37, no. 1, pp. 71–79, 2001.

[30] L. F. Randolph, "Chromosome numbers in *Zea mays* L.," *Cornell University Agricultural Experiment Station Mem*, vol. 117, pp. 1–44, 1928.

[31] H. W. Li, W. K. Pao, and C. H. Li, "Desynapsis in the common wheat," *American Journal of Botany*, vol. 32, pp. 92–101, 1945.

[32] R. Rieger, A. Michaelis, and M. M. Green, *Glossary of Genetics and Cytogenetics*, Springer, New York, NY, USA, 4th edition, 1976.

[33] B. N. Peirson, S. E. Bowling, and C. A. Makaroff, "A defect in synapsis causes male sterility in a T-DNA-tagged *Arabidopsis thaliana* mutant," *Plant Journal*, vol. 11, no. 4, pp. 659–669, 1997.

[34] M. L. H. Kaul and T. G. K. Murthy, "Mutant genes affecting higher plant meiosis," *Theoretical and Applied Genetics*, vol. 70, no. 5, pp. 449–466, 1985.

[35] R. J. Singh, *Plant Cytogenetics*, CRC Press, Boca Raton, Fla, USA, 2nd edition, 2002.

[36] P. Kumar, V. K. Singhal, and D. Kaur, "Impaired male meiosis due to irregular synapsis coupled with cytomixis in a new diploid cytotype of *Dianthus angulatus* (Caryophyllaceae) from Indian Cold Deserts," *Folia Geobotanica*. In press.

[37] S. K. Sharma, M. S. Bisht, and M. K. Pandit, "Synaptic mutation-driven male sterility in *Panax sikkimensis* Ban. (Araliaceae) from Eastern Himalaya, India," *Plant Systematics and Evolution*, vol. 287, no. 1, pp. 29–36, 2010.

[38] S. K. Sharma, S. Kumaria, P. Tandon,, and S. R. Rao, "Synaptic variation derived plausible cytogenetical basis of rarity and endangeredness of endemic *Mantisia spathulata* Schult," *The Nucleus*, vol. 54, no. 2, pp. 85–93, 2011.

[39] S. A. Henderson, "Temperature and chiasma formation in *Schistocerca gregaria*—II. Cytological effects at 40°C and the mechanism of heat-induced univalence," *Chromosoma*, vol. 13, no. 4, pp. 437–463, 1962.

[40] F. W. J. Havekes, J. H. De Jong, C. Heyting, and M. S. Ramanna, "Synapsis and chiasma formation in four meiotic mutants of tomato (*Lycopersicon esculentum*)," *Chromosome Research*, vol. 2, no. 4, pp. 315–325, 1994.

[41] R. C. V. Verma and S. N. Raina, "NMU induced translocation and inversion in *Phlox drummondii*," *Cytologia*, vol. 47, no. 3-4, pp. 609–614, 1982.

[42] R. B. Singh, B. D. Singh, V. Laxmi, and R. M. Singh, "Meiotic behaviour of spontaneous and mutagen induced partial desynaptic plants in pearl millet," *Cytologia*, vol. 42, no. 1, pp. 41–47, 1977.

[43] S. R. Rao and A. Kumar, "Cytological investigations in a synaptic variant of *Anogeissus sericea* var. *sericea* Brandis (Combretaceae), an important hardwood tree of Rajasthan," *Botanical Journal of the Linnean Society*, vol. 142, no. 1, pp. 103–109, 2003.

[44] R. Prakken, "Studies of asynapsis in rye," *Hereditas*, vol. 29, pp. 475–495, 1943.

[45] B. S. Ahloowalia, "Effect of temperature and barbiturates on a desynaptic mutant of ryegrass," *Mutation Research*, vol. 7, no. 2, pp. 205–213, 1969.

[46] R. G. Stanley and H. F. Linskens, *Pollen: Biology, Biochemistry, Management*, Springer, Berlin, Germany, 1974.

[47] E. L. Camadro and S. J. Peloquin, "The occurrence and frequency of *2n* pollen in three diploid solanums from Northwest Argentina," *Theoretical and Applied Genetics*, vol. 56, no. 1-2, pp. 11–15, 1980.

[48] G. Orjeda, R. Freyre, and M. Iwanaga, "Production of *2n* pollen in diploid *Ipomoea trifida*, a putative wild ancestor of sweet potato," *Journal of Heredity*, vol. 81, no. 6, pp. 462–467, 1990.

[49] N. O. Maceira, A. D. Haan, R. Lumaret, M. Billon, and J. Delay, "Production of *2n* gametes in diploid subspecies of *Dactylis glomerata* L. 1. Occurrence and frequency of *2n* pollen," *Annals of Botany*, vol. 69, no. 4, pp. 335–343, 1992.

[50] R. C. Jansen and A. P. M. Den Nijs, "The statistical analysis of *2n* and *4n* pollen formation in *Lolium perenne* using pollen diameters," in *Proceedings of the 16th Meeting of the Fodder Crop Section of Eucarpia*, Wageningen, The Netherlands, November 1993.

[51] L. A. Becerra Lopez-Lavalle and G. Orjeda, "Occurrence and cytological mechanism of *2n* pollen formation in a tetraploid accession of *Ipomoea batatas* (sweet potato)," *Journal of Heredity*, vol. 93, no. 3, pp. 185–192, 2002.

[52] S. M. Ghaffari, "Occurrence of diploid and polyploid microspores in *Sorghum bicolor* (Poaceae) is the result of cytomixis,"

African Journal of Biotechnology, vol. 5, no. 16, pp. 1450–1453, 2006.

[53] B. Peng, *Study on* 2n *pollen in Chinese Jujube*, Dissertation, Agricultural University of Hebei, 2008.

[54] R. P. S. Pundir, N. K. Rao, and L. J. G. van der Maesen, "Induced autotetraploidy in chickpea (*Cicer arietinum* L.)," *Theoretical and Applied Genetics*, vol. 65, no. 2, pp. 119–122, 1983.

[55] J. D. Berdhal and R. E. Barker, "Characterization of autotetraploid Russian wild rye produced with nitrous oxide," *Crop Science*, vol. 31, pp. 1153–1155, 1991.

[56] C. Humbert-Droz and F. Felber, "Etude biometrique des stomates et des grains de pollen comme indicateurs du degre de polyploidie chez *Anthoxanthum alpinum* Love et Love," *Bulletin de la Société des Sciences Naturelles de Neuchatel*, vol. 115, pp. 31–45, 1992.

[57] D. Southworth and P. Pfahler, "The effects of genotype and ploidy level on pollen surface sculpturing in maize (*Zea mays* L.)," *American Journal of Botany*, vol. 79, pp. 1418–1422, 1992.

[58] J. A. Fortescue and D. W. Turner, "Reproductive biology," in *Banana Breeding and Production*, M. Pillay and A. Tenkouano, Eds., pp. 145–180, Boca Raton, Fla, USA, 2011.

[59] M. M. Rhoades, "Cross-over chromosomes in unreduced gametes of asynaptic maize," *Genetics*, vol. 32, p. 101, 1947.

[60] J. S. Kim, K. Oginuma, and H. Tobe, "Syncyte formation in the microsporangium of *Chrysanthemum* (asteraceae): a pathway to infraspecific polyploidy," *Journal of Plant Research*, vol. 122, no. 4, pp. 439–444, 2009.

[61] V. K. Singhal and P. Kumar, "Variable sized pollen grains due to impaired male meiosis in the cold desert plants of Northwest Himalayas (India)," in *Pollen: Structure, Types and Effects*, B. J. Kaiser, Ed., pp. 101–126, Nova Science Publishers, New York, NY, USA, 2010.

[62] P. Kumar and V. K. Singhal, "Meiotic aberrations and chromosomal variation in the plants of Lahaul-Spiti and adjoining high hills in Himachal Pradesh," in *Biodiversity Evaluation-Botanical Perspective*, N. S. Atri, R. C. Gupta, M. I. S. Saggoo, and V. K. Singhal, Eds., Bishen Singh Mahendra Pal Singh, Dehra Dun, India, 2011.

[63] F. Bretagnolle and J. D. Thompson, "Gametes with the somatic chromosome number: mechanisms of their formation and role in the evolution of autopolyploid plants," *New Phytologist*, vol. 129, no. 1, pp. 1–22, 1995.

[64] J. R. Harlan and J. M. J. deWet, "On Ö. Winge and a Prayer: the origins of polyploidy," *The Botanical Review*, vol. 41, no. 4, pp. 361–390, 1975.

[65] R. Veilleux, "Diploid and polyploid gametes in crop plants: mechanisms of formation and utilization in plant breeding," *Plant Breeding Reviews*, vol. 3, pp. 253–288, 1985.

[66] A. Dewitte, T. Eeckhaut, J. Van Huylenbroeck, and E. Van Bockstaele, "Meiotic aberrations during 2n pollen formation in *Begonia*," *Heredity*, vol. 104, no. 2, pp. 215–223, 2010.

[67] N. Silva, A. B. Mendes-Bonato, J. G.C. Sales, and M. S. Pagliarini, "Meiotic behavior and pollen viability in *Moringa oleifera* (Moringaceae) cultivated in southern Brazil," *Genetics and Molecular Research*, vol. 10, no. 3, pp. 1728–1732, 2011.

[68] S. J. Peloquin, L. S. Boiteux, P. W. Simon, and S. H. Jansky, "A chromosome-specific estimate of transmission of heterozygosity by 2n gametes in potato," *Journal of Heredity*, vol. 99, no. 2, pp. 177–181, 2008.

[69] R. K. Soost, "Comparative cytology and genetics of asynaptic mutants in *Lycopersicon esculentum* Mill," *Genetics*, vol. 36, pp. 410–434, 1951.

[70] J. Sjödin, "Induced asynaptic mutants in *Vicia faba* L.," *Hereditas*, vol. 66, pp. 215–232, 2003.

[71] M. K. Pandit and C. R. Babu, "The effects of loss of sex in clonal populations of an endangered perennial *Coptis teeta* (Ranunculaceae)," *Botanical Journal of the Linnean Society*, vol. 143, no. 1, pp. 47–54, 2003.

[72] S. Bala, B. Kaushal, H. Goyal, and R. C. Gupta, "A case of synaptic mutant in *Erigeron karvinskianus* DC. (Latin American Fleabane)," *Cytologia*, vol. 75, no. 3, pp. 299–304, 2010.

[73] A. Levan, "The cytology of Allium amplectens and the occurrence in nature of its asynapsis," *Hereditas*, vol. 26, pp. 353–394, 1940.

[74] V. K. Singhal, P. K. Rana, and P. Kumar, "Syncytes during male meiosis resulting in 2n pollen grain formation in *Lindelofia longiflora* var. *falconeri*," *Journal of Systematics and Evolution*, vol. 49, pp. 406–410, 2011.

[75] J. Ramsey and D. W. Schemske, "Pathways, mechanisms, and rates of polyploid formation in flowering plants," *Annual Review of Ecology and Systematics*, vol. 29, pp. 467–501, 1998.

12

NADP-Dependent Isocitrate Dehydrogenase from *Arabidopsis* Roots Contributes in the Mechanism of Defence against the Nitro-Oxidative Stress Induced by Salinity

Marina Leterrier,[1] Juan B. Barroso,[2] Raquel Valderrama,[2] José M. Palma,[1] and Francisco J. Corpas[1]

[1] *Departamento de Bioquímica, Biología Celular y Molecular de Plantas, Estación Experimental del Zaidín, CSIC, Apartado 419, 18080 Granada, Spain*
[2] *Grupo de Señalización Molecular y Sistemas Antioxidantes en Plantas, Unidad Asociada al CSIC (EEZ), Departamento de Bioquímica y Biología Molecular, Universidad de Jaén, 23071 Jaén, Spain*

Correspondence should be addressed to Francisco J. Corpas, javier.corpas@eez.csic.es

Academic Editors: S. Cuzzocrea and H. Verhoeven

NADPH regeneration appears to be essential in the mechanism of plant defence against oxidative stress. Plants contain several NADPH-generating dehydrogenases including isocitrate dehydrogenase (NADP-ICDH), glucose-6-phosphate dehydrogenase (G6PDH), 6-phosphogluconate dehydrogenase (6PGDH), and malic enzyme (ME). In *Arabidopsis* seedlings grown under salinity conditions (100 mM NaCl) the analysis of physiological parameters, antioxidant enzymes (catalase and superoxide dismutase) and content of superoxide radical ($O_2^{\bullet-}$), nitric oxide (NO), and peroxynitrite ($ONOO^-$) indicates a process of nitro-oxidative stress induced by NaCl. Among the analysed NADPH-generating dehydrogenases under salinity conditions, the NADP-ICDH showed the maximum activity mainly attributable to the root NADP-ICDH. Thus, these data provide new insights on the relevance of the NADP-ICDH which could be considered as a second barrier in the mechanism of response against the nitro-oxidative stress generated by salinity.

1. Introduction

In higher plants, salinity can provoke alterations in the metabolism of proteins and nucleic acids, photosynthesis and respiration [1–3]. In addition, the production and participation of reactive oxygen species (ROS) during different plant stress conditions including salinity is also well documented [4–8], and more recently the involvement of nitric oxide (NO) and related molecules designated as reactive nitrogen species (RNS) seems to be also a complementary part of the mechanism of response of plants against environmental stresses [9] which can participate in a nitro-oxidative stress situation.

NADPH is a key cofactor in the cellular redox homeostasis, being an indispensable electron donor in numerous enzymatic reactions, biosynthetic pathways, and detoxification processes [10, 11]. In this sense, NADPH is necessary in the metabolism of ROS and RNS; for example, it is a reducing equivalent for the regeneration of reduced glutathione (GSH) by glutathione reductase (component of ascorbate-glutathione cycle) and for the activity of the NADPH-dependent thioredoxin system, two important cell antioxidants against oxidative damage. Moreover, NADPH is also required for the generation of superoxide radical by the NADPH oxidase (NOX) [12], but is also a necessary cofactor for the generation of nitric oxide (NO) by the L-arginine-dependent nitric oxide synthase activity [13]. The most important enzymes which have the capacity to generate reducing power in the form of NADPH in plants are the ferredoxin-NADP reductase as a component of photosystem I [14] and a group of NADP-dehydrogenases located in different subcellular compartments which includes the NADP-isocitrate dehydrogenase (NADP-ICDH), the glucose-6-phosphate dehydrogenase (G6PDH) and 6-phosphogluconate dehydrogenase (6PGDH) (both belonging to the pentose phosphate pathway), and the NADP-malic enzyme (ME) [15–17]. Among the different NADP-ICDH isoforms present in higher plants, it has been shown that the cytosolic NADP-ICDH represents more than 90% of the total cellular NADP-ICDH activity [18–21], and very recently *in vitro* assays have shown that the *Arabidopsis* cytosolic NADP-ICDH activity from *Arabidopsis* roots and leaves is differentially regulated by molecules involved in ROS and RNS metabolism [22] including H_2O_2, NO, and $ONOO^-$ indicating a metabolic interconnection among this enzyme and these molecules.

In the present work, using *Arabidopsis* as model plant, it is shown that under salinity (100 mM NaCl) stress there is a concomitant nitro-oxidative imbalance that is accompanied by a general induction of NADP-dehydrogenase activities being the NADP-ICDH from roots, the enzyme with the most prominent activity. The present data support that the recycling of NADPH is important as a mechanism against cellular nitro-oxidative damage produced by salinity.

2. Material and Methods

2.1. Plant Material and Growth Conditions.

Arabidopsis thaliana ecotype Columbia seeds were surface sterilized for 5 min in 70% (v/v) ethanol containing 0.1% (w/v) SDS, then placed for 20 min in sterile water containing 20% (v/v) bleach and 0.1% (w/v) SDS, and washed four times in sterile water. The seeds were sown for 2 days at 4°C in the dark for vernalization on the basal growth medium composed of 4.32 g/L commercial Murashige and Skoog medium (Sigma) with a pH of 5.5, containing 1% (w/v) sucrose and 0.8% (w/v) phyto agar. The Petri plates containing the *Arabidopsis* seeds were then grown at 22°C/18°C (16 h light/8 h dark, long-day conditions) under a light intensity of 100 μE m^{-2} s^{-1}. For the experiments with NaCl stress, 6-day-old seedlings were transferred to MS medium plates both with and without 100 mM NaCl for another 7 days under long-day conditions [23].

2.2. Crude Extracts of Plant Tissues.

Arabidopsis seedlings were frozen in liquid N_2 and ground in a mortar with a pestle. The powder was suspended in a homogenizing medium containing 50 mM Tris-HCl, pH 7.8, 0.1 mM EDTA, 0.2% (v/v) Triton X-100, and 10% (v/v) glycerol. Homogenates were centrifuged at 27,000 g for 20 min, and the supernatants were used for the assays.

2.3. Histochemical Analyses.

Histochemical detection of plasma membrane loss integrity in *Arabidopsis* root apexes was performed by the method described by Yamamoto et al. [24]. For this analysis, the *Arabidopsis* seedlings were incubated in 15 mL of Evans blue solution [0.2% (w/v) in water] for 10 min, and then they were washed three times in distilled water for 10 min each. Blue color indicates damage to the plasma membrane.

2.4. Enzymatic Activity Assays.

Catalase activity (EC 1.11.1.6) was determined by measuring the disappearance of H_2O_2, as described by Aebi [25]. Glycolate oxidase (GOX; EC 1.1.3.1) was assayed as described previously [26] by measuring the formation of glyoxylate-phenylhydrazone. Hydroxypyruvate reductase (HPR) was assayed according to Schwitzguébel and Siegenthaler [27].

Glucose-6-phosphate dehydrogenase (G6PDH; EC 1.1.1.49) activity was determined spectrophotometrically by recording the reduction of NADP at 340 nm. Assays were performed at 25°C in a reaction medium (1 mL) containing 50 mM HEPES, pH 7.6, 2 mM $MgCl_2$, and 0.8 mM NADP, and the reaction was initiated by the addition of 5 mM glucose-6-phosphate. For the determination of 6-phosphogluconate dehydrogenase (6PGDH; EC 1.1.1.44) activity, the reaction mixture was similar to that described for G6PDH, but the substrate was 5 mM 6-phosphogluconate [28]. NADP-isocitrate dehydrogenase (NADP-ICDH, EC 1.1.1.42) activity was also measured by following the NADP reduction according to Corpas et al. [29]. Thus, the assay was performed at 25°C in a reaction medium (1 mL) containing 50 mM HEPES, pH 7.6, 2 mM $MgCl_2$ and 0.8 mM NADP, and the reaction was initiated by the addition of 10 mM 2R,3S-isocitrate. NADP-malic enzyme (NADP-ME; EC 1.1.1.40) activity was also determined spectrophotometrically by recording the reduction of NADP at 340 nm using the same reaction mixture (1 mL) indicated above for other dehydrogenases, but in this case, the reaction was initiated by the addition of 1 mM L-malate [30].

2.5. Superoxide Dismutase Isozymes.

Superoxide dismutase (SOD; EC 1.15.1.1) isozymes were separated by native PAGE on 12% acrylamide gels and visualized by a photochemical NBT (nitroblue tetrazolium) reduction method [31]. To identify the type of SOD isozymes, gels were preincubated separately at 25°C for 30–45 min in 50 mM K-phosphate, pH 7.8, in the presence or absence of either 2 mM KCN or 5 mM H_2O_2. CuZn-SODs are inhibited by CN^- and H_2O_2, Fe-SODs are inhibited by H_2O_2 but not by CN^-, whilst Mn-SODs are not inhibited by either CN^- or H_2O_2 [32].

2.6. RNA Isolation and Semiquantitative RT-PCR.

Total RNA was extracted with Trizol according to Gibco BRL, Life Technologies. Two μg of total RNA were used to produce cDNA by RT-PCR. Semiquantitative reverse transcription-PCR amplification of actin cDNA from *Arabidopsis* was chosen as control. *NADP-ICDH* and *actin* cDNAs were amplified by the PCR as follows: 1 μL of each cDNA (30 ng) was added to 250 mM dNTPs, 1.5 mM $MgCl_2$, 1 × PCR buffer, 0.5 U of Hot Master TaqTM DNA polymerase (Eppendorf), and 0.5 mM of each primer (cytosolic *ICDH*: 5′-TTGTGG-AGAGGAGTGTTGAG-3′ and 5′-CCTAAAAGACCCTAA-TACCA-3′; mitochondrial/chloroplastic *ICDH* 5′-GGG-AATTGGGAACAATACA-3′ and 5′-TGTTGGATACGA-AACTGAA-3′; peroxisomal *ICDH*: 5′-CAGCGTGATGTT-TGATTTG-3′ and 5′-TAGCCA TTTCTGTTGATTGG-3′; *actin II*: 5′-TCCCTCAGCACATTCCAGCAGAT-3′ and 5′-AACGATTCCTGGACCTGCCTCATC-3′) in a final volume

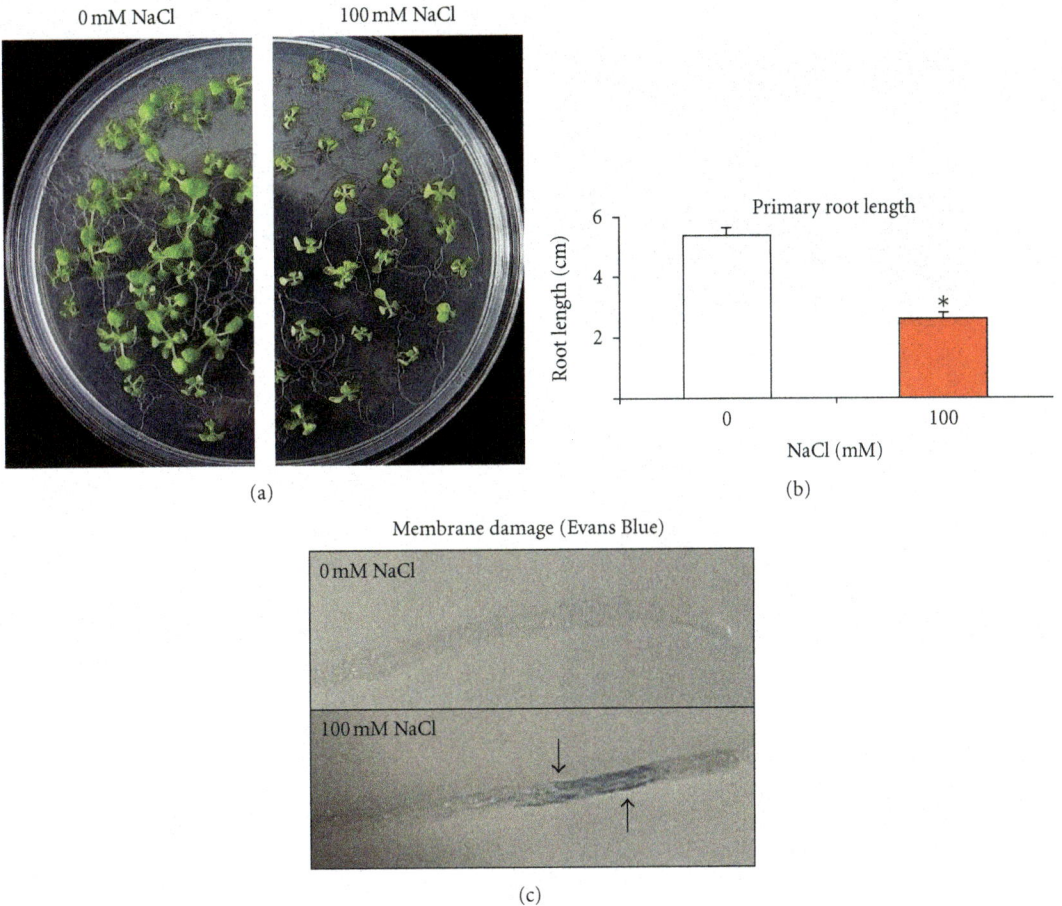

FIGURE 1: Effect of salinity in *Arabidopsis* seedlings growth. (a) Appearance of 14-day-old *Arabidopsis* seedling growth in MS medium supplemented and nonsupplemented with 100 mM NaCl. (b) Primary root length. Results are the mean of three different experiments ± SEM. *Differences in relation to control values were significant at $P < 0.05$. (c) Histochemical detection of plasma membrane integrity by staining with Evan blue solutions. Blue colour (arrows) indicates the root area where the membrane integrity is affected by salinity.

of 20 μL. Reactions were carried out in a Hybaid thermocycler. A first step of 2 min at 95°C was followed by 28 cycles of 20 s at 94°C, 20 s at 55°C, and 30 s at 65°C plus a final step of 10 min at 65°C. Then, PCR products were detected by electrophoresis in 1% (w/v) agarose gels and staining with ethidium bromide. Quantification of the bands was performed using a Gel Doc system (Bio-Rad Laboratories) coupled with a high-sensitive charge-coupled device (CCD) camera.

2.7. Detection of Superoxide Radical ($O_2^{\bullet-}$), Nitric Oxide (NO), and Peroxynitrite ($ONOO^-$) by Confocal Laser Scanning Microscopy (CLSM).

Detection of superoxide radicals ($O_2^{\bullet-}$) in roots of *Arabidopsis* seedlings was carried out using 10 μM dihydroethidium (DHE) [33] by incubation of *Arabidopsis* seedlings with this fluorescent probe for 1 h at 37°C in darkness.

Nitric oxide (NO) and peroxynitrite ($ONOO^-$) were detected using the fluorescent reagents 10 μM of 4-aminomethyl-2′,7′-difluorofluorescein diacetate (DAF-FM DA, Calbiochem) and 10 μM 3′-(p-aminophenyl) fluorescein (APF, Invitrogen), respectively, according to Corpas et al. [34].

In all cases, the images obtained by CLSM system (Leica TCS SL; Leica Microsystems, Wetzlar, Germany) from control and treated *Arabidopsis* seedlings were maintained constant during the course of the experiments in order to produce comparable data. The images were processed and analyzed using statistical Leica-Lite software.

2.8. Other Assays.

Protein concentration was determined with the Bio-Rad Protein Assay (Hercules, CA) using bovine serum albumin as standard. To estimate the statistical significance between means, the data was analyzed by Student's *t* test.

3. Results

3.1. Effect of Salinity in Physiological Parameters and in the Metabolism of Reactive Oxygen Species (ROS).

Previous studies have shown that *Arabidopsis* seedlings grown with 100 mM NaCl underwent salinity stress [23, 35], and therefore this concentration was chosen for the salinity treatment. Figure 1(a) shows the appearance of *Arabidopsis* seedlings grown with 100 mM NaCl. These seedlings had a smaller size,

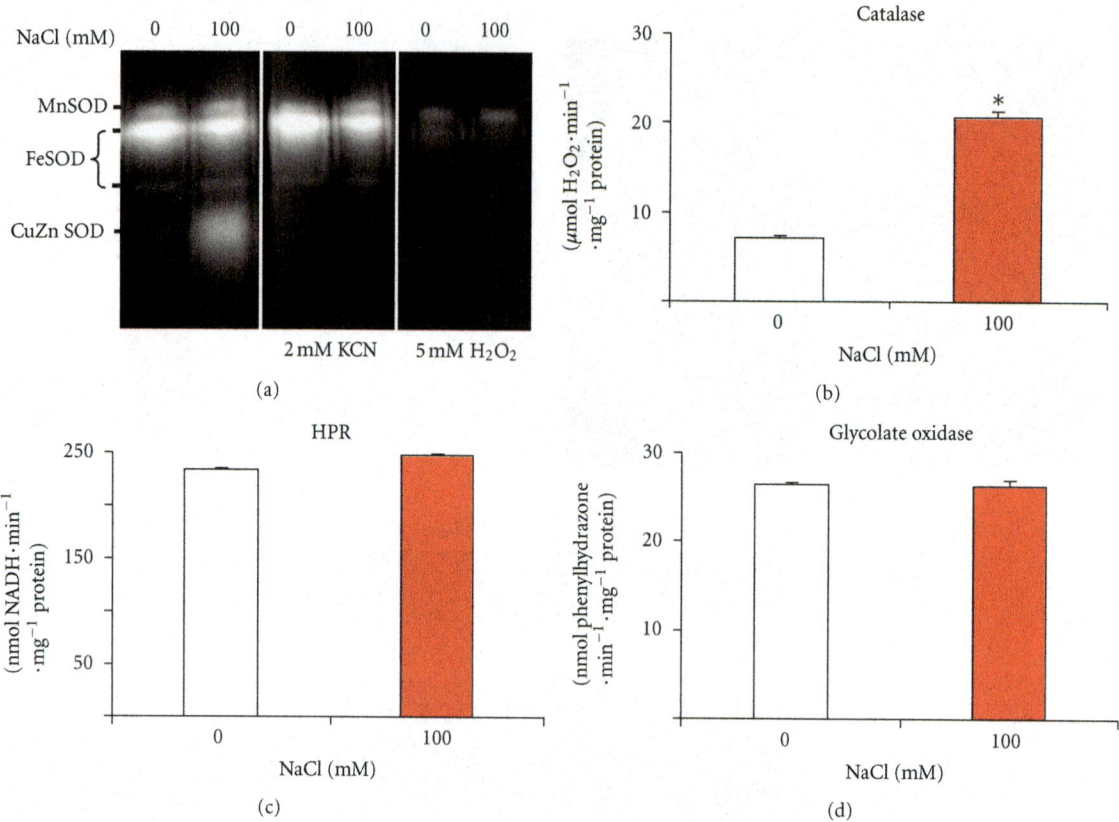

FIGURE 2: Enzyme activities in *Arabidopsis* seedlings exposed to 100 mM NaCl. (a) Superoxide dismutase (SOD) isoenzymes activities in native gels incubated in the absence and the presence of specific inhibitors, either cyanide or hydrogen peroxide. (b) Catalase activity. (c) Hydroxypruvate reductase (HPR) activity. (d) Glycolate oxidase activity. Results are the mean of three different experiments ± SEM. *Differences in relation to control values were significant at $P < 0.05$.

leaves with chlorotic symptoms, and a root length reduced by 24% (Figure 1(b)). To determine whether NaCl could affect the cell-membrane integrity of the root cells, a histochemical method based on the Evans Blue staining was used. Thus, an intense blue color appeared in roots of seedlings exposed to 100 mM NaCl, indicating the loss of cell-membrane integrity (Figure 1(c)).

To establish whether our experimental salinity conditions affect ROS metabolism, the activity of the first line of anti-oxidant enzymes was analyzed, including superoxide dismutase (SOD) and catalase, and some key enzymes of the photorespiratory pathway (NADH-hydroxypyruvate reductase and glycolate oxidase). The analysis of SOD activity in native gel showed the presence of four isozymes which differed according to their susceptibility to the inhibitor, whether cyanide or hydrogen peroxide: one Mn-SOD, two Fe-SODs, and one CuZn-SOD, which displayed increasing electrophoretic mobility (Figure 2). As can also be seen, only the CuZn-SOD isozyme was strongly induced by salinity without significantly affecting the other SOD isozymes (Figure 2(a)). On the other hand, the catalase activity increased 2.9-fold under salinity conditions (Figure 2(b)). However, the hydroxypyruvate reductase (HPR) and the glycolate oxidase activities were not affected after 100 mM NaCl treatment (Figures 2(c) and 2(d), resp.).

3.2. Cellular Analysis of Superoxide Radical ($O_2^{•-}$), NO, and ONOO$^-$ Production Induced by Salinity.

Figure 3(a) to 3(f) shows the analysis by confocal laser scanning microscope (CLSM) of the content of $O_2^{•-}$, NO, and ONOO$^-$ in the root tips of *Arabidopsis* seedlings exposed to 100 mM NaCl. The cellular production of $O_2^{•-}$ was analyzed using the fluorescent probe DHE, which is specific for this radical. In control seedlings, the green fluorescence corresponding to $O_2^{•-}$ was slightly detected in the root tips (Figure 3(a)). However, in roots from NaCl-treated seedlings, the green fluorescence was intensified in the root tips (Figure 3(b)). When NO generation was analyzed using DAF-FM DA as the fluorescence probe, a significant increase in NO production (green color) was noted in the roots under salt stress with a homogenous distribution throughout the root (Figures 3(c) and 3(d)), whereas in control plants, labeling was detected only in root tips. On the other hand, ONOO$^-$, which results from the reaction between $O_2^{•-}$ and NO was also analyzed in roots by CLSM using the fluorescence probe APF. Figure 3(e) shows the location of ONOO$^-$ in the control roots of *Arabidopsis* seedlings with very slight fluorescent signal. However, this RNS significantly increased in roots under salinity stress with a homogeneous distribution throughout the root (Figure 3(f)), similar to the distribution of the NO.

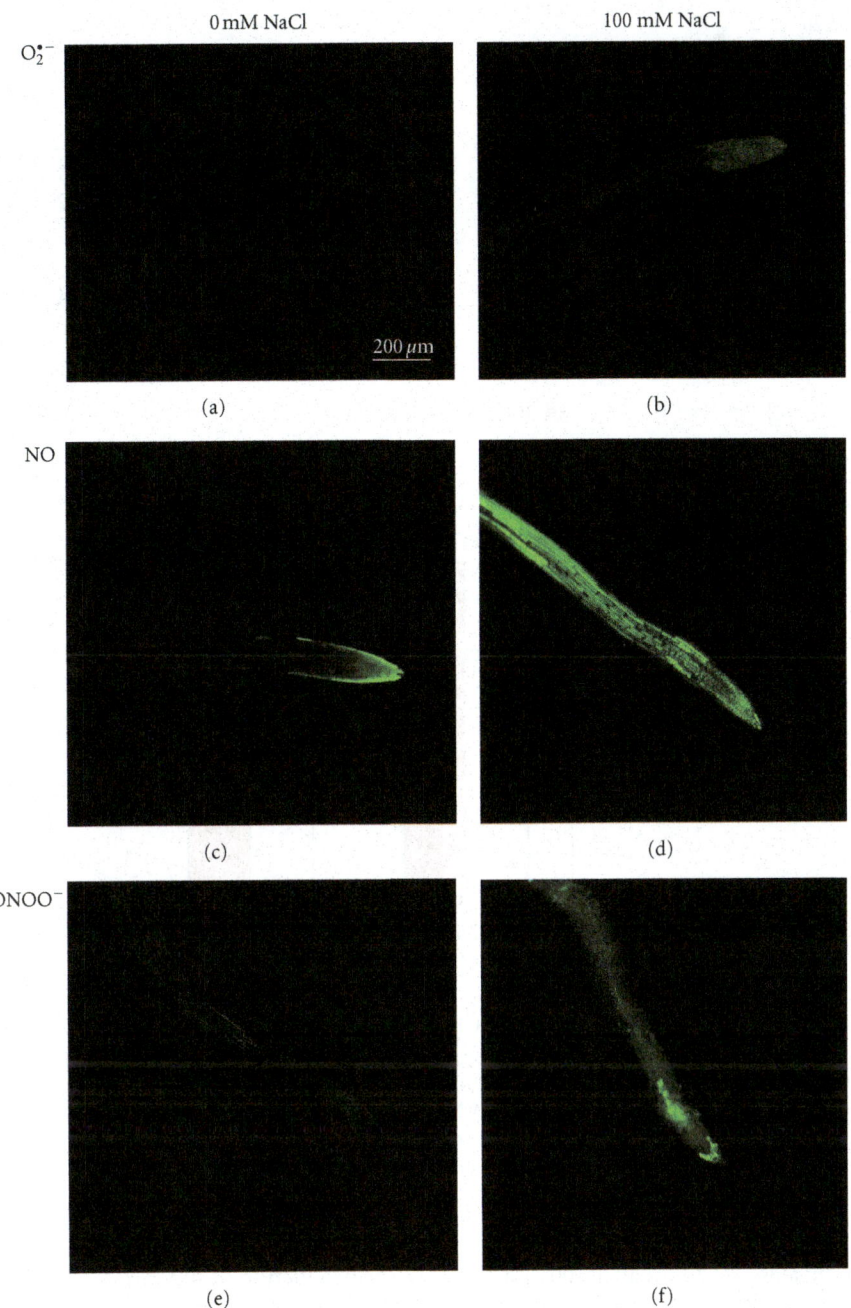

Figure 3: Representative images illustrating the CLSM *in vivo* detection of superoxide radical ($O_2{}^{\bullet-}$) ((a) and (b)), nitric oxide (NO) ((c) and (d)), and peroxynitrite ($ONOO^-$) ((e) and (f)) in root tips of *Arabidopsis* seedlings exposed to 100 mM NaCl.

3.3. Effect of Salinity on NADP-Dehydrogenase Activities. The analysis of the activity of the main NADP dehydrogenases is shown in Figure 4(a). The activity of NADP-ICDH, ME, and G6PDH increased 1.6-, 1.5-, and 1.9-fold, respectively, with respect to control seedlings. However, the 6PGDH activity was not affected by salinity treatment. Considering that the NADP-ICDH showed a higher relative specific activity under salinity conditions in comparison to the other NADP dehydrogenases, further analyses were focused on this enzyme.

Arabidopsis has several NADP-ICDH isoforms localized in different subcellular compartments including the cytosol, chloroplasts/mitochondria and peroxisomes [20]. For an evaluation of the potential contribution of each isoform under salinity stress, its gene expression was analyzed. Figure 4(b) showed the gene expression of the cytosolic (At1g65930), chloroplastic/mitochondrial (At5g14590) and peroxisomal (At1g54340) NADP-ICDH evaluated by semi-quantitative RT-PCR. Contrary to what happened in the activity analysis, none of the genes appeared to undergo significant changes under salinity stress. With the goal of gaining fuller knowledge of the potential function of the NADP-ICDH activity, its activity was investigated independently in

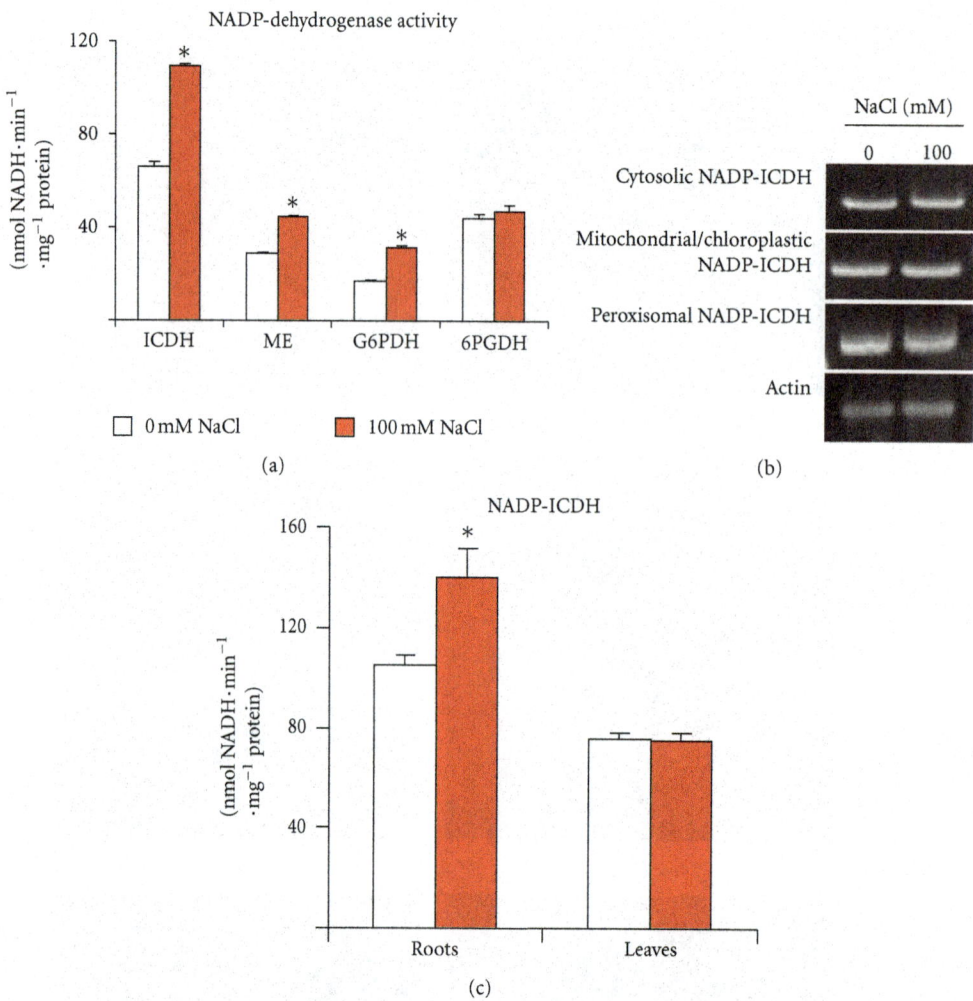

FIGURE 4: (a) Activity of NADP-isocitrate dehydorgenase (ICDH), malic enzymes (ME), glucose-6-phosphate dehydrogenase and 6-phosphogluconate dehydrogenase in *Arabidopsis* seedlings exposed to 100 mM NaCl. (b) Representative agarose electrophoresis gel of the semiquantitative RT-PCR analysis of the cytosolic (At1g65930), mitochondrial/chloroplastic (At5g14590), and peroxisomal (At1g54340) *NADP-ICDH* genes in *Arabidopsis* seedlings exposed to 100 mM NaCl. Gel was visualized by ethidium bromide staining, and *actin* was used as internal control. (c) NADP-ICDH activity in roots and leaves of *Arabidopsis* seedlings exposed to 100 mM NaCl. Results are the mean of three different experiments ± SEM. *Differences in relation to control values were significant at $P < 0.05$.

roots and leaves of *Arabidopsis* seedlings after 100 mM NaCl treatment (Figure 4(c)). NADP-ICDH activity was found to be higher in roots than in leaves from control plants. Also, it was observed that, under salinity conditions, the activity significantly increased (by 39%) in roots whereas the activity in leaves showed no change.

4. Discussion

Salinity is recognized to influence plant productivity due to its negative effects on plant growth, ion balance, and water relations. In addition, in many plant species such as pea [4, 5], tomato [36, 37], or olive [30], the salinity stress is usually accompanied by an oxidative stress. In this sense, the data gathered in our *Arabidopsis in vitro* model system corroborate that salinity (100 mM NaCl) significantly reduces root growth, damages root plasma-membrane integrity, boosts

the production of superoxide radical, and significantly raises catalase and CuZn-SOD activities, although photorespiration appears not to be affected. The remarkable induction of a CuZn-SOD in salt-treated *Arabidopsis* seedlings, enhances the relevance of this enzymatic system in the response of plants to salinity stress, as has been found earlier [5, 30, 36]. On the other hand, the analysis of some RNS such as nitric oxide (NO) and peroxynitrite ($ONOO^-$) also showed a higher content under salinity stress, which also agrees with previous data in different plant species [33, 34, 38, 39]. Therefore, in this context, where the ROS and RNS metabolism is affected under salinity stress, the analysis of NADPH-generating dehydrogenase activity was studied, considering that NADPH is necessary for the metabolism of these species because it occurs in some antioxidant systems such as the ascorbate-glutathione cycle, the generation of superoxide radical ($O_2^{\bullet-}$) by the NADPH oxidase [12], and

NO generation by a L-arginine nitric oxide synthase [13, 14]. Thus, the general increase in the activity of these NADP-dehydrogenases is reasonable considering the increase of peroxynitrite observed in roots. This molecule, being a strong oxidant which results from the interaction of $(O_2^{\bullet-})$ and NO, must provoke cellular damage. Consequently, the general increase of the NADPH-generating dehydrogenases, with the exception of the 6PGDH, suggests the participation of these enzymes in the mechanism of response against the nitro-oxidative stress prompted by the salinity treatment. Accordingly, in dune reed (*Phragmites communis*) callus under 50–150 mM NaCl treatments, the G6PDH activity was induced, being necessary for GSH maintenance and H_2O_2 accumulation under salt stress [40]. Furthermore, in *Carex moorcroftii* callus under salt stress (100 mM NaCl), G6PDH was also involved in the regulation of plasma membrane H^+-ATPase [41]. These results also agree with the behavior of these NADP dehydrogenases under other kinds of environmental stress such as cadmium [42] or low temperature [43] where the activity of some of these NADP-dehydrogenases was induced.

Among these NADP dehydrogenases, special attention was placed on NADP-ICDH, since this activity was higher than that of other NADPH-generating dehydrogenases. In previous works, it has been reported that the NADP-ICDH was significantly greater in oxidative stress situation promoted after paraquat treatment in pea nodule [44], biotic stress in *Arabidopsis* [21], mechanical wounding, high and low temperature in pea leaves [26], and low temperature in pepper leaves [43], thus indicating the contribution of NADP-ICDH to the redox state of the cell. In the facultative halophyte *Mesembryanthemum crystallinum* adapted to high salinity (400 mM), the NADP-ICDH activity increased in leaves and decreased in roots [45]. However, in our experimental model of *Arabidopsis*, the comparison of NADP-ICDH activity between the two organs (roots and leaves) points to a significant role of this enzyme in roots. This difference in NADP-ICDH activity between the two organs in *M. crystallinum* and *A. thaliana* must be related to the degree of resistance to salinity in each plant species, and thus the increase of the NADP-ICDH activity must be related to the NADPH requirement in each organ. Thus, in the case of *M. crystallinum* under salt stress the excess Na^+ is transported very efficiently to the leaves whereas only a minor part is accumulated in root tissue [46]; however, in *Arabidopsis* the situations is totally different, considering the sensitivity of this plant to salinity in comparison to *M. crystallinum* [2].

In summary, these data suggest that the activities of the NADPH-generating dehydrogenases, especially the NADP-ICDH in roots, contributed to maintaining the cellular redox status as a mechanism to support the antioxidative system during the nitro-oxidative stress generated by salinity stress in *Arabidopsis*. Thus, it is proposed that NADP-ICDH dehydrogenase acts in *Arabidopsis* seedlings as a second barrier in the response mechanism of salinity stress, but they could also have a protective function in other types of abiotic stress.

Acknowledgments

This work was supported by ERDF-cofinanced grants from the Ministry of Science and Innovation (BIO2009-12003-C02-01 and BIO2009-12003-C02-02), Spain. Confocal laser scanning microscopy analyses were carried out at the Technical Services of the University of Jaén. It is also acknowledged the excellent technical support provided by Mr. Carmelo Ruíz-Torres.

References

[1] P. M. Hasegawa, R. A. Bressan, J. K. Zhu, and H. J. Bohnert, "Plant cellular and molecular responses to high salinity," *Annual Review of Plant Biology*, vol. 51, pp. 463–499, 2000.

[2] R. Munns and M. Tester, "Mechanisms of salinity tolerance," *Annual Review of Plant Biology*, vol. 59, pp. 651–681, 2008.

[3] M. M. Chaves, J. Flexas, and C. Pinheiro, "Photosynthesis under drought and salt stress: regulation mechanisms from whole plant to cell," *Annals of Botany*, vol. 103, no. 4, pp. 551–560, 2009.

[4] J. A. Hernández, F. J. Corpas, M. Gómez, L. A. del Río, and F. Sevilla, "Salt-induced oxidative stress mediated by activated oxygen species in pea leaf mitochondria," *Plant Physiology*, vol. 89, pp. 103–110, 1993.

[5] J. A. Hernández, E. Olmos, F. J. Corpas, F. Sevilla, and L. A. del Río, "Salt-induced oxidative stress in chloroplasts of pea plants," *Plant Science*, vol. 105, no. 2, pp. 151–167, 1995.

[6] T. Demiral and I. Türkan, "Comparative lipid peroxidation, antioxidant defense systems and proline content in roots of two rice cultivars differing in salt tolerance," *Environmental and Experimental Botany*, vol. 53, no. 3, pp. 247–257, 2005.

[7] S. S. Gill and N. Tuteja, "Reactive oxygen species and antioxidant machinery in abiotic stress tolerance in crop plants," *Plant Physiology and Biochemistry*, vol. 48, no. 12, pp. 909–930, 2010.

[8] G. Miller, N. Suzuki, S. Ciftci-Yilmaz, and R. Mittler, "Reactive oxygen species homeostasis and signalling during drought and salinity stresses," *Plant, Cell and Environment*, vol. 33, no. 4, pp. 453–467, 2010.

[9] F. J. Corpas, M. Leterrier, R. Valderrama et al., "Nitric oxide imbalance provokes a nitrosative response in plants under abiotic stress," *Plant Science*, vol. 181, no. 5, pp. 604–611, 2011.

[10] J. B. Barroso, J. Peragón, C. Contreras-Jurado et al., "Impact of starvation-refeeding on kinetics and protein expression of trout liver NADPH-production systems," *American Journal of Physiology*, vol. 274, no. 6, pp. R1578–R1587, 1998.

[11] G. Noctor, "Metabolic signalling in defence and stress: the central roles of soluble redox couples," *Plant, Cell and Environment*, vol. 29, no. 3, pp. 409–425, 2006.

[12] M. Sagi and R. Fluhr, "Production of reactive oxygen species by plant NADPH oxidases," *Plant Physiology*, vol. 141, no. 2, pp. 336–340, 2006.

[13] F. J. Corpas, J. M. Palma, L. A. del Río, and J. B. Barroso, "Evidence supporting the existence of l-arginine-dependent nitric oxide synthase activity in plants," *New Phytologist*, vol. 184, no. 1, pp. 9–14, 2009.

[14] A. K. Arakaki, E. A. Ceccarelli, and N. Carrillo, "Plant-type ferredoxin-$NADP^+$ reductases: a basal structural framework and a multiplicity of functions," *FASEB Journal*, vol. 11, no. 2, pp. 133–140, 1997.

[15] M. F. Drincovich, P. Casati, and C. S. Andreo, "NADP-malic enzyme from plants: a ubiquitous enzyme involved in different

metabolic pathways," *FEBS Letters*, vol. 490, no. 1-2, pp. 1–6, 2001.

[16] M. Hodges, V. Flesch, S. Gálvez, and E. Bismuth, "Higher plant NADP$^+$-dependent isocitrate dehydrogenases, ammonium assimilation and NADPH production," *Plant Physiology and Biochemistry*, vol. 41, no. 6-7, pp. 577–585, 2003.

[17] R. M. Mateos, D. Bonilla-Valverde, L. A. del Río, J. M. Palma, and F. J. Corpas, "NADP-dehydrogenases from pepper fruits: effect of maturation," *Physiologia Plantarum*, vol. 135, no. 2, pp. 130–139, 2009.

[18] S. Fieuw, B. Muller-Rober, S. Gálvez, and L. Willmitzer, "Cloning and expression analysis of the cytosolic NADP$^+$-dependent isocitrate dehydrogenase from potato. Implications for nitrogen metabolism," *Plant Physiology*, vol. 107, no. 3, pp. 905–913, 1995.

[19] F. Gallardo, S. Gálvez, P. Gadal, and F. M. Cánovas, "Changes in NADP$^+$-linked isocitrate dehydrogenase during tomato fruit ripening. Characterization of the predominant cytosolic enzyme from green and ripe pericarp," *Planta*, vol. 196, no. 1, pp. 148–154, 1995.

[20] M. Leterrier, L. A. del Río, and F. J. Corpas, "Cytosolic NADP-isocitrate dehydrogenase of pea plants: genomic clone characterization and functional analysis under abiotic stress conditions," *Free Radical Research*, vol. 41, no. 2, pp. 191–199, 2007.

[21] A. Mhamdi, C. Mauve, H. Gouia, P. Saindrenan, M. Hodges, and G. Noctor, "Cytosolic NADP-dependent isocitrate dehydrogenase contributes to redox homeostasis and the regulation of pathogen responses in Arabidopsis leaves," *Plant, Cell and Environment*, vol. 33, no. 7, pp. 1112–1123, 2010.

[22] M. Leterrier, J. B. Barroso, J. M. Palma, and F. J. Corpas, "Biochemical and pharmacological characterization of cytosolic NADP-isocitrate dehydrogenase in Arabidopsis leaves and roots," *Biologia Plantarum*. In press.

[23] J. Sun, H. Jiang, Y. Xu et al., "The CCCH-type zinc finger proteins AtSZF1 and AtSZF2 regulate salt stress responses in Arabidopsis," *Plant and Cell Physiology*, vol. 48, no. 8, pp. 1148–1158, 2007.

[24] Y. Yamamoto, Y. Kobayashi, and H. Matsumoto, "Lipid peroxidation is an early symptom triggered by aluminum, but not the primary cause of elongation inhibition in Pea roots," *Plant Physiology*, vol. 125, no. 1, pp. 199–208, 2001.

[25] H. Aebi, "Catalase *in vitro*," *Methods in Enzymology*, vol. 105, pp. 121–126, 1984.

[26] M. W. Kerr and D. Groves, "Purification and properties of glycollate oxidase from *Pisum sativum* leaves," *Phytochemistry*, vol. 14, no. 2, pp. 359–362, 1975.

[27] J. P. Schwitzguébel and P. A. Siegenthaler, "Purification of peroxisomes and mitochondria from spinach leaf by percoll gradient centrifugation," *Plant Physiology*, vol. 75, pp. 670–674, 1984.

[28] F. J. Corpas, J. B. Barroso, L. M. Sandalio et al., "A dehydrogenase-mediated recycling system of NADPH in plant peroxisomes," *Biochemical Journal*, vol. 330, no. 2, pp. 777–784, 1998.

[29] F. J. Corpas, J. B. Barroso, L. M. Sandalio, J. M. Palma, J. A. Lupiáñez, and L. A. Del Río, "Peroxisomal NADP-dependent isocitrate dehydrogenase. Characterization and activity regulation during natural senescence," *Plant Physiology*, vol. 121, no. 3, pp. 921–928, 1999.

[30] R. Valderrama, F. J. Corpas, A. Carreras et al., "The dehydrogenase-mediated recycling of NADPH is a key antioxidant system against salt-induced oxidative stress in olive plants," *Plant, Cell and Environment*, vol. 29, no. 7, pp. 1449–1459, 2006.

[31] C. Beauchamp and I. Fridovich, "Superoxide dismutase: improved assays and an assay applicable to acrylamide gels," *Analytical Biochemistry*, vol. 44, no. 1, pp. 276–287, 1971.

[32] A. Fernández-Ocaña, M. Chaki, F. Luque et al., "Functional analysis of superoxide dismutases (SODs) in sunflower under biotic and abiotic stress conditions. Identification of two new genes of mitochondrial Mn-SOD," *Journal of Plant Physiology*, vol. 168, pp. 1303–1308, 2011.

[33] R. Valderrama, F. J. Corpas, A. Carreras et al., "Nitrosative stress in plants," *FEBS Letters*, vol. 581, no. 3, pp. 453–461, 2007.

[34] F. J. Corpas, M. Hayashi, S. Mano, M. Nishimura, and J. B. Barroso, "Peroxisomes are required for in vivo nitric oxide accumulation in the cytosol following salinity stress of arabidopsis plants," *Plant Physiology*, vol. 151, no. 4, pp. 2083–2094, 2009.

[35] A. Nishizawa, Y. Yabuta, and S. Shigeoka, "Galactinol and raffinose constitute a novel function to protect plants from oxidative damage," *Plant Physiology*, vol. 147, no. 3, pp. 1251–1263, 2008.

[36] V. Mittova, M. Tal, M. Volokita, and M. Guy, "Salt stress induces up-regulation of an efficient chloroplast antioxidant system in the salt-tolerant wild tomato species *Lycopersicon pennellii* but not in the cultivated species," *Physiologia Plantarum*, vol. 115, no. 3, pp. 393–400, 2002.

[37] V. Mittova, M. Tal, M. Volokita, and M. Guy, "Up-regulation of the leaf mitochondrial and peroxisomal antioxidative systems in response to salt-induced oxidative stress in the wild salt-tolerant tomato species *Lycopersicon pennellii*," *Plant, Cell and Environment*, vol. 26, no. 6, pp. 845–856, 2003.

[38] G. Tanou, A. Molassiotis, and G. Diamantidis, "Hydrogen peroxide- and nitric oxide-induced systemic antioxidant prime-like activity under NaCl-stress and stress-free conditions in citrus plants," *Journal of Plant Physiology*, vol. 166, no. 17, pp. 1904–1913, 2009.

[39] K. Gémes, P. Poór, E. Horváth et al., "Cross-talk between salicylic acid and NaCl-generated reactive oxygen species and nitric oxide in tomato during acclimation to high salinity," *Physiologia Plantarum*, vol. 142, no. 2, pp. 179–192, 2011.

[40] X. Wang, Y. Ma, C. Huang, Q. Wan, N. Li, and Y. Bi, "Glucose-6-phosphate dehydrogenase plays a central role in modulating reduced glutathione levels in reed callus under salt stress," *Planta*, vol. 227, no. 3, pp. 611–623, 2008.

[41] J. Li, G. Chen, X. Wang, Y. Zhang, H. Jia, and Y. Bi, "Glucose-6-phosphate dehydrogenase-dependent hydrogen peroxide production is involved in the regulation of plasma membrane H$^+$-ATPase and Na$^+$/H$^+$ antiporter protein in salt-stressed callus from Carex moorcroftii," *Physiologia Plantarum*, vol. 141, no. 3, pp. 239–250, 2011.

[42] A. M. León, J. M. Palma, F. J. Corpas et al., "Antioxidative enzymes in cultivars of pepper plants with different sensitivity to cadmium," *Plant Physiology and Biochemistry*, vol. 40, no. 10, pp. 813–820, 2002.

[43] M. Airaki, M. Leterrier, R. M. Mateos et al., "Metabolism of reactive oxygen species and reactive nitrogen species in pepper (*Capsicum annuum* L.) plants under low temperature stress," *Plant, Cell and Environment*, vol. 35, no. 2, pp. 281–295, 2012.

[44] D. Marino, E. M. González, P. Frendo, A. Puppo, and C. Arrese-Igor, "NADPH recycling systems in oxidative stressed pea nodules: a key role for the NADP$^+$-dependent isocitrate dehydrogenase," *Planta*, vol. 225, no. 2, pp. 413–421, 2007.

[45] O. V. Popova, S. F. Ismailov, T. N. Popova, K. J. Dietz, and D. Golldack, "Salt-induced expression of NADP-dependent

NADP-Dependent Isocitrate Dehydrogenase from Arabidopsis Roots Contributes in the Mechanism of Defence against the
Nitro-Oxidative Stress Induced by Salinity

103

isocitrate dehydrogenase and ferredoxin-dependent glutamate
synthase in *Mesembryanthemum crystallinum*," *Planta*, vol.
215, no. 6, pp. 906–913, 2002.

[46] D. Golldack and K. J. Dietz, "Salt-induced expression of the
vacuolar H$^+$-ATPase in the common ice plant is developmentally controlled and tissue specific," *Plant Physiology*, vol. 125,
no. 4, pp. 1643–1654, 2001.

Biological Effects of Weak Electromagnetic Field on Healthy and Infected Lime (*Citrus aurantifolia*) Trees with Phytoplasma

Fatemeh Abdollahi,[1] **Vahid Niknam,**[1] **Faezeh Ghanati,**[2]
Faribors Masroor,[3] **and Seyyed Nasr Noorbakhsh**[3]

[1] *Department of Plant Sciences, School of Biology and Center of Excellence in Phylogeny of Living Organisms, College of Sciences, University of Tehran, Tehran 14155-6455, Iran*
[2] *Department of Plant Science, Faculty of Biological Science, Tarbiat Modares University, Tehran 14115-154, Iran*
[3] *Department of Chemistry, Engineering Research Institute, Sooliran Street, 16 km Tehran-Karaj Old Road, Tehran 13455-754, Iran*

Correspondence should be addressed to Vahid Niknam, vniknam@khayam.ut.ac.ir

Academic Editor: Mehmet Zulkuf Akdag

Exposure to electromagnetic fields (EMF) has become an issue of concern for a great many people and is an active area of research. Phytoplasmas, also known as mycoplasma-like organisms, are wall-less prokaryotes that are pathogens of many plant species throughout the world. Effects of electromagnetic fields on the changes of lipid peroxidation, content of H_2O_2, proline, protein, and carbohydrates were investigated in leaves of two-year-old trees of lime (*Citrus aurantifolia*) infected by the Candidatus *Phytoplasma aurantifoliae*. The healthy and infected plants were discontinuously exposed to a 10 KHz quadratic EMF with maximum power of 9 W for 5 days, each 5 h, at 25°C. Fresh and dry weight of leaves, content of MDA, proline, and protein increased in both healthy and infected plants under electromagnetic fields, compared with those of the control plants. Electromagnetic fields decreased hydrogen peroxide and carbohydrates content in both healthy and infected plants compared to those of the controls.

1. Introduction

During the past years considerable evidence has been accumulated with notice to the biological effects of low-frequency electromagnetic fields (EMF), such as those bringing from modern world such as power lines and household electrical wiring [1, 2]. The effects of electromagnetic fields of much lower frequency than visible light on plant growth and development have rarely been studied until relatively recently, and knowledge is still limited. Several studies have been showed that low-frequency EMFs may influence plant growth and development [3–5]. Also the international discussion about the biological effects of electromagnetic fields, in which we were involved in the past [6], led us to examine the possibility of using such fields to inhibit phytoplasmas growth on plants such as lime. Phytoplasmas are endocellular prokaryotes without cell wall associated with more than 600 diseases in at least 300 plant species [7]. Moreover, knowledge about phytoplasmas has been limited by inability to isolate them in pure culture.

Reactive oxygen substances (ROS), such as singlet oxygen, superoxide anion, and hydroxyl radical, are produced by a free radical chain reaction and may contribute to tissue damage. To mitigate the oxidative damage initiated by ROS, plants have developed a complex antioxidative defense system, including production of low-molecular mass antioxidants as well as antioxidative enzymes, such as superoxide dismutase (SOD), catalase (CAT), ascorbate peroxidase (APX), guaiacol peroxidase (GPX), and glutathione reductase (GR) [8]. Moreover, the level of malondialdehyde (MDA), a product of lipid peroxidation, has also been considered an indicator of oxidative damage under magnetic field [9].

Lime (family Rutaceae) is one of the most important and most economic horticulture products in the south part of Iran. Lime is susceptible to a large number of diseases caused by plant pathogens [10]. Witches' broom disease of lime (WBDL) associated with "*Candidatus* Phytoplasma aurantifolia" is one of the most destructive diseases of lime in the southern provinces of Iran [11]. Witches' broom disease of lime was first observed in the Sultanate of Oman and later was found to be present in United Arab Emirates [12], India [13], and Iran [14].

Studies on physiological relationships between phytoplasmas and some host plants have been reported [10, 15] but so far none have focused on the responses of phytoplasmas-infected lime plants to electromagnetic fields. Thus, the objective of the present work was to study some biochemical aspects related to lipid peroxidation, content of H_2O_2, proline, protein, and carbohydrates in phytoplasmas-infected lime plant under electromagnetic fields.

2. Materials and Methods

2.1. Plant Materials and Electromagnetic Treatment. Lime plants (*Citrus aurantifolia* L. Swingle cv. Keyline) were infected with the Witches' broom disease of lime (WBDL) *Phytoplasma* by graft transmission and were grown in plastic pots (10×10 cm) under greenhouse condition in Engineering Research Institute, 16 km Tehran-Karaj Old Road, Iran. Infected lime plants used for grafting were collected from *Takht, Bandar-e-Abbas*, south of Iran and transported to the greenhouse. Shoots showing typical symptoms were grafted on two-years-old lime plants.

Exposure to EMF was performed by a locally designed electromagnetic wave generator able to generate different wave shapes including sinusoidal, triangular, and quadratic. The system could generate EMF in range of 0.1 Hz–10 KHz with a continuous fine control in stable conditions and maximum consuming power density of 9 W. It was consisted of two vertical coils each 28 turns of 0.3 mm copper wire rounded around a quadratic frame of 48×34 cm. One coil was oriented in vertical plane (XOZ) with pointing vector in horizontal direction (antiparallel with the gravity), while the other one was oriented in horizontal plane (XOY) with pointing vector in vertical direction (perpendicular with the gravity). Impedance of each coil was 8 ohm. The calibration of the system was performed at different frequencies. For the present experiment the healthy and infected plants were discontinuously exposed to a quadratic EMF for 5 days, each 5 h, at 25°C. The applied frequency was 10 KHz with consuming power of 8.3. The average electric and magnetic strength were 168.5 ± 4 (KV/m) and 3400 ± 43 (mA/m), respectively [16]. All measurements were conducted either in fresh harvested tissue ground immediately after excision or from leaves quickly deep frozen in liquid nitrogen and kept at −80°C until the assay.

2.2. Plant Water Relations. Leaf water content (WC) was calculated based on [17]

$$\text{WC (\%)} = \left[\frac{(\text{fresh mass} - \text{dry mass})}{\text{fresh mass}} \right] \times 100. \quad (1)$$

2.3. Protein Content. For determination of protein content, 500 mg fresh leaf was homogenized in a chilled (4°C) mortar using a 50 mM Tris-HCl buffer (pH 7.0) containing 10 mM EDTA, 2 mM $MgSO_4$, 20 mM dithiotreitol, 10% (v/v) glycerol, and 2% (m/v) polyvinylpyrrolidone. After centrifugation at 13000 g for 45 min at 4°C, the supernatant was filtered and then transferred to Eppendorf tubes and the sample kept on ice at 4°C. A portion of eluent was stored at −70°C. Total protein content was measured by the spectrophotometric method of Bradford [18] using bovine serum albumin (BSA) as the standard.

2.4. Proline Content. Free proline content was determined according to Bates et al. [19] using L-proline as a standard. High-speed centrifuge (Beckman J2-21M, Palo Alto, USA) and UV-visible spectrophotometer (Shimadzu UV-160, Tokyo, Japan) with 10 mm matched quartz cells were used for centrifugation of the extracts and determination of the absorbance, respectively.

2.5. Malondialdehyde Content. The level of lipid peroxidation was measured in terms of thiobarbituric acid reactive substances (TBARS), following the method of Heath and Packer [20]. The plant materials (0.5 g) were homogenized in 5 mL of 0.1% (w/v) trichloroacetic acid (TCA) and centrifuged at 10,000 g for 20 min. To 1 mL aliquot of the supernatant, 4 mL of 0.5% thiobarbituric acid (TBA) in 20% TCA was added. The mixture was heated at 95°C for 30 min and quickly cooled in an ice bath. After centrifugation at 10,000 g for 15 min, the absorbance of the supernatant was recorded at 532 and 600 nm. The value for nonspecific absorption at 600 nm was subtracted. The concentration of MDA was calculated using extinction coefficient of $155 \text{ mM}^{-1} \text{ cm}^{-1}$.

2.6. Total Carbohydrates Content. For determination of carbohydrates content, 50 mg of dry powder was extracted using 10 mL of ethanol : distilled water (8 : 2; v/v), and supernatant was collected after twice centrifugation at 1480 g. The residue from ethanol extraction was subsequently used for polysaccharide extraction by boiling water [21]. Total carbohydrates content was estimated by the method of Dubois et al. [22].

2.7. Hydrogen Peroxide Content. The content of hydrogen peroxide was determined according to Sergiev et al. [23]. The plant materials (0.5 g) were homogenized in 5 mL of 0.1% (w/v) trichloroacetic acid (TCA) on ice and centrifuged at 12,000 g for 15 min. To 0.5 mL aliquot of the supernatant, 1 mL potassium phosphate buffer and 1 mL KI was added. The absorbance of the supernatant was recorded at 390 nm.

2.8. Statistical Analysis. All analyses were performed based on a completely randomized design. The data determined in triplicate were analyzed by analysis of variance (*ANOVA*) using *SPSS* (version *9.05*). Each data point was the mean of three replicates ($n = 3$). The significance of differences was determined according to Duncan's multiple range test (DMRT). *P* values < 0.05 are considered to be significant.

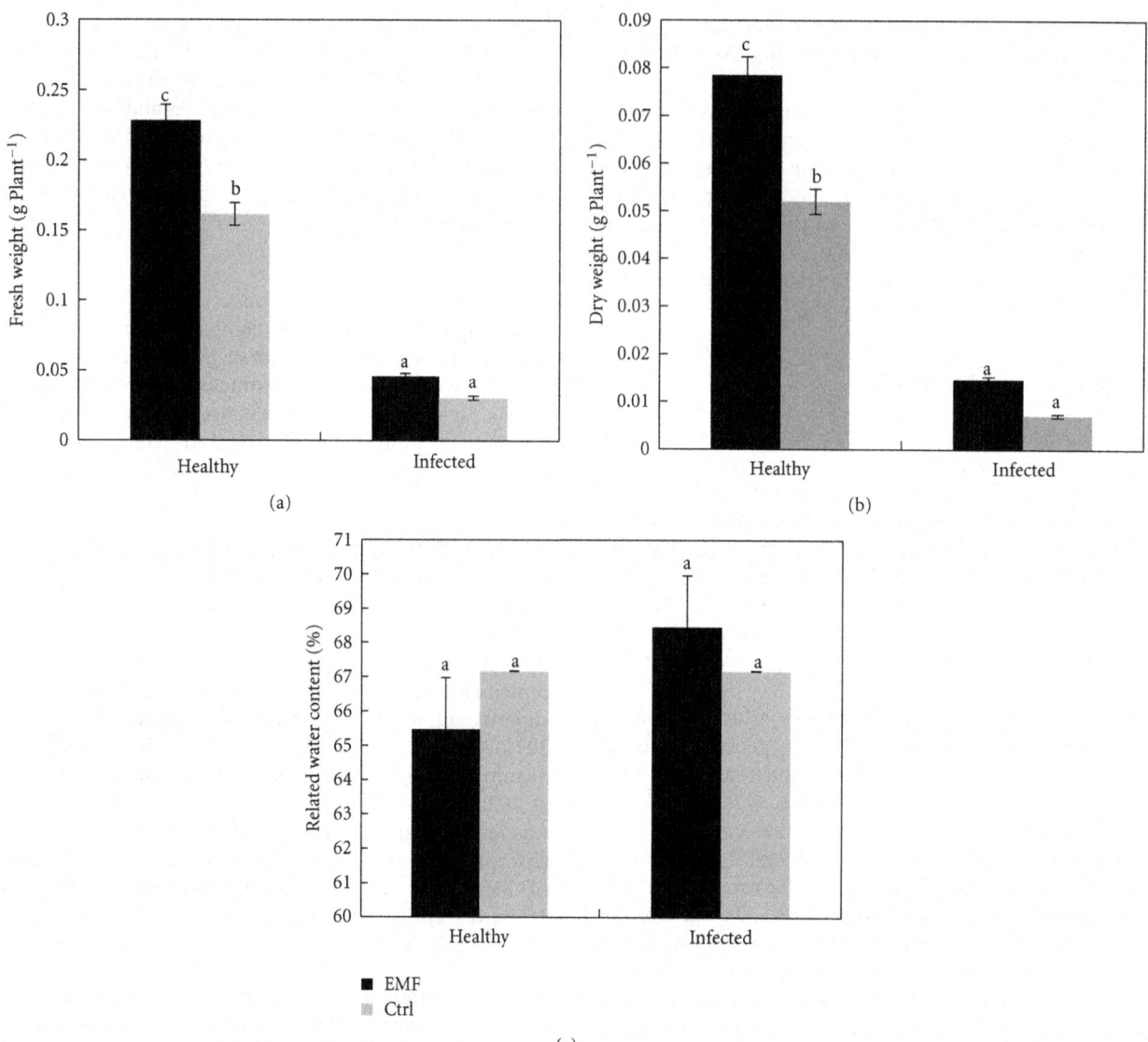

FIGURE 1: Effect of EMF on the fresh weight (a), dry weight (b), and relative water content (c) in healthy and infected plants of *Citrus aurantifolia*. Vertical bars indicate mean ± SE of three replicates. Different letters indicate significant differences ($P < 0.05$).

3. Result and Discussion

In order to determine the effect of electromagnetic field on leaf growth and biochemistry in *Citrus aurantifolia*, we treated the healthy and infected lime plants with electromagnetic field. The treatment of healthy and infected plants with electromagnetic field affected significantly the growth of the lime plants. Fresh and dry weight of leave in both healthy and infected plants increased under electromagnetic field (Figure 1). The present data agree with the previous results reported on *Prunus maritime*, *Cucumis sativus*, *Raphanus sativus*, and *Helianthus annuus* [24–26]. Relative water content (RWC) decreased in healthy plants and increased in infected plants under EMF stress (Figure 1(c)).

Protein content in leaves of both healthy and infected plants increased significantly under EMF (Figure 2). EMF decreased slightly the content of total carbohydrates in both healthy and infected plants (Figure 3). The reduction in carbohydrates content is more prominent in healthy plants than that of infected ones.

Free proline content increased significantly under EMF exposure (Figure 4). However, enhanced levels of proline accumulation may not be enough to maintain water balance of the healthy plants under EMF treatment (Figure 1(a)). Many plants accumulated proline as nontoxic and protective osmolyte under stress conditions [27–29]. Higher accumulation of proline in lime plants under EMF may afford it much protection against electromagnetic field. Although the precise role of proline accumulation is still debated, proline is also considered to be involved in the preservation of cellular structures, enzymes, and to exploit as a free radical scavenger [30, 31].

Changes in lipid peroxidation serve as an indicator of the extent of oxidative damage under stress, with an unchanged

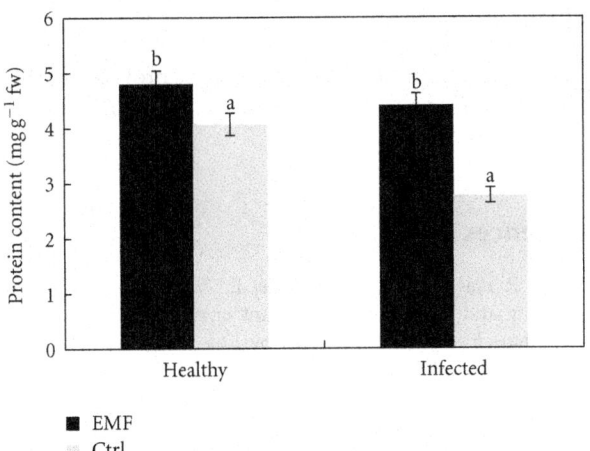

FIGURE 2: Effect of EMF on the protein content in healthy and infected plants of *Citrus aurantifolia*. Vertical bars indicate mean ± SE of three replicates. Different letters indicate significant differences ($P < 0.05$).

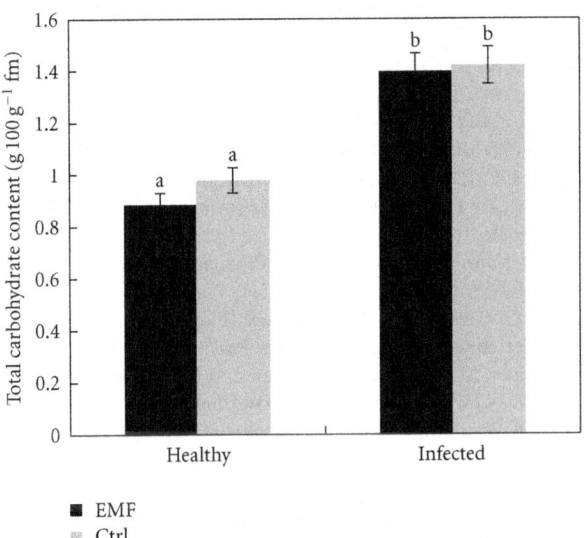

FIGURE 3: Effect of EMF on carbohydrates content in healthy and infected plants of *Citrus aurantifolia*. Vertical bars indicate mean ± SE of three replicates. Different letters indicate significant differences ($P < 0.05$).

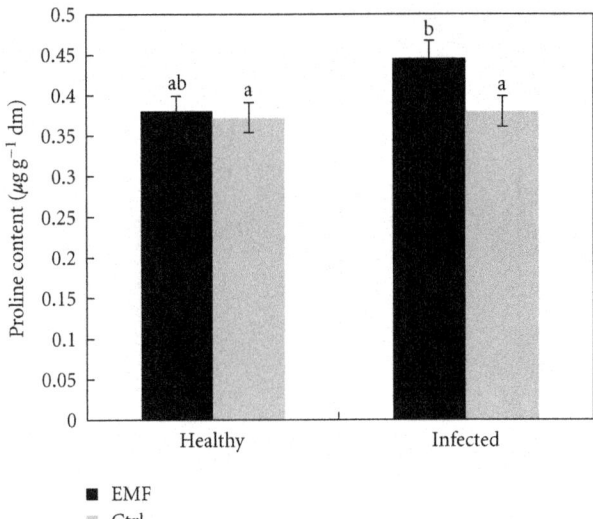

FIGURE 4: Effect of EMF on proline content in healthy and infected plants of *Citrus aurantifolia*. Vertical bars indicate mean ± SE of three replicates. Different letters indicate significant differences ($P < 0.05$).

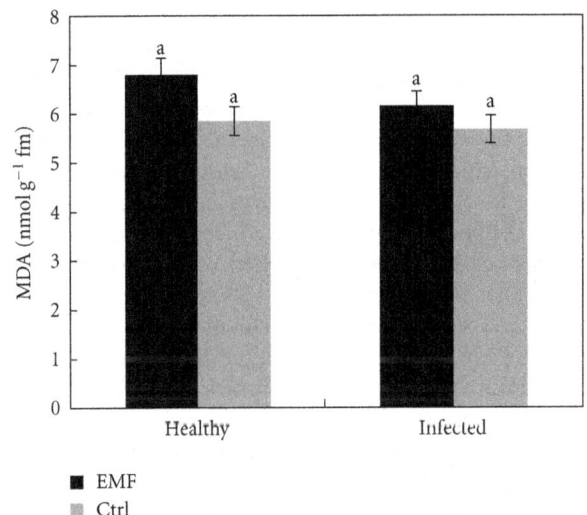

FIGURE 5: Effect of EMF on MDA content in healthy and infected plants of *Citrus aurantifolia*. Vertical bars indicate mean ± SE of three replicates. Different letters indicate significant differences ($P < 0.05$).

lipid peroxidation level seeming to be a characteristic of tolerant plants coping with elevated levels of stress. MDA content was higher in healthy and infected plants under EMF exposure (Figure 5). However these differences are not significant. Lipid peroxidation is mostly triggered by hydroxyl radicals, although the peroxidation can be caused by other reactive oxygen species as well. Treatment with the EMF did not cause any significant changes in the rate of peroxidation of the membrane lipids in lime plants. This result is against some results published on the detrimental effects of EMF [32, 33] and ELF-MF on membranes [9]. However, our result regarding MDA content is in accordance with the obtained result in maize under EMF [34] and rats

under ELF magnetic field [35]. Damaging effects of magnetic field on DNA in animals are also reported [36].

For determination of ROS scavenging capacity, the H_2O_2 content of lime plants under EMF stress was estimated. H_2O_2 content was always significantly lower under EMF stress throughout the experiments performed here (Figure 6). This result is in contrast to the obtained result on MDA content (Figure 5). Lower content of H_2O_2 might be a result of the significantly higher induction of the activities of antioxidant enzymes in the lime plants under EMF stress. Several authors have described the overproduction of toxic

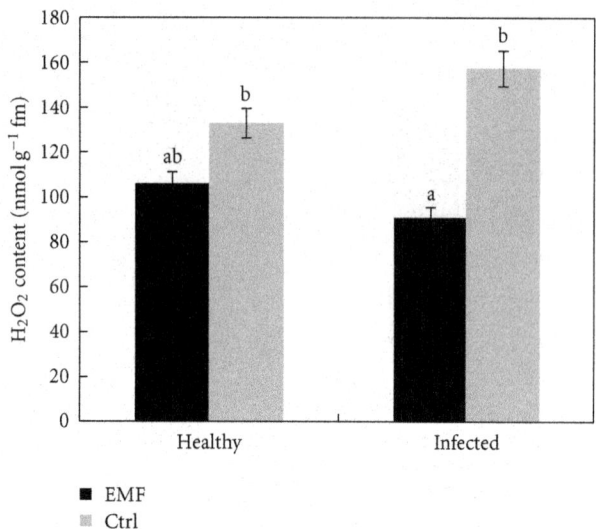

FIGURE 6: Effect of EMF on H_2O_2 content in healthy and infected plants of *Citrus aurantifolia*. Vertical bars indicate mean ± SE of three replicates. Different letters indicate significant differences ($P <$ 0.05).

oxygen forms with aging in plants [37, 38]. It is known that ROS can cause peroxidation of membrane fatty acids [39]. In turn, these oxidized fatty acids may give rise to the propagation of peroxidation resulting in membrane damage. In a recent report, Lacan and Baccou [40, 41], besides supporting that ROS are involved in the ripening and senescence events in nonnetted muskmelon fruit, also showed that the observed delay in senescence in the long-storage life variety Clipper is closely linked to the very low free radical induced membrane lipid peroxidation in comparison with that of the short-storage life variety Jerac. These data explain the EMF-improved ability to scavenge free radicals, leading to a delay in senescence and alterations in membrane integrity, as demonstrated by the growth and survival responses. Furthermore, these results suggest that, under specific conditions, it takes the combined action of more than one antioxidant to provide an increased resistance to oxidative stress and lengthened survival in plants. This is in agreement with some previous reports [42, 43].

4. Conclusions

The obtained result show that 10 KHz EMF field can simulate the growth rate of healthy and infected lime plants. Moreover, it seems that EMF field could reduce the intensity of Witches' broom disease in infected trees and this effect could probably be due to the reduction of phytoplasma in plant tissues. This study provides an initial understanding of the response of infected lime plants to EMF stress, which is important for future studies aimed at developing strategies for struggle against phytoplasma and Witches' broom disease in lime plants.

Acknowledgments

The financial support of this research was provided partly by College of Science, University of Tehran and partly by Engineering Research Institute.

References

[1] M. P. Piacentini, D. Fraternale, E. Piatti et al., "Senescence delay and change of antioxidant enzyme levels in Cucumis sativus L. etiolated seedlings by ELF magnetic fields," *Plant Science*, vol. 161, no. 1, pp. 45–53, 2001.

[2] A. Ubeda, M. Díaz-Enriquez, M. A. Martínez-Pascual, and A. Parreño, "Hematological changes inr ats exposed to weak electromagnetic fields," *Life Sciences*, vol. 61, no. 17, pp. 1651–1656, 1997.

[3] A. Yano, Y. Ohashi, T. Hirasaki, and K. Fujiwara, "Effects of a 60 Hz magnetic field on photosynthetic CO2 uptake and early growth of radish seedlings," *Bioelectromagnetics*, vol. 25, no. 8, pp. 572–581, 2004.

[4] R. Ruzic and I. Jerman, "Weak magnetic field decreases heat stress in cress seedlings," *Electromagnetic Biology and Medicine*, vol. 21, no. 1, pp. 69–80, 2002.

[5] B. C. Stange, R. E. Rowland, B. I. Rapley, and J. V. Podd, "ELF Magnetic Fields Increase Amino Acid Uptake into *Vicia faba* L. Roots and alter Ion movement across the plasma membrane," *Bioelectromagnetics*, vol. 23, no. 5, pp. 347–354, 2002.

[6] E. Piatti, M. Albertini, W. Baffone et al., "Antibacterial effect of a magnetic field on *Serratia marcescens* and related virulence to *Hordeum vulgare* and *Rubus fruticosus* callus cells," *Comparative Biochemistry and Physiology B*, vol. 132, no. 2, pp. 359–365, 2002.

[7] B. C. Kirkpatrick, "Mycoplasma-like organisms: plant and invertebrate pathogens," in *The Prokaryotes*, A. Balows, H. G. Truper, M. Dworkin, W. Harder, and K. H. Schleifer, Eds., pp. 4050–4067, Springer-Verlag, New York, NY, USA, 1992.

[8] G. Noctor and C. H. Foyer, "Ascorbate and glutathione: keeping active oxygen under control," *Annual Review of Plant Biology*, vol. 49, pp. 249–279, 1998.

[9] M. Z. Akdag, S. Dasdag, E. Ulukaya, A. K. Uzunlar, M. A. Kurt, and A. Taşkln, "Effects of extremely low-frequency magnetic field on caspase activities and oxidative stress values in rat brain," *Biological Trace Element Research*, vol. 138, no. 1–3, pp. 238–249, 2010.

[10] S. Zafari, V. Niknam, R. Musetti, and S. N. Noorbakhsh, "Effect of phytoplasma infection on metabolite content and antioxidant enzyme activity in lime (*Citrus aurantifolia*)," *Acta Physiologiae Plantarum*, vol. 34, no. 2, pp. 561–568, 2012.

[11] M. Mirzai, J. Heydarnejad, M. Salehi, A. Hosseini-Pour, H. Massumi, and M. Shaabanian, "Production of polycolonal antiserum against causal agent of lime witches' broom," *Iranian Journal of Plant Pathology*, vol. 45, no. 2, pp. 40–41, 2009.

[12] M. Garnier, L. Zreik, and J. Bové, "Witches' broom, a lethal mycoplasmal disease of lime in the Sultanate of Oman and the United Arab Emirates," *Plant Disease*, vol. 75, pp. 546–555, 1991.

[13] D. K. Ghosh, A. K. Das, S. Singh, S. J. Singh, and Y. A. Ahlawat, "Occurrence of witches' broom, a new phytoplasma disease of acid lime (*Citrus aurantifolia*) in India," *Plant Disease*, vol. 83, no. 3, p. 302, 1999.

[14] J. M. Bové, J. L. Danet, K. Bananej et al., "Witches' broom disease of lime in Iran," in *Proceedings of the 14th Conference*

of the International Organization of Citrus Virolo (IOCV '00), pp. 207–212, 2000.

[15] R. Musetti, "Biochemical changes in plants infected by phytoplasmas," in Phytoplasmas: Genomes, Plant Hosts and Vectors, P. G. Weintraub and P. Jones, Eds., pp. 135–149, CABI Publishing, Wallingford, UK, 2009.

[16] F. Ghanati, E. Rajabbeigi, and P. Abdolmaleki, "Influence of electromagnetic field exposure on the growth of Ocimum basilicum and its essential oil," in Proceedings of the 4th International Workshop on Biological Effects of ElectroMagnetic Fields and the participant's, pp. 125–128, NCSR Demokritos, 2006.

[17] A. N. Molassiotis, T. Sotiropoulos, G. Tanou, G. Kofidis, G. Diamantidis, and E. Therios, "Antioxidant and anatomical responses in shoot culture of the apple rootstock MM 106 treated with NaCl, KCl, mannitol or sorbitol," Biologia Plantarum, vol. 50, no. 1, pp. 61–68, 2006.

[18] M. M. Bradford, "A rapid and sensitive method for the quantitation of microgram quantities of protein utilizing the principle of protein dye binding," Analytical Biochemistry, vol. 72, no. 1-2, pp. 248–254, 1976.

[19] L. S. Bates, R. P. Waldren, and I. D. Teare, "Rapid determination of free proline for water-stress studies," Plant and Soil, vol. 39, no. 1, pp. 205–207, 1973.

[20] R. L. Heath and L. Packer, "Photoperoxidation in isolated chloroplasts. I. Kinetics and stoichiometry of fatty acid peroxidation," Archives of Biochemistry and Biophysics, vol. 125, no. 1, pp. 189–198, 1968.

[21] V. Niknam, M. Bagherzadeh, H. Ebrahimzadeh, and A. Sokhansanj, "Effect of NaCl on biomass and contents of sugars, proline and proteins in seedlings and leaf explants of Nicotiana tabacum grown in vitro," Biologia Plantarum, vol. 48, no. 4, pp. 613–615, 2004.

[22] M. Dubois, K. A. Gilles, J. K. Hamilton, P. A. Rebers, and F. Smith, "Colorimetric method for determination of sugars and related substances," Analytical Chemistry, vol. 28, no. 3, pp. 350–356, 1956.

[23] L. Sergiev, V. Alexieva, and E. Karanov, "Effect of spermine, atrazin and combination between them on some endogenous protective systems and stress markers in plants," Comptes Rendus de l'Academie Bulgare des Sciences, vol. 51, pp. 121–124, 1997.

[24] Y. Dao-liang, G. Yu-qi, Z. Xue-ming, W. Shu-wen, and P. Qin, "Effects of electromagnetic fields exposure on rapid micropropagation of beach plum (Prunus maritima)," Ecological Engineering, vol. 35, no. 4, pp. 597–601, 2009.

[25] M. D. Potts, W. C. Parkinson, and L. D. Noodén, "Raphanus sativus and electromagnetic fields," Bioelectrochemistry and Bioenergetics, vol. 44, no. 1, pp. 131–140, 1997.

[26] A. Vashisth and S. Nagarajan, "Effect on germination and early growth characteristics in sunflower (Helianthus annuus) seeds exposed to static magnetic field," Journal of Plant Physiology, vol. 167, no. 2, pp. 149–156, 2010.

[27] C. A. Jaleel, A. Kishorekumar, P. Manivannan, B. Sankar, M. Gomathinayagam, and R. Panneerselvam, "Salt stress mitigation by calcium chloride in Phyllanthus amarus," Acta Botanica Croatica, vol. 67, no. 1, pp. 53–62, 2008.

[28] M. H. Siddiqui, F. Mohammad, and M. N. Khan, "Morphological and physio-biochemical characterization of Brassica juncea L. Czern. & Coss. genotypes under salt stress," Journal of Plant Interactions, vol. 4, no. 1, pp. 67–80, 2009.

[29] M. N. Khan, M. H. Siddiqui, F. Mohammad, M. Naeem, and M. M. A. Khan, "Calcium chloride and gibberellic acid protect linseed (Linum usitatissimum L.) from NaCl stress by inducing antioxidative defence system and osmoprotectant

accumulation," Acta Physiologiae Plantarum, vol. 32, no. 1, pp. 121–132, 2010.

[30] L. van Resenburg, G. H. J. Kruger, and H. Kruger, "Proline accumulation as drought tolerance selection criterion: its relationship to membrane integrity and chloroplast ultrastructure in Nicotiana tabacum L.," Journal of Plant Physiology, vol. 141, no. 2, pp. 188–194, 1993.

[31] A. Solomon, S. Beer, Y. Waisel, G. P. Jones, and L. G. Paleg, "Effects of NaCl on the carboxylating activity of rubisco from tamarix jordanis in the presence and absence of proline-related compatible solutes," Physiologia Plantarum, vol. 90, no. 1, pp. 198–204, 1994.

[32] B. C. Seref, A. K. Baltaci, R. Mogulkoc, and E. Öztekin, "Zinc supplementation ameliorates electromagnetic field-induced lipid peroxidation in the rat brain," Tohoku Journal of Experimental Medicine, vol. 208, no. 2, pp. 133–140, 2006.

[33] H. Sahebjamei, P. Abdolmaleki, and F. Ghanati, "Effects of magnetic field on the antioxidant enzyme activities of suspension-cultured tobacco cells," Bioelectromagnetics, vol. 28, no. 1, pp. 42–47, 2007.

[34] A. Hajnorouzi, M. Vaezzadeh, F. Ghanati, H. jamnezhad, and B. Nahidian, "Growth promotion and a decrease of oxidative stress in maize seedlings by a combination of geomagnetic and weak electromagnetic fields," Journal of Plant Physiology, 2011.

[35] M. Z. Akdag, S. Dasdag, F. Aksen, B. Isik, and F. Yilmaz, "Effect of ELF magnetic fields on lipid peroxidation, sperm count, p53, and trace elements," Medical Science Monitor, vol. 12, no. 11, pp. BR366–BR371, 2006.

[36] B. Yokus, M. Z. Akdag, S. Dasdag, D. U. Cakir, and M. Kizil, "Extremely low frequency magnetic fields cause oxidative DNA damage in rats," International Journal of Radiation Biology, vol. 84, no. 10, pp. 789–795, 2008.

[37] M. J. Droillard, A. Paulin, and J. C. Massot, "Free radical production, catalase and superoxide dismutase activities and membrane integrity during senescence of petals of cut carnations (Dianthus caryophyllus)," Physiologia Plantarum, vol. 71, no. 2, pp. 197–202, 1987.

[38] A. Borochov, A. H. Halevy, and M. Shinitzky, "Senescence and fluidity of rose petal plasma membranes: the effect of phospholipid metabolism," Plant Physiology, vol. 69, no. 2, pp. 296–299, 1982.

[39] S. Mayak, R. L. Legge, and J. E. Thompson, "Superoxide radical production by microsomal membranes from senescing carnation flowers: an effect on membrane fluidity," Phytochemistry, vol. 22, no. 6, pp. 1375–1380, 1983.

[40] D. Lacan and J. C. Baccou, "Changes in lipids and electrolyte leakage during nonnetted muskmelon ripening," Journal of the American Society for Horticultural Science, vol. 121, no. 3, pp. 554–558, 1996.

[41] D. Lacan and J. C. Baccou, "High levels of antioxidant enzymes correlate with delayed senescence in nonnetted muskmelon fruits," Planta, vol. 204, no. 3, pp. 377–382, 1998.

[42] G. M. Pastori and V. S. Trippi, "Antioxidative protection in a drought-resistant strain during leaf senescence," Plant Physiology, vol. 87, no. 2, pp. 227–231, 1993.

[43] A. S. Gupta, R. P. Webb, A. S. Holaday, and R. D. Allen, "Overexpression of superoxide dismutase protects plants from oxidative stress. Induction of ascorbate peroxidase in superoxide dismutase-overexpressing plants," Plant Physiology, vol. 103, no. 4, pp. 1067–1073, 1993.

Identification of Xylem Occlusions Occurring in Cut *Clematis* (*Clematis* L., fam. *Ranunculaceae* Juss.) Stems during Their Vase Life

Agata Jedrzejuk,[1] Julia Rochala,[1] Jacek Zakrzewski,[2] and Julita Rabiza-Świder[1]

[1] *Department of Ornamental Plants, Faculty of Horticulture and Landscape Architecture, Warsaw University of Life Sciences, Nowoursynowska 166, 02-787 Warsaw, Poland*
[2] *Department of Forest Botany, Faculty of Forestry, Warsaw University of Life Sciences, Nowoursynowska 166, 02-787 Warsaw, Poland*

Correspondence should be addressed to Agata Jedrzejuk, agata.jedrzejuk@wp.pl

Academic Editors: M. Edery and D. Granot

During the vase life of cut stems obstruction of xylem vessels occurs due to microbial growth, formation of tyloses, deposition of materials in the lumen of xylem vessels and the presence of air emboli in the vascular system. Such obstructions may restrict water uptake and its transport towards upwards thus lowering their ornamental value and longevity of cut flowers. *Clematis* is a very attractive plant material which may be used as cut flower in floral compositions. Nothing is known about the histochemical or cytological nature of xylem blockages occurring in cut stems of this plant. This study shows that in *clematis*, tyloses are the main source of occlusions, although bacteria and some amorphic substances may also appear inside the vessels. A preservative composed of 200 mg dm^{-3} 8-HQC (8-hydroxyquinolin citrate) and 2% sucrose arrested bacterial development and the growth of tyloses. This information can be helpful in the development of new treatments to improve keeping qualities of cut *clematis* stems.

1. Introduction

Clematis is used in Europe mostly as a climber plant, but, because of its beautiful flowers, this genus may also provide cut flowers for floral compositions. It is used as such in the United States; the European flower market still lacks suitable *clematis* cultivars and methods allowing to control thier postharvest quality. This creates a broad opportunity for the European growers and breeders of ornamental plants. Several Polish cultivars are proving themselves to be potential sources of a good cut ornamental material.

The postharvest life of *clematis* ranges between 2 and 14 days, and it depends on a cultivar. The standard preservative to effectively prolong the vase life of *clematis* is a solution of 200 mg dm^{-3} 8-hydroxyquinolin citrate (8HQC) with 2% sucrose [1], but more advanced studies are needed to develop preservatives and treatments suitable for *clematis* during all steps of the market chain. Proper water balance in cut stems is crucial for the flower postharvest longevity, and

blockages occurring in vessels disturb it by limiting water uptake and transport to the flower. The main cause of reduced water uptake in cut stems is obstruction of xylem vessels by microbial growth, formation of tyloses, deposition of materials in the lumen of xylem vessels, and the presence of air emboli in the vascular system [2, 3].

The invasion of the dead lumens of tracheary elements by living parenchyma cells (formation of tyloses) is a well-known response to infection by pathogens and to wounding [4]. It is often accompanied or followed by the transformation of gums and tannins which add to the strength and durability of the composite polymers. The nature of such material was investigated cytologically, revealing the presence of pectic elements, callose, or lignin-like molecules [5–9]. Such material is produced by the plant in response to invasion by the bacteria [10–14] or in response to phytotoxins produced by bacteria [15].

This study was conducted to provide cytochemical and immunohistochemical information on vessel occlusions and the involvement of tyloses, gels, or gums in their formation in cut *clematis* stems kept in different vase solutions.

2. Material and Methods

2.1. Plant Material. The study was done on flowering stems of *clematis* (*Clematis* L.) kindly provided by Mr. Szczepan Marczyński and Władysław Piotrowski from the plant nursery in Duchnice near Warsaw. The choice of the cultivar was based on observations of the vase life length made by Skutnik and Rabiza-Świder [1, 16] and previous anatomical observations of stems of five different cultivars, of which two were short-lasting ("Andromeda" and "Viola"), one medium lasting ("Isago"), and two long-lasting ("Solidarność" and "Silver moon"). Additionally, anatomical studies of stem blockage formation were done in all five cultivars, while the histochemical, immunohistochemical and cytological identification of the nature of blockages was done in only one, cv. "Solidarność."

Flowering stems were harvested at the same stage of development, immediately transferred to laboratory and trimmed to 20 cm. Shoots were placed in distilled water or the standard preservative composed of the bactericide 8HQC + sucrose (SUC) which was tested as the most effective preservative [1, 16]. There were eight shoots in each treatment, individually tagged and treated as individual replications. The experiments were conducted at 18–20°C and a 12 h photoperiod, provided by luminescence light with a quantum irradiance of $25 \, mol \, m^{-2} \, s^{-1}$. The relative air humidity was maintained at 60%.

2.2. Anatomy, Histochemistry, and Immunolocalization. Stem ends *ca* 5 mm long were sampled on three dates: just after harvest (control, day 0), after 7 days (wilting of flowers kept in distilled water, term I), and after 12 days, when wilting and loss of a decorative value occurred in flowers placed into the preservative (term II). On terms I and II, the stem fragments were collected from both treatments (distilled water, preservative).

The specimens were fixed in the PFA fixative: 4% paraformaldehyde (Sigma), 0.4% DMSO (Sigma), 0.05M phosphate-buffered saline (PBS) (pH 7.0), DEPC-treated water (Sigma) for 12 h under 0.6 atm. Fixed samples were washed twice for 30 min. each in the phosphate-buffered saline (PBS), dehydrated in the graded ethanol series (30%, 50%, 70%, 80%, 95%, 100%), each series for 1 h in RT (room temperature), and twice in Histoclear (Histochoice Clearing Agent, Sigma) for 30 min each. Paraplast pellets (Sigma) were added to the last series of Histoclear in the paraffin oven, twice a day for 5–7 days, in temperature 56–58°C, until the Histoclear evaporated completely. In the last step, specimens were embedded in clear Paraplast (Sigma). Semi-thin sections ($10 \, \mu m$) were sectioned on a rotary microtome (Reichert Jung). All preparations were made on the RNase, DNase-free objective slides (Thermo Scientific MenzelGläser, Superfrost Plus), and dried at 42°C for 2–4 days.

For general anatomical identification of xylem occlusions in *clematis* stems, permanent slides were stained using the

TABLE 1: Histological tests used for detection of occlusions in *Clematis* stems.

Type of reaction	Color of product	Detection target
Periodic acid-Schiff reaction (PAS) [36]	Red	Aldehydes are created by oxidative cleavage of saccharides with H6IO5. Coloration is produced by the reaction
Bradford reagent [37]	Blue	Proteins
Azur B [38]	Blue	Lignins
Phloroglucinol-HCL [39]	Red	Lignins
Sudan IV [40]	Orange	Suberins

safranin—fast green method. For the histochemical identification of xylem occlusions, slides were stained as listed in Table 1.

For some cell wall components, sections were incubated with monoclonal primary antibodies Jim 5, Jim 7 (detection of homogalacturonans), Jim 11, Jim 12, Jim 20 (detection of extensins) synthesized by Dr Knox, Centre of Plant Sciences, University of Leeds, Leeds, UK (details at http://www.plant-probes.net/). Primary antibodies diluted 1 : 20 in PBS were applied for 2 h at 37°C. Secondary, antiRat IgG antibody labeled with the alkaline phosphatase (SIGMA) was applied for 2 h in 37°C, and slides were incubated with the nitroblue tetrazolium chloride and 5-bromo-4-chloro-3′-indolyphosphate p-toluidine salt (NBT/BCIP) (Sigma) diluted in 100 mM Tris, 100 mM NaCl, 50 mM MgCl$_2$, for 2 h in the dark. All observations were made using olympus BX41 bright field microscope.

2.3. Electron Microscopy (EM). For conventional EM observations, stem fragments were fixed for 6 h in 2.5% glutaraldehyde (Sigma) buffered with 0.1 M cacodylate buffer, pH 7.2, rinsed in the same buffer and postfixed for 2 h in 1% osmium tetroxide (Merck). Samples were dehydrated in a graded series of alcohol followed by dehydrated acetone and embedded in Epon (Fluka). After thin sectioning, samples were stained with 3% uranyl acetate and Reynold's lead citrate and examined under a JEOL JEM100C transmission electron microscope.

2.4. Statistical Analyses. The xylem vessel data were tested using analysis of variance (Anova 1) with the Stagraphics 4.1. program. Means were compared using the Duncan's multiple range test at $P = 0.95$.

3. Results and Discussion

3.1. Anatomical Organization of Clematis Stems. Stems of different cultivars of *clematis* contain between six to twelve primary vascular bundles with the diameter in the metaxylem between 17.9 and $110.5 \, \mu m$ (see Table 2). Stems of cv. "Solidarność" contain six primary vascular bundles with well visible cambium between the phloem and xylem zones (Figure 1). Cambium consists of 3-4 meristematic cells in

TABLE 2: Diameter of xylem lumen in 5 different cultivars of clematis stems.

Cultivar	Diameter of xylem lumen (μm)			
	Minimum	Maximum	Mean	Standard error
Andromeda	17.9	53.9	34.8	±2.32
Viola	20.4	53.4	37.4	±7.02
Isago	25.0	66.7	44.7	±2.69
Solidarność	27.0	69.9	46.4	±2.26
Silver moon	28.9	110.5	59.0	±1.48

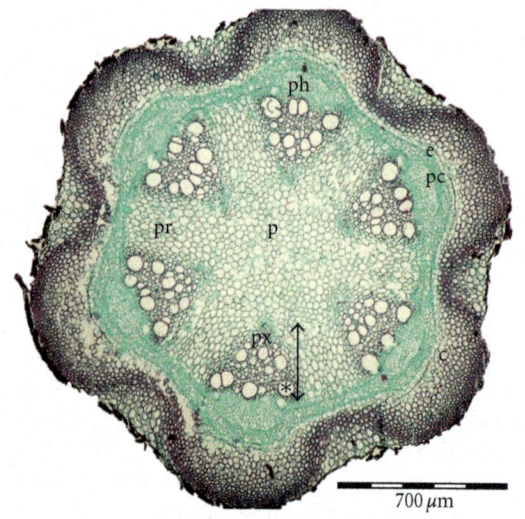

FIGURE 1: Transverse section of *Clematis* "Solidarność" stem, p: pith, pr: pith rays, ph: phloem, px: protoxylem, *: vessel of metaxylem, ↕: xylem, e: endodermis, c: cortex with lamellar collenchymas, pc: pericycle.

a radial arrangement. At the inner side, the primary xylem contains small vessels of protoxylem and, on the outside, a rather wide region of metaxylem with large diameter vessels (Figure 1). The primary rays have parenchymatous cells with 3–5 layers in tangential direction. The pith region is wide and parenchymatic and occupies mostly the center part of the stem. At the outer side of the vascular bundle, there is a multicellular, parenchymatic pericycle. The cortex contains a parenchymatous endodermis, a layer of parenchyma cells, and a ring of lamellar collenchyma cells. The diameter of vessels in metaxylem ranges from 26.4 to 66.7 μm.

According to van Meeteren et al. [17], the vessel diameter may affect the duration of the postharvest life. This is based on the fact that wide vessels are more efficient in water transport [2, 18], and the xylem occlusions do not block the entire lumen of the vessel. Our preliminary research (Table 2) showed that diameter of primary xylem vessels seems to be associated with the length of postharvest life. Observations on all five cultivars showed a similar architecture of stem anatomy, except for the number of primary xylem vessels and their diameters. According to Skutnik and Rabiza-Świder [1, 17] the cut stems of cv. "Solidarność" are long-lasting and their wilting in distilled water occurs after 10 days.

3.2. *Cytological Identification of Xylem Occlusions.* In freshly harvested control stems, the xylem vessels were free of occlusions. The thickness of cell walls ranged from 1.0 to 1.2 μm (Figure 2(a)). After seven days in distilled water (date I), xylem vessels were blocked primarily by tyloses and to a lesser extent by bacteria. Tyloses contained mostly amyloplasts, and ther nuclei showed well-advanced fragmentation (Figure 2(b)). On this collection date, no amorphic or jelly substances were observed in the vessel lumen, but, in 5–7% of the specimen examined, some bacteria were present. On the second collection date, 12 days of vase life in water, tyloses filled nearly the entire volume of xylem vessels. Tyloses contained all components of parenchymatic cell matrix, for example, plastids (Figure 2(c)), mitochondria, plastids, degenerated lipid bodies (data not shown), and plenty of amorphic substances probably originating from degenerating cytoplasm (data not shown). Bacteria were responsible for 27–30% of blocked vessels. In most cases, the vessels blocked by bacteria were free of tyloses.

After seven days of vase life, the stems kept in the preservative solution contained only several tyloses (5.6–7.8% of all observed specimen) with very thin membranes (0.03 μm) and big vacuoles, filling most of the tyloses' spaces. Tyloses also contained mitochondria, autophagosomal vacuoles and intact cytoplasm near the cell wall (Figure 2(d)). Tyloses did not fill the entire lumen of xylem vessel. On the second collection date (after 12 days of vase life), the thickness of the cell membranes was *ca* 0.1 μm. Tyloses contained big amyloplasts (Figure 2(e)) and traces of lipid bodies which preserved their shape and did not degenerate (Figure 2(f)) as they did in stems placed in distilled water (data not shown). Large vacuoles were also observed, similarly to flowers kept in water (Figure 2(f)). Apart from tyloses, some bacteria were present in 10–12% of blocked vessels as well as tubular material at the border of the xylem cell wall (Figure 2(g)).

In both treatments, young tyloses (those from collection date I) usually were globular in shape and elongated during the tylose development. According to Clérivet et al. [15], globular-shaped tyloses are outgrowths of the vessel-associated parenchyma cells, which balloon through pit cavities into adjacent vessel elements. They are generally considered as a primary defense mechanism during vascular attacks and hamper the pathogen transportation within xylem vessels [18–21]. Our observations show that the main source of xylem blockage in cut *clematis* stems is tyloses and that their development is delayed when stems are kept in a standard preservative containing 8HQC + SUC. Even

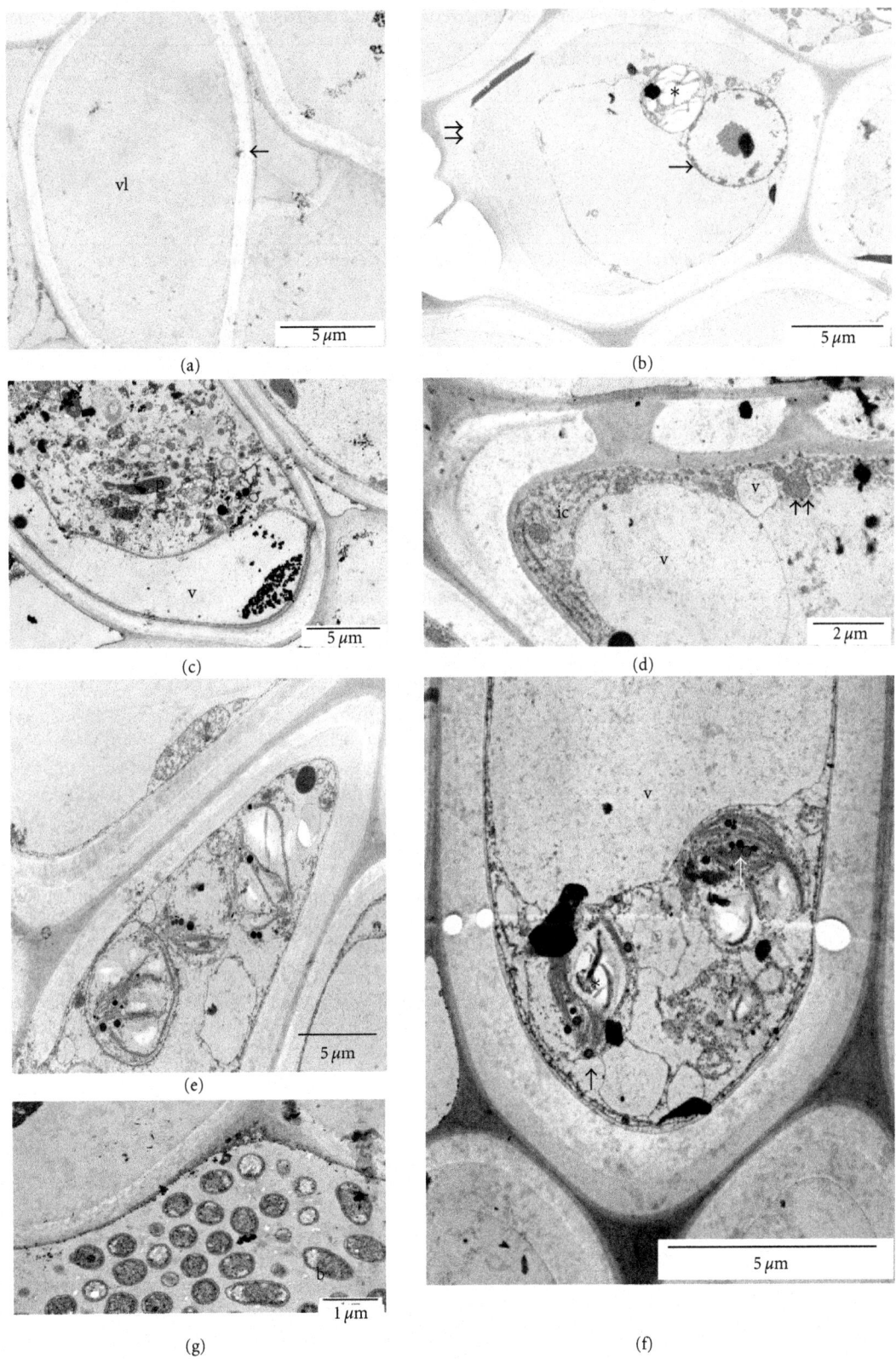

FIGURE 2: Ultrastructural identification of xylem occlusions in stems of *Clematis* "Solidarność" (a) empty vessel, control, (b) young tylose, stems kept in distilled water, date I, (c) mature tyloses, stems kept in distilled water, date II, (d) young tylose, stems kept in 8HQC with 2% sucrose, date I, (e) and (f) mature tylose, stems kept in 8HQC with 2% sucrose, date II, (g) bacteria in blocked vessels, stems kept in 8HQC with 2% sucrose, date II. Vl: vessel lumen, ←: cell wall, *: amyloplast, →: degrading nuclei, ⇉: tylose cell membrane, P: plastid, ic: intact cytoplasm, v: vacuole, ↑↑: mitochondria, ↑: lipid bodies, b: bacteria, Control: term 0, Term I: after 7 days of postharvest life (wilting of flowers kept in distilled water), Term II: and after 12 days, when wilting and loss of a decorative value occurred in flowers placed into the preservative.

TABLE 3: Average number of blocked vessels (in percent) in *Clematis* stems stored in distilled water and 8HQC with 2% sucrose, term I.

Cultivar	Completely blocked vessels		Half-blocked vessels	
	Distilled water	8HQC + SUC	Distilled water	8HQC + SUC
Andromeda	7.88g	5.7e	8.64b	33.84g
Viola	5.14d	0.74b	8.72b	18.58d
Isago	7.06f	3.0c	11.0c	21.4f
Solidarność	5.66e	0.1a	9.34b	20.24e
Silver moon	7.14f	0.1a	11.36c	6.2a

Numbers followed by the same letter do not differ significantly at $P = 0.95$, according to Duncan's multiple range test, $P \leq 0.05$.
P: the least significant difference.

TABLE 4: Average number of blocked vessels (in percent) in *Clematis* stems stored in distilled water and 8HQC with 2% sucrose, term II.

Cultivar	Completely blocked vessels		Half-blocked vessels	
	Distilled water	8HQC + SUC	Distilled water	8HQC + SUC
Andromeda	10.2c	28.28h	15.88b	37.54e
Viola	14.18f	11.7d	11.26a	35.48e
Isago	18.94g	11.4d	16.6b	22.36c
Solidarność	12.7e	3.74b	9.52a	28.64d
Silver moon	13.74f	2.9a	15.24b	22.6c

Numbers followed by the same letter do not differ significantly at $P = 0.95$, according to Duncan's multiple range test, $P \leq 0.05$.
P: the least significant difference.

though some bacteria and amorphous material were present in observed specimens, their spread was reduced by the preservative by nearly one half relative to distilled water. The amount of amorphous material in the xylem lumen was localized only near the secondary walls of xylem vessels and was probably connected with the activity of bacteria and a degradation of cell components.

3.3. Anatomical, Histochemical, and Immunohistochemical Identification of Xylem Occlusions. In the freshly harvested stems (collection date 0), no traces of mechanical vessel blockage or gel occlusions were observed in any of the five cultivars studied (Figures 3(a) and 3(b)). After seven days of vase life (collection date I), in stems kept in water, completely blocked vessels presented *ca* 5.06%–7.88% and half- blocked vessels *ca* 8.64–11.36% of the total vessel number in all observed cultivars (Figures 3(c) and 3(d), Table 3) Blocked vessels were observed only in the metaxylem. After 12 days of the vase life in distilled water (collection date II), completely blocked vessels represented about 10–18.94% and the half-blocked vessels about 9.5–16.6% of the total vessel number (Figures 3(e) and 3(f), Table 4). Completely and half-blocked vessels were present both in the proto and metaxylem (Figures 3(e) and 3(f)).

In stems kept in the standard preservative on the collection date I, completely blocked vessels were blocked in 5.7% in the short-lasting cv. "Andromeda" and in 3% in cv. "Isago." The remaining three cultivars had less than 1% vessels blocked (Table 3); half blocked vessels were present in 6.2–33,8%, and they were seen in the proto and metaxylem (Table 3). On the collection date II, completely blocked and half-blocked vessels represented 3%–29% and 22.3–37.5% of

the total number of vessels observed, respectively, and they were present both in the proto- and in metaxylem (Figures 3(g) and 3(h), Table 4).

There was no correlation between the vessel diameter and proportion of blocked or half-blocked vessels. In stems stored in distilled water (collection date I), the short-lasting cv. "Andromeda" had the highest number of completely blocked vessels (7.8%), but cv. "Viola," another short-lasting cultivar had the lowest number of completely blocked vessels (5.1%). For the collection date II, the highest number of completely blocked vessels was in the mid-lasting cv. "Isago," around 19%, and the lowest number of completely blocked vessels was in the short-lasting cv. "Andromeda" and the long-lasting cultivar "Solidarność" (12.7%). Stems stored in $200 \, mg \, dm^{-3}$ 8HQC with 2% sucrose in sampling date II showed a clear effect of the preservative: the short-lasting cv. "Andromeda" had 28.3% of completely blocked vessels while the long-lasting cv. "Silver Moon" had 2.9% of such completely blocked vessels. Skutnik and Rabiza-Świder [1, 17] rate cv. "Andromeda" and "Viola" as short-lasting, and both had the lowest diameters of the metaxylem vessels, 17–54 μm. Long-lasting cultivars had larger diameters of xylem vessels, 27–110 μm. Their better postharvest longevity may be associated with a better water hydraulic conductivity through wider lumen of the vessels. We have observed that on sampling date II, the number of completely blocked vessels was higher in stems kept in 8HQC + SUC than in stems kept in distilled water in only one, short-lasting cv. "Andromeda" (Table 4). In cv. "Viola" and "Isago," the number of completely blocked vessels in the stems kept in 8HQC + SUC was lower when compared with stems kept in distilled water and it was significantly lower in the long-lasting cv. "Solidarność" and "Silver Moon" (Table 4). 8HQC

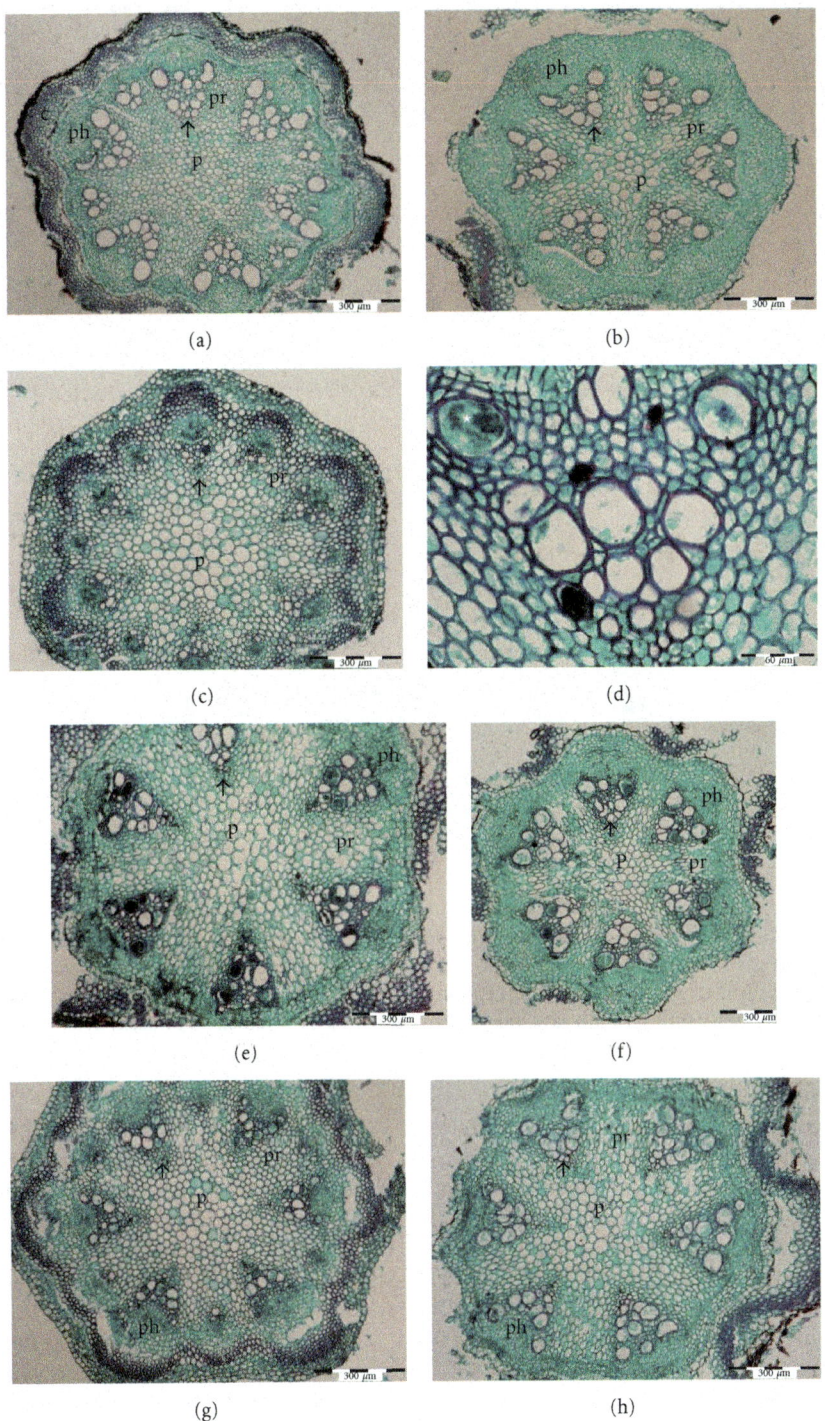

FIGURE 3: Identification of xylem occlusions in short-lasting (Andromeda) and long-lasting (Solidarność) cultivars of *Clematis* by staining with fast green and safranin. (a) and (b) Control stems with xylem vessels free of any occlusions in Andromeda (a) and Solidarność (b). (c) and (d) Stems kept in distilled water, term I with xylem vessels half and completely blocked by the occlusions in Andromeda (c) and Solidarność (d). (e) and (f) Stems kept in distilled water, term II with xylem vessels half and completely blocked by the occlusions in Andromeda (e) and Solidarność (f). (g) and (h) Stems kept in 8HQC + 2% sucrose, term II with xylem vessels half and completely blocked by the occlusions in Andromeda (g) and Solidarność (h). c: cortex with lamellar collenchyma, p: pith, pr: pith rays, ph: phloem, arrow: xylem vessel, *: blocked xylem lumen. The choice of the cultivar was based on observations of the vase life length made by Skutnik and Rabiza-Świder [1, 16] and previous anatomical observations of stems of 5 different cultivars. Two of them were characterized as short-lasting ("Andromeda", "Viola"), one medium ("Isago"), and two long-lasting ("Solidarność" and "Silver moon") cultivars. Control: term 0, term I: after 7 days of postharvest life (wilting of flowers kept in distilled water), term II: after 12 days, when wilting and loss of a decorative value occurred in flowers placed into the preservative.

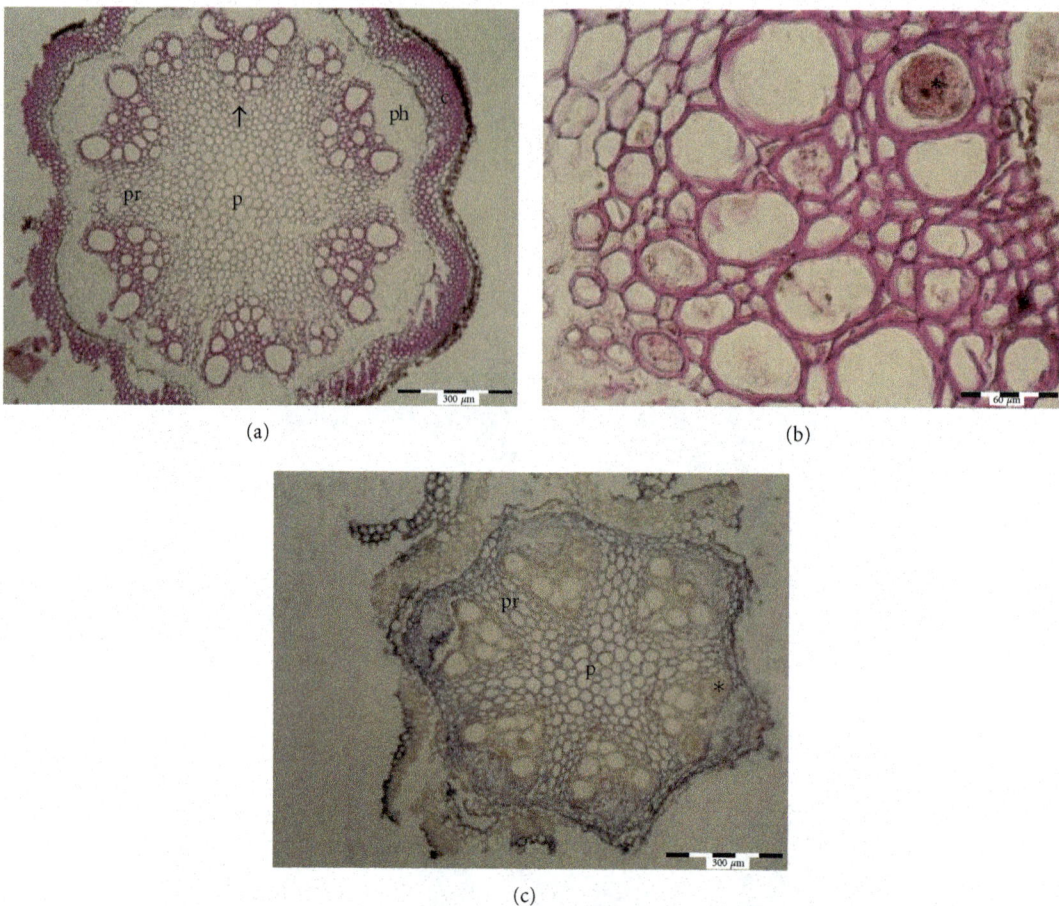

(a)

(b)

(c)

FIGURE 4: Histochemical identification of polysaccharide components and homogalacturonan epitopes in xylem occlusions in stems of *Clematis* "Solidarność" by using PAS reaction and Jim5 antibody. (a) and (b) Identification of polysaccharide components in xylem of control stems (a), stems kept in distilled water (b), (c) identification of homogalacturonan epitopes in xylem of stems kept in distilled water. C: cortex with lamellar collenchyma, p: pith, pr: pith rays, ph: phloem, arrow: xylem vessel, ∗: blocked xylem lumen.

+ SUC significantly reduced the numbers of completely blocked vessels for the sampling date in all studied cultivars, when flowers were still decorative and at the onset of wilting (Table 3). The number of half-blocked vessels in the stems treated by 8HQC + SUC was higher than in stems stored in distilled water in all observed cultivars in sampling dates I and II (Tables 3 and 4). This suggests that 8HQC + SUC delays the development of tyloses compared to distilled water but does not stop it completely.

3.4. Histological and Immunohistological Identification of Xylem Occlusions in cv. "Solidarność"

3.4.1. Polysaccharides. In controls (sampling date 0) free of any xylem pollutants, polysaccharides were detected in cortex, proto- and metaxylem cell walls with parenchymatic cells of the primary rays. The red color of the PAS reaction was also visible in pith rays, but the color intensity suggested lower accumulation of polysaccharides in pith cells. No color reaction was observed in the phloem and cambium. The epidermis stained brown which is the natural color of the tissue and not the test reaction (Figure 4(a)).

In distilled water after 7 days of vase life (collection date I), the PAS reaction produced clear, red color of blocked vessels. The color intensity ranged from light in the half-blocked vessels to intense in the completely blocked vessels (Figure 4(b)). On the second collection date (after 12 days), strongly red stained vessels were present also in protoxylem (data not shown).

On both collection dates (after 7 and 12 days), in stems kept in the standard preservative, only light red color was visible in the blocked and half-blocked vessels. In this treatment, only metaxylem showed weak color coming from the reaction (data not shown).

Homogalacturonan epitopes were recognized by JIM 5 and JIM 7 antibodies [22]. Histological immunolocalization of homogalacturonans showed strong, blue coloration in the phloem, cambium, pith, and particular cells of primary rays when specimens were incubated with the Jim 5 antibody. No evidence of homogalacturonans was present in xylem occlusions observed in stems placed either in distilled water (Figure 4(c)) or in 8HQC + SUC on any collection date. The Jim 7 antibody gave a much weaker signal than Jim 5 antibody (data not shown).

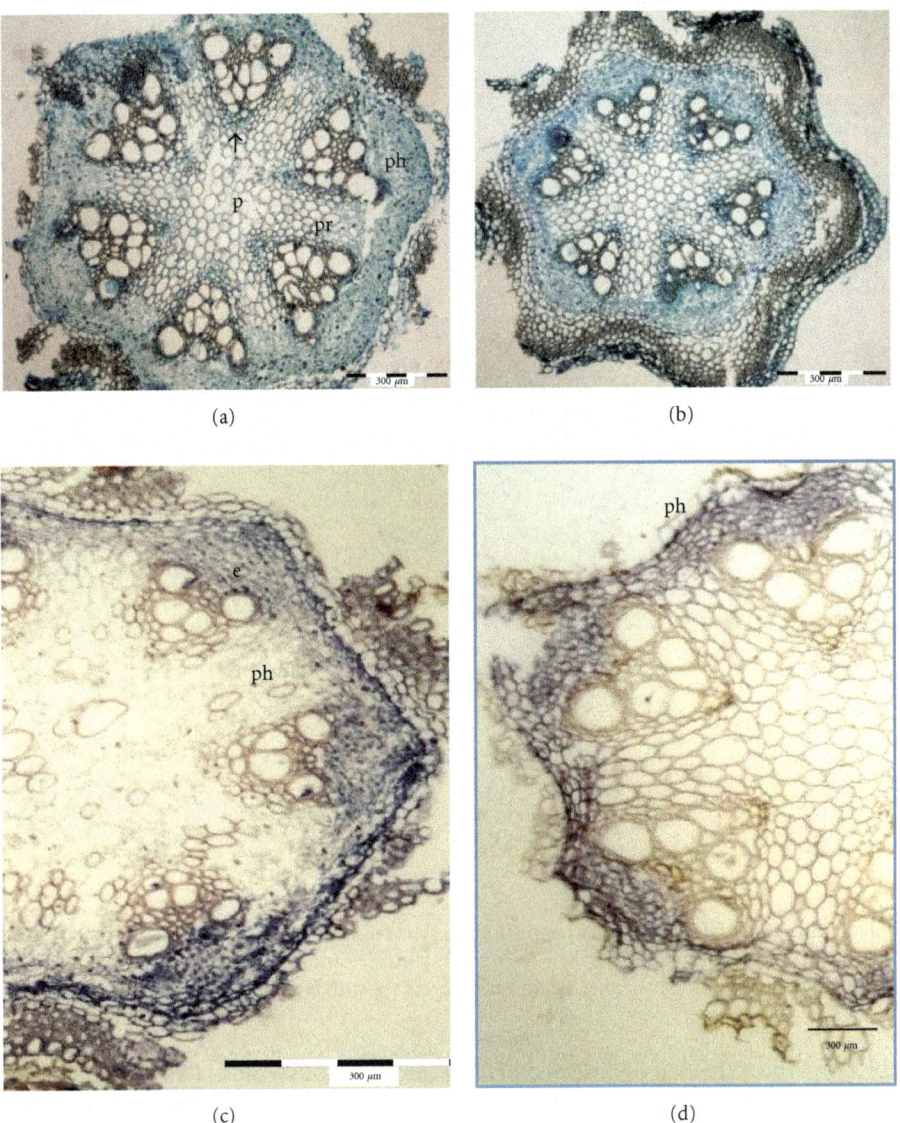

FIGURE 5: Histochemical identification of protein components and extensins in xylem occlusions in stems of *Clematis* "Solidarność" by using Bradford reagent and Jim11 antibody. (a) and (b) Identification of protein components in xylem of control stems (a), stems kept in distilled water (b), (c) and (d) identification of extensins in xylem of stems kept in distilled water (c) and 8HQC + 2% sucrose (d). p: pith, pr: pith rays, ph: phloem, arrow: xylem vessel, ∗: blocked xylem lumen, e: endodermis.

Strong red color of the xylem occlusions after the PAS reaction in the stems kept in distilled water confirmed a high concentration of polysaccharides in the tyloses. Histoimmunochemistry did not localize any homogalacturonans in tyloses nor in the tubular material blocking xylem vessels. According to the literature [8, 15, 23], pectins concentrate in gels which may occlude xylem vessels, but their presence in tyloses is rather rare. In our study, no pectins were detected on the histochemical level, neither in tyloses nor in amorphous, extracellular material occluding xylem vessels. However, in some plants, tylose differentiation may correlate with the accumulation of pectins in parenchymatic cells of pith [15, 24]. In the stems of *clematis*, accumulation of pectin epitopes in pith rays was quite evident, but they were not observed in tyloses.

3.4.2. Proteins. The presence of proteins in xylem occlusions was checked using the Bradford reagent. In control stems, an intensive blue color was visible in phloem, cambium, and the parenchymatic cells surrounding the primary xylem. No evidence of proteins was observed in pith rays, xylem vessels, and collenchyma (Figure 5(a)). On both collection dates during the vase life, in stems kept in distilled water, proteins were visible in all blocked vessels (the half- and completely blocked) both in the proto- and metaxylem (Figure 5(b)).

In stems kept in the standard preservative blocked vessels colored light blue on the sampling date I and strongly blue on the sampling date II. The coloration intensity was equally strong in the completely and half-blocked vessels. Proteins are very often present in the gels or gums occluded into the lumen of the vessel. Tyloses mostly include polyphenols with

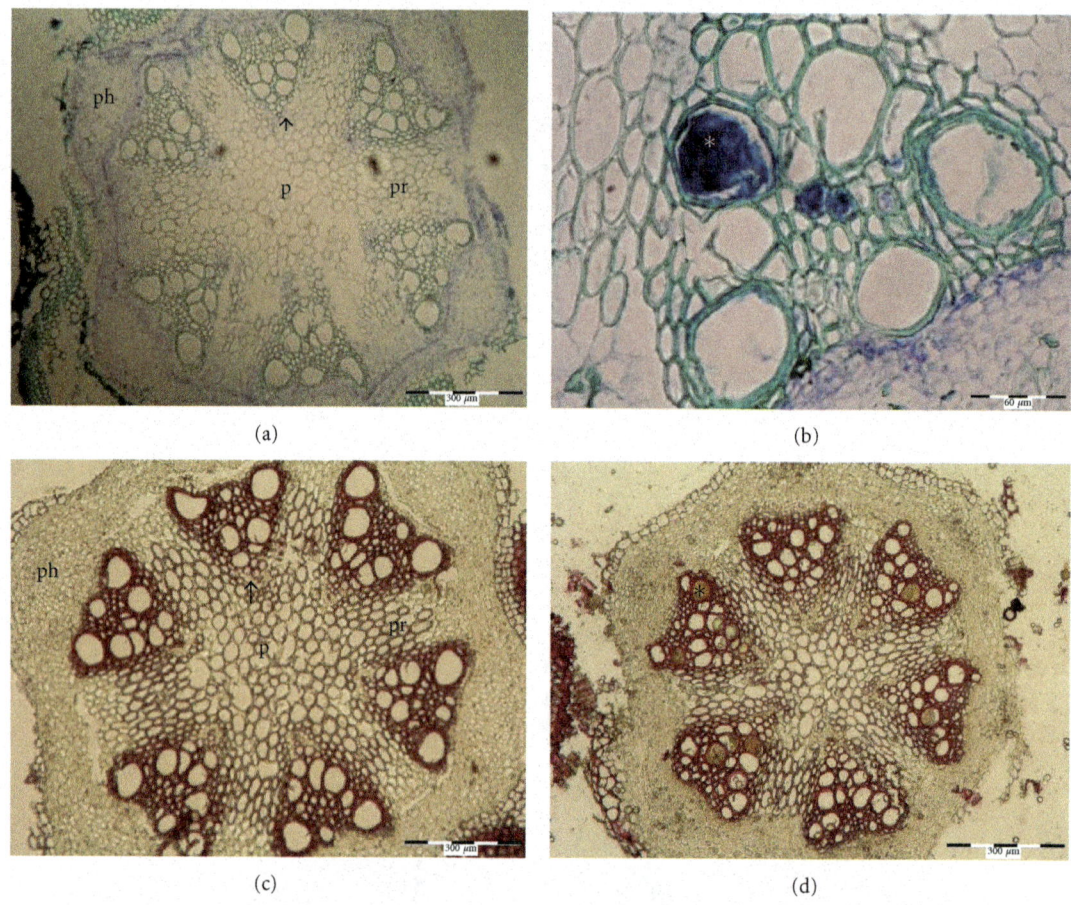

FIGURE 6: Histochemical identification of lignins in xylem occlusions in stems of *Clematis* "Solidarność" by using Azur B and phloroglucinol-HCl methods. (a) and (b) Identification of lignins in xylem of control stems; (c) and (d) Identification of lignins in xylem of stems kept in distilled water; p: pith, pr: pith rays, ph: phloem, arrow: xylem vessel, c: cortex with lamellar collenchyma, ∗: blocked xylem lumen.

their serious antiseptic properties. The tyloses observed here also included proteins probably transported as an extracellular material from the parenchymatic cells surrounding the primary xylem.

Extensins were identified with the extensin-specific antibodies (Jim 11, Jim 12, Jim 20). None of the antibodies bound to the xylem occlusions. When Jim11 was used, the dark blue color indicating the presence of extensin was not detected in the phloem and endodermis (Figures 5(c) and 5(d)). Jim 12 and Jim 20 did not detect any extensins at the histological level. The synthesis of extracellular structural proteins after an injury or a pathogen attack, as well as their subsequent incorporation into cell walls *via* oxidative cross-links, has frequently been reported [25–28]. During their immobilization in the cell wall, these proteins can be linked to other extracellular compounds [29, 30] or to other proteins [31–33]. In this study, even though proteins were detected in blocked xylem vessels, extensins were not localized either in tyloses or in the amorphous material occluding xylem vessels.

3.4.3. Lignins. Azur B showed intensive blue, and the HCL phloroglucinol showed intensive red coloration in control xylem cell walls (Figures 6(a) and 6(b)) indicating the

presence of lignins. On both sampling dates, the occlusions present in stems kept in distilled water gave strong, blue coloration in metaxylem when stained with Azur B (Figure 6(c)). In phloroglucinol staining, xylem occlusions preserved their natural, orange color in contrast to strongly stained vessels (Figure 6(d)). This phenomenon appeared in xylem vessels from stems kept both in distilled water and in the preservative solution. Azur B staining of xylem blockage in stems placed in the preservative produced blue coloration of the occlusions, but the intensity was lower than in stems kept in distilled water. These confirms cytological observations that tyloses are the main cause of xylem blockage in *clematis* stems. According to Soukup and Votrubova [34] and preceding authors [15, 35], tyloses contain plenty of polyphenols which are main components of lignins and their role is mainly antibacterial. In *clematis* here, staining with Azur B was more specific for lignins relative to the HCL phloroglucinol test. The test on lignin's presence in blocked vessels clearly shows that 8HQC + SUC arrests the development of tyloses in cut stems of *clematis*, as shown by weaker color of stained occlusions.

3.4.4. Suberins. Sections prepared from freshly harvested stems (sampling date 0) and from those kept in the solution

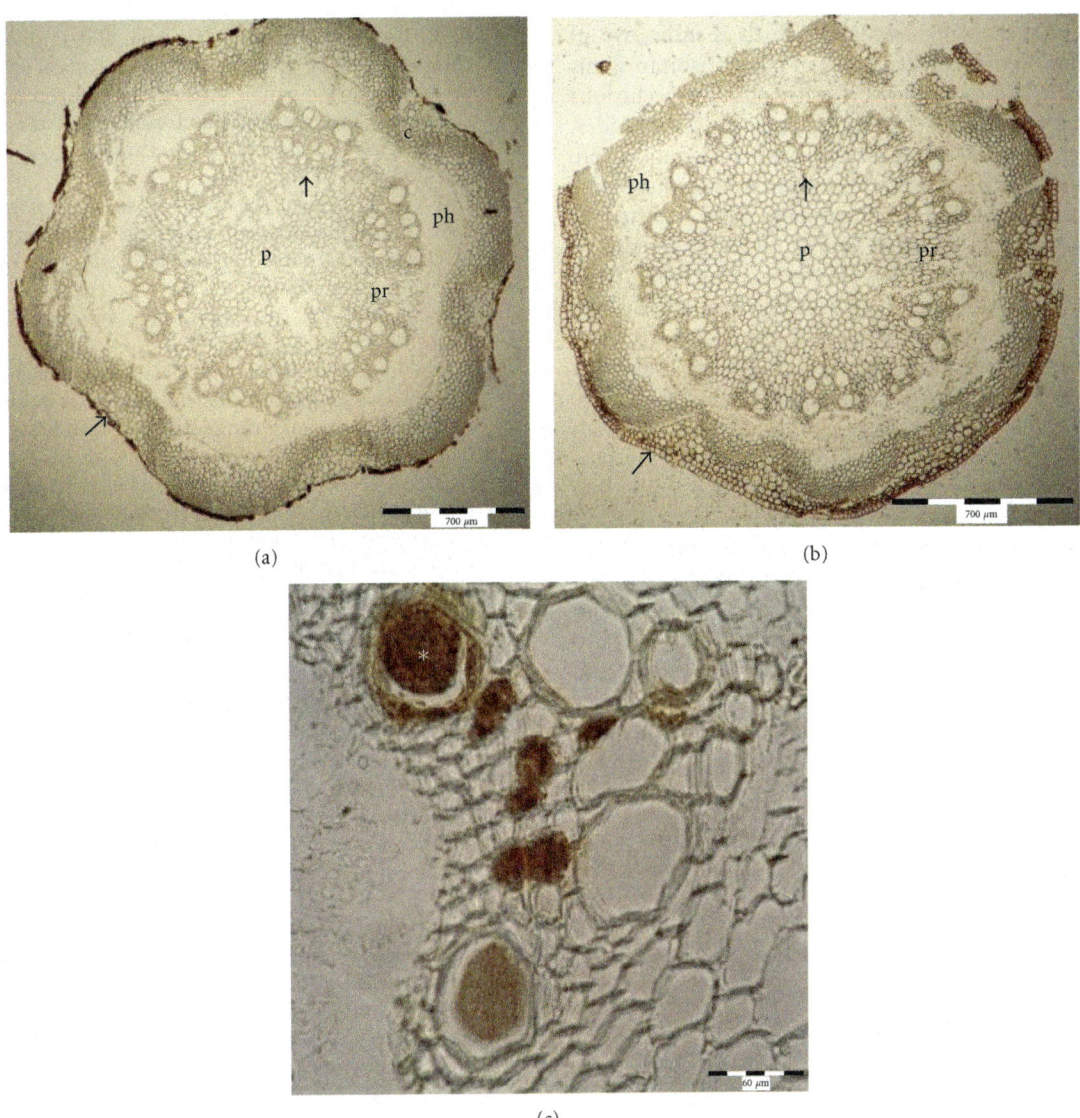

(a)   (b)

(c)

FIGURE 7: Histochemical identification of suberins in xylem occlusions in stems of *Clematis* "Solidarność" by using Sudan IV. (a) Identification of suberins in xylem of control stems; (b) identification of suberins in xylem of stems kept in 8HQC + 2% sucrose; (c) identification of suberins in stems kept in distilled water. P: pith, pr: pith rays, ph: phloem, arrow: xylem vessel, c: cortex with lamellar collenchyma, ∗: blocked xylem lumen, ╱: epidermis.

of 8HQC + SUC, did not show any evidence of suberins (Figures 7(a) and 7(b)). Suberins were only present in xylem occlusions from stems kept in distilled water. On sampling dates I and II, suberins were present only in blocked vessels (Figure 7(c)).

4. Conclusions

The main reasons of xylem blockage in cut flowers are air embolism, tyloses, plant and soil microparticles present in the water, bacteria developing in old, dirty water and spreading into vessels, and gums and gels formed in response to cutting [41, 42]. In this study, we have focused mainly on the identification of xylem occlusions appearing in the *clematis* stems during their postharvest life, depending on the keeping

solution. According to Skutnik and Rabiza-Świder [1, 17], solution composed of 8HQC + SUC efficiently prolongs *clematis* vase life. 8HQC acts as a bactericide, and it may be responsible for the reduced formation of tyloses and other artifacts. According to Van Doorn et al. [43], in some woody ornamentals such as common lilac for cut flower, tylose formation is suppressed by 8HQC (8-hydroxyquinoline citrate) and AVG (aminoethoxyvinylglycine). These results were confirmed by Jedrzejuk and Zakrzewski [44] on stems of common lilac stored in distilled water, $200\,\mathrm{mg\,dm^{-3}}$ 8HQC, and Chrysal Professional. This study shows that tyloses, when they appear in *clematis* stem kept in s8HQC, they never occlude the entire lumen of a vessel as they do in stems kept in water. Tyloses produced in stems treated by 8HQC + 2% sucrose did not show the presence of degraded lipids or autophagosomal vacuoles, which is indicative of less degeneration of stem structures in comparison to stems kept in water. Histological tests revealed that the occlusions are mainly composed of proteins, lipids, polysaccharides, phenolics, and lignin-like material. These compounds were reported to appear both in woody and in herbaceous plants [24]. In *clematis*, all these components were detected in occluded vessels of the stems kept both in water and in the preservative solution. However, when stems were placed in distilled water, the percentage of blocked vessels was higher than in the stems kept in the preservative. This is demonstrated by relative differences in staining intensities in quantitative color reactions employed in this study. The 8HQC + SUC combination also significantly arrested bacterial proliferation: on the second sampling date, the percentage of vessels containing bacteria was only one half that of stems kept in distilled water (27–30%).

This study confirms that the preservative composed of $200\,mg\,dm^{-3}$ 8HQC, and 2% sucrose arrests the development of bacteria in the vessels of cut *Clematis* stems and to some extent also reduces the growth of tyloses. We also identify the origin of the xylem occlusions in *clematis* and compare their development in water and in the preservative solution. Such information can be useful to develop new treatments aiming to improve keeping qualities of cut *clematis* stems.

List of Abbreviations

8HQC:	8-Hydroxyquinoline citrate
8HQC + SUC:	$200\,mg\,dm^{-3}$ + 2% sucrose
DEPC:	Diethylpyrocarbonate
PFA:	Paraformaldehyde
DMSO:	Dimethylsulfoxide
NBT/BCIP:	Nitroblue tetrazolium and 5-bromo-4-chloro-3′-indolyphosphate.

Acknowledgment

This research was supported by Grant of Ministry of Science and Higher Education no. 0893/B/P01/2009/36.

References

[1] E. Skutnik and J. Rabiza-Świder, "Przydatnosc kwiatow cietych wybranych odmian powojnika (*Clematis* L.) do wykorzystania we florystyce," *Zeszyty Problemowe Postępów Nauk Rolniczych*, vol. 504, pp. 507–513, 2005.

[2] W. G. Van Doorn, "Water relations of cut flowers," *Horticultural Reviews*, vol. 18, pp. 1–85, 1997.

[3] P. Twumasi, W. van Ieperen, E. J. Woltering et al., "Effects of water stress during growth on xylem anatomy, xylem functioning and vase life in three *Zinnia elegans* cultivars," in *Proceedings of the 8th International Symposium on Postharvest Physiology of Ornamental Plants*, N. Marissen, W. VanDoorn, and U. VanMeeteren, Eds., vol. 669, pp. 303–311, Acta Horticulturae, 2005.

[4] M. J. Canny, "Tyloses and the maintenance of transpiration," *Annals of Botany*, vol. 80, no. 4, pp. 565–570, 1997.

[5] K. E. Bretschneider, M. P. Gonella, and D. J. Robeson, "A comparative light and electron microscopical study of compatible and incompatible interactions between Xanthomonas campestris pv. campestris and cabbage (*Brassica oleracea*)," *Physiological and Molecular Plant Pathology*, vol. 34, no. 4, pp. 285–297, 1989.

[6] B. Boher, I. Brown, M. Nicole et al., "Histology and cytochemistry of interactions between plants and Xanthomonas," in *Histology, Ultrastructure and Molecular Cytology of Plant-Microorganism Interactions*, M. Nicole and V. Gianinazzi-Pearson, Eds., pp. 193–210, Kluwer, Dordecht, The Netherlands, 1996.

[7] K. Kpémoua, B. Boher, M. Nicole, P. Calatayud, and J. P. Geiger, "Cytochemistry of defense responses in cassava infected by Xanthomonas campestris pv. manihotis," *Canadian Journal of Microbiology*, vol. 42, no. 11, pp. 1131–1143, 1996.

[8] D. Rioux, M. Nicole, M. Simard, and G. B. Ouellette, "Immunocytochemical evidence that secretion of pectin occurs during gel (gum) and tylosis formation in trees," *Phytopathology*, vol. 88, no. 6, pp. 494–505, 1998.

[9] S. Espino and H. J. Schenk, "Mind the bubbles: achieving stable measurements of maximum hydraulic conductivity through woody plant samples," *Journal of Experimental Botany*, vol. 62, no. 3, pp. 1119–1132, 2011.

[10] D. S. Teackle, P. M. Smith, and D. R. L. Steindl, "Ratoon stunting disease of sugarcane: possible correlation of resistance with vascular anatomy," *Phytopathology*, vol. 65, pp. 138–141, 1975.

[11] D. L. Hopkins and H. H. Mollenhauer, "*Rickettsia*-like bacterium associated with Pierce's disease of grapes," *Science*, vol. 179, no. 4070, pp. 298–300, 1973.

[12] P. Y. Huang, R. D. Milholland, and M. E. Daykin, "Structural and morphological changes associated with the Pierce's disease bacterium in bunch and muscadine grape tissues," *Phytopathology*, vol. 76, pp. 1232–1238, 1986.

[13] S. M. Fry and R. D. Milholland, "Response of resistant tolerant and susceptible grapevine tissues to invasion by the Pierce's disease bacterium *Xylella fastidiosa*," *Phytopatology*, vol. 80, pp. 66–69, 1990.

[14] J. F. Stevenson, M. A. Matthews, L. C. Greve, J. M. Labavitch, and T. L. Rost, "Grapevine susceptibility to Pierce's disease II: progression of anatomical symptoms," *American Journal of Enology and Viticulture*, vol. 55, no. 3, pp. 238–245, 2004.

[15] A. Clérivet, V. Déon, I. Alami, F. Lopez, J. P. Geiger, and M. Nicole, "Tyloses and gels associated with cellulose accumulation in vessels are responses of plane tree seedlings (*Platanus x acerifolia*) to the vascular fungus *Ceratocystis fimbriata* f. sp platani," *Trees*, vol. 15, no. 1, pp. 25–31, 2000.

[16] E. Skutnik and J. Rabiza-Świder, "Wplyw chlodzenia na pozbiorcza trwalosc wybranych odmian powojnika (*Clematis* L.)," *Zeszyty Problemowe Postępów Nauk Rolniczych*, vol. 510, pp. 587–592, 2006.

[17] U. van Meeteren, L. Arévalo-Galarza, and W. G. van Doorn, "Inhibition of water uptake after dry storage of cut flowers: role of aspired air and wound-induced processes in Chrysanthemum," *Postharvest Biology and Technology*, vol. 41, no. 1, pp. 70–77, 2006.

[18] C. H. Beckman and P. W. Talboys, "Anatomy of resistance," in *Fungal Wilt Diseases of Plants*, M. E. Mace, A. A. Bell, and C. H. Beckman, Eds., pp. 487–521, Academic Press, New York, NY, USA, 1981.

[19] G. E. VanderMolen, C. H. Beckman, and E. Rodehorst, "The ultrastructure of tylose formation in resistant banana following inoculation with *Fusarium oxysporum* f.sp. *Cubense*,"

Physiological and Molecular Plant Pathology, vol. 31, no. 2, pp. 185–200, 1987.

[20] A. A. Bell, "Verticillium wilt," in *Cotton Diseases*, R. J. Hillocks, Ed., pp. 87–126, CAB International, Wallingford, UK, 1992.

[21] G.B. Ouellette and D. Rioux, "Anatomical and physiological aspects of resistance to Dutch elm disease," in *Defense Mechanisms of Woody Plants against Fungi*, A. Blanchette and A. Biggs, Eds., pp. 257–307, Springer, New York, NY, USA, 1992.

[22] M. H. Clausen, W. G. T. Willats, and J. P. Knox, "Synthetic methyl hexagalacturonate hapten inhibitors of anti-homogalacturonan monoclonal antibodies LM7, JIM5 and JIM7," *Carbohydrate Research*, vol. 338, no. 17, pp. 1797–1800, 2003.

[23] G. J. Niemann, R. P. Baayen, and J. J. Boon, "Localization of phytoalexin accumulation and determination of changes in lignin and carbohydrate composition in carnation (*Dianthus caryophyllus* L.) xylem as a consequence of infection with *Fusarium oxysporum* f. sp. dianthi, by pyrolysis-mass spectrometry," *Netherlands Journal of Plant Pathology*, vol. 96, no. 3, pp. 133–153, 1990.

[24] K. S. Rajput, G. V. Sanghvi, R. D. Koyani, and K. S. Rao, "Anatomical changes in the stems of *Azadirachta indica* (meliaceae) infected by Pathogenic Fungi," *IAWA Journal*, vol. 30, no. 1, pp. 27–36, 2009.

[25] G. I. Cassab and J. E. Varner, "Cell wall proteins," *Annual Review of Plant Biology*, vol. 39, pp. 321–353, 1988.

[26] D. J. Bradley, P. Kjellbom, and C. J. Lamb, "Elicitor- and wound-induced oxidative cross-linking of a proline-rich plant cell wall protein: a novel, rapid defense response," *Cell*, vol. 70, no. 1, pp. 21–30, 1992.

[27] M. T. Tyree, S. D. Davis, and H. Cochard, "Biophysical perspectives of xylem evolution: is there a tradeoff of hydraulic efficiency for vulnerability to dysfunction?" *IAWA Journal*, vol. 15, no. 4, pp. 335–360, 1994.

[28] G. Merkouropoulos and A. H. Shirsat, "The unusual *Arabidopsis* extensin gene atExt1 is expressed throughout plant development and is induced by a variety of biotic and abiotic stresses," *Planta*, vol. 217, no. 3, pp. 356–366, 2003.

[29] K. Iiyama, Lam Thi Bach Tuyet, and B. A. Stone, "Covalent cross-links in the cell wall," *Plant Physiology*, vol. 104, no. 2, pp. 315–320, 1994.

[30] L. Saulnier, C. Marot, E. Chanliaud, and J. F. Thibault, "Cell wall polysaccharide interactions in maize bran," *Carbohydrate Polymers*, vol. 26, no. 4, pp. 279–287, 1995.

[31] S. C. Fry, "Isodityrosine, a new cross-linking amino acid from plant cell-wall glycoprotein," *Biochemical Journal*, vol. 204, no. 2, pp. 449–455, 1982.

[32] K. J. Biggs and S. C. Fry, "Solubilization of covalently bound extensin from Capsicum cell walls," *Plant Physiology*, vol. 92, no. 1, pp. 197–204, 1990.

[33] J. D. Brady, I. H. Sadler, and S. C. Fry, "Di-isodityrosine, a novel tetrameric derivative of tyrosine in plant cell wall proteins: a new potential cross-link," *Biochemical Journal*, vol. 315, no. 1, pp. 323–327, 1996.

[34] A. Soukup and O. Votrubová, "Wound-induced vascular occlusions in tissues of the reed *Phragmites australis*: their development and chemical nature," *New Phytologist*, vol. 167, no. 2, pp. 415–424, 2005.

[35] M. T. Tyree and M. H. Zimmermann, *Xylem Structure and the Ascent of Sap*, Springer, Berlin, Germany, 2002.

[36] A. G. Pearse, *Histochemistry (Theoretical and Applied)*, J. & A. Churchill Ltd, London, UK, 1968.

[37] M. M. Bradford, "A rapid and sensitive method for the quantitation of microgram quantities of protein utilizing the principle of protein dye binding," *Analytical Biochemistry*, vol. 72, no. 1-2, pp. 248–254, 1976.

[38] W. A. Jensen, *Botanical Histochemistry*, W. H. Freeman, San Francisco, Calif, USA, 1962.

[39] M. N. Clifford, "Specificity of acidic phloroglucinol reagents," *Journal of Chromatography A*, vol. 94, pp. 321–324, 1974.

[40] M. C. Brundrett, B. Kendrick, and C. A. Peterson, "Efficient lipid staining in plant material with Sudan Red 7B or fluoral yellow 088 in polyethylene glycol-glycerol," *Biotechnic and Histochemistry*, vol. 66, no. 3, pp. 133–142, 1988.

[41] W. G. Van Doorn and P. Cruz, "Evidence for a wounding-induced xylem occlusion in stems of cut chrysanthemum flowers," *Postharvest Biology and Technology*, vol. 19, no. 1, pp. 73–83, 2000.

[42] M. Loubaud and W. G. Van Doorn, "Wound-induced and bacteria-induced xylem blockage in roses, *Astilbe*, and *Viburnum*," *Postharvest Biology and Technology*, vol. 32, no. 3, pp. 281–288, 2004.

[43] W. Van Doorn, H. Harkema, and E. Otma, "Is vascular blockage in stems of cut lilac flowers mediated by ethylene?" *Acta Horticulturae*, vol. 298, pp. 177–182, 1991.

[44] A. Jedrzejuk and J. Zakrzewski, "Xylem occlusions in the stems of common lilac during postharvest life," *Acta Physiologiae Plantarum*, vol. 31, no. 6, pp. 1147–1153, 2009.

Carrizo citrange Plants Do Not Require the Presence of Roots to Modulate the Response to Osmotic Stress

Rosa M. Pérez-Clemente, Almudena Montoliu, Sara I. Zandalinas, Carlos de Ollas, and Aurelio Gómez-Cadenas

Department of Agricultural Sciences, Universitat Jaume I, Campus Riu Sec, 12071 Castelló de la Plana, Spain

Correspondence should be addressed to Aurelio Gómez-Cadenas, aurelio.gomez@uji.es

Academic Editors: G. Galiba, H. Verhoeven, and B. Vyskot

The study of the effects of a specific stress condition on the performance of plants grown under field conditions is difficult due to interactions among multiple abiotic and biotic factors affecting the system. *In vitro* tissue-culture-based techniques allow the study of each adverse condition independently and also make possible to investigate the performance of genotypes of interest under stress conditions avoiding the effect of the root. In this paper, the response of Carrizo citrange, a commercial citrus rootstock, to osmotic stress was evaluated by culturing *in vitro* intact plants and micropropagated shoots. The osmotic stress was generated by adding two different concentrations of polyethyleneglycol to the culture media. Different parameters such as plant performance, organ length, antioxidant activities, and endogenous contents of proline, malondialdehyde, and hormones were determined. Differently to that observed under high salinity, when subjected to osmotic stress conditions, Carrizo citrange showed increased endogenous levels of MDA, proline, and ABA. These results evidence that the mechanisms of response of Carrizo citrange to saline or osmotic stress are different. The presence of roots was not necessary to activate any of the plant responses which indicates that the organs involved in the stress perception and signaling depends on the type of adverse condition to which plants are subjected.

1. Introduction

Water is essential for plant growth and a necessary component of most physiological processes. Drought and other abiotic factors often decrease soil water potential and difficult plant water uptake. Water deficiency in a plant, which results from an imbalance between water uptake and loss, leads to the disturbance of various physiological processes [1].

Citrus is one of the most important horticultural crops globally and is considered as being salt sensitive [2, 3]. In the Mediterranean area, low rainfall and high temperatures in summer along with the high concentrations of salts found in the irrigation water often result in agricultural crops suffering simultaneous water and salt stress [4]. Water stress in citrus reduces growth and metabolism, leading to a reduction in fruit yield and quality [5, 6].

The study of the effects of a specific stress condition on the performance of plants grown under field conditions is difficult due to different uncontrolled situations such as interactions among multiple abiotic and biotic stress factors.

Moreover, the measurement of root traits in field-grown plants is laborious, time-consuming, and not always possible [7].

Development of tools to separately investigate specific components of the multiple factors affecting plant development can be useful for both academic and applied research. The *in vitro* tissue culture techniques can overcome some of the above described limitations and allow growing clonal plants under controlled climatic and nutritional conditions to carry out experiments in identical conditions all year round [8]. In a previous work, we used this approach to study the performance of different citrus genotypes cultured under salt stress conditions, avoiding the effect of the root by culturing shoots of those genotypes without the root system. The method proved to be a good tool for studying biochemical processes involved in the response of citrus to salt stress [9].

Addition of sorbitol or polyethylene glycol (PEG) decreases medium water potential, inducing water stress that adversely affects both shoot and root growth of the plantlets [10]. Polymers of PEG have been used for many years, mainly because PEG molecules with a molecular weight of 6000 g/mol (PEG-6000) cannot penetrate the cell wall pores [11]. Because PEG does not enter the apoplast, water is with drawn not only from the cell but also from the cell wall. Therefore, PEG solutions mimic dry soil more closely than solutions of compounds with low molecular weights, which infiltrate the cell wall with solute [12].

Plants grown under osmotic stress conditions are seriously affected by oxidative stress. One of the earliest responses aiming water loss avoidance involves stomatal closure, which subsequently downregulates the photosynthetic machinery due to a decrease in CO_2 uptake [13]. As a consequence, the photosynthetic electron transport chain becomes overreduced, resulting in the generation of reactive oxygen species (ROS) [14]. In plant cells, the excessive production of ROS is potentially harmful to lipids, proteins, and nucleic acids [15], whose oxidation may, in turn, lead to detrimental effects such as enzyme inhibition, chlorophyll degradation, disruption of membranes, among others [16]. Malondialdehyde (MDA), a decomposition product of polyunsaturated fatty acids in biomembranes, is often used as indicator of peroxidation, and this compound accumulates under several abiotic stress conditions [17, 18]. Several enzymatic systems and antioxidant molecules are responsible to counteract the deleterious effects of ROS. The first enzymatic step in the detoxifying process is the superoxide dismutase activity (SOD), which catalyses the conversion of $O_2^{\bullet-}$ to H_2O_2. The H_2O_2 is then reduced to water by the enzymatic activities ascorbate peroxidase (APX), which utilizes ascorbate as the specific electron donor, and catalase (CAT), which does not require any reducing equivalent [16].

Besides the activation of different detoxification sys tems, plant tissues accumulate compatible solutes [19] such as the amino acid proline which is one of the most widespread osmoprotectants [20] to neutralize osmotic stress; in addition to its role for osmotic adjustment under stress conditions, its capacity for quenching reactive oxygen radicals may help cells to overcome oxidative damage caused by water deficit [21]. In this respect, proline has been proposed to stabilize DNA, membranes, and proteins [22]. Moreover, the role of proline in plant responses to oxidative stress has been extensively shown by experiments utilizing exogenous application of this amino acid [23] or by genetic manipulation of its metabolism; it has been reported that transgenic "Swingle" citrumelo plants overexpressing Δ1-pyrroline-5-carboxylate synthetase gene (P5CS), involved in the first two steps of proline biosynthesis, exhibited better osmotic adjustment, and tolerated longer period in severe conditions of drought stress than wild-type plants [18].

The present study was designed to evaluate the response of Carrizo citrange, a citrus rootstock widely used in citriculture worldwide, to moderate and severe osmotic stress conditions. To depict the specific effect of this abiotic stress on plant physiology without the interferences of other factors, an in vitro tissue culture system was used. The osmotic stress was generated by using two different concentrations of PEG. Stress impact on the plant was evaluated by measuring growth and levels of two metabolites, proline and MDA, as stress markers. Physiological responses to osmotic stress were evaluated by measuring plant hormonal concentration and different antioxidant activities.

2. Materials and Methods

2.1. Plant Materials and Treatments. Carrizo citrange (*C. sinensis* L. Osb. x *P. trifoliata* L. Raf.) seeds were peeled, and after removing their both seed coats, they were disinfected for 10 min in a 0.5% (vol/vol) sodium hypochlorite solution containing 0.1% (vol/vol) Tween-20 wetting agent and rinsed three times with sterile distilled water. Seeds were sown individually in 25×150 mm culture tubes with 25 mL of germination medium (GM) consisting of Murashige and Tucker (MT) salt solution [24], 100 mg/L *i*-inositol, 1 mg/L pyridoxine-HCl, 0.2 mg/L thiamine-HCl, 1 mg/L nicotinic acid, and 30 g/L sucrose. The pH was set at 5.7 ± 0.1 with 0.1 N NaOH before autoclaving. The medium was solidified by the addition of agar (Pronadisa, Madrid, Spain). The cultures were maintained at 26°C, first in darkness for 2 weeks and then with a 16 h photoperiod and illumination of 45 mmol m^{-2} s^{-1}. Once root system reached 3 cm length, plants were used as plant material to carry out the osmotic stress experiments.

Polyethylene glycol (PEG-6000, Panreac, Barcelona, Spain) was added to GM at two concentrations to achieve medium osmotic potential of −0.75 and −1.5 MPa, according to Michel and Kaufmann [25]. All the experiments were performed using liquid GM medium supplemented with PEG when necessary: control medium (GM), moderate-stress condition (osmotic potential of −0.75 MPa, PEG075), and severe-stress condition (osmotic potential of −1,50 MPa, PEG150). Plants were maintained in the culture conditions described above throughout experimental period.

Another set of experiments was performed using Carrizo citrange micropropagated shoots. Both, establishment of cultures from greenhouse growing plants and the micro-propagation, they were performed following the protocol described in Montoliu et al. [9]. When shoots reached 15 mm length, they were excised and used as plant material in successive experiments.

The culture medium (MT), consisted of inorganic salts of Murashige and Skoog [26], 100 mg/L *i*-inositol, 1 mg/L pyridoxine-HCl, 0.2 mg/L thiamine-HCl, 1 mg/L nicotinic acid, and 30 g/L sucrose supplemented with 0.2 mg/L of gibberellic acid and 0.2 mg/L 6-benzylaminopurine. The pH was set at 5.7 ± 0.1 with 0.1 N NaOH before autoclaving.

As described above, for the osmotic stress treatments, different concentrations of PEG-6000 were added to MT medium to achieve osmotic potentials of −0.50 (PEG050, moderate-stress condition) and −0.75 MPa (PEG075, severe-stress condition). Shoots grown on MT were used as controls.

Samples of shoots and roots were taken for analyses at 10, 20, and 30 days after the plants were transferred to treatment media. Shoots and roots were collected, rinsed with distilled water to eliminate any residue, and frozen in liquid nitrogen. Plant material was kept at −80°C until further analyses.

2.2. Visible Symptoms of Plant Damage. The presence of yellowish spots at the leaf tip that progressively led to severe burning injuries was considered to be a good visible estimate of osmotic stress-induced damage to leaves. The number of damaged leaves was regularly recorded during the experimental period and expressed as a percentage of the total number of leaves. Plants showing a percentage of damaged leaves equal to or over 50% and or damage in the root (considered this as necrosis at the root apex) were considered "affected" for the osmotic treatment. Shoot and root length were also recorded after 10, 20, and 30 days of treatment.

2.3. Malondialdehyde Concentration. Malondialdehyde concentration was measured following the procedure described in Hodges et al. [27]. Plant material (root or shoot tissue) was homogenized in 5 mL of 80% cold ethanol (Panreac, Barcelona, Spain) using a tissue homogenizer (Ultra-Turrax; IKA-Werke, Staufen, Germany). Homogenates were centrifuged at 4°C to pellet debris and different aliquots of the supernatant were mixed either with 20% trichloroacetic acid (TCA) (Panreac) or a mixture of 20% TCA and 0.5% thiobarbituric acid (Sigma-Aldrich, Madrid, Spain). Both mixtures were allowed to react in a water bath at 90°C for 1 h. After this time, samples were cooled down in an ice bath and centrifuged. Absorbance at 440, 534, and 600 nm was read in the supernatant against a blank. The MDA concentration in the extracts was calculated as in Arbona et al. [17].

2.4. Antioxidant Enzyme Activity. Enzyme assays were performed as described in Arbona et al. [2]. Briefly, 0.5 g of frozen plant material was extracted in 2.5 mL of PBS using sea sand as an abrasive. The homogenate was filtered through two layers of muslin cloth. The different buffers used for enzyme extraction were for APX, 50 mM PBS pH 7.1 supplemented with 1 mM sodium ascorbate, 0.1 mM EDTA, and two drops of Triton X-100 (Panreac) and for CAT was 50 mM PBS pH 6.8. The APX activity (EC 1.11.1.11) was determined following the depletion in absorbance at 290 nm because of ascorbate consumption. CAT (EC 1.11.1.6) was assayed using the hydrogen peroxide-dependent reduction of titanium chloride. Protein content in extracts was assessed by means of the protein-dye binding method using Coomassie blue G-250 (Sigma-Aldrich). Enzyme activity was expressed as arbitrary units per mg protein.

2.5. Proline Content. Proline content was analyzed following the procedure described in Arbona et al. [2]. Briefly, 0.05 g of frozen plant tissue (shoot or root) was homogenized in 5% sulphosalicylic acid (Panreac) using a tissue homogenizer (Ultra-Turrax). After extraction, homogenates were centrifuged to pellet cell debris at 4°C and 1 mL aliquot of the supernatant was combined with an equal volume of glacial acetic acid (Panreac) and ninhydrin reagent. This mixture was boiled in a water bath for 1 h and then cooled in an ice bath. The solution was partitioned against 2 mL of toluene (ACS grade; Panreac) and absorbance at 520 nm measured in this organic layer. A calibration curve was performed using commercial proline as a standard (Sigma-Aldrich).

2.6. Hormone Analyses. Plant hormones were analyzed by HPLC coupled to tandem mass spectrometry as described in Durgbanshi et al. [28] and Arbona and Gómez-Cadenas [29]. Briefly, frozen plant material (shoot or root) was ground to a fine powder with a prechilled mortar and a pestle and then 0.5 g of powdered tissue was extracted in ultrapure water using a tissue homogenizer (Ultra-Turrax). Before extraction, samples were spiked with 100 ng of $[^2H_6]$-ABA, 100 ng of $[^2H_4]$-SA, and 100 ng of dihydrojasmonic acid to assess recovery and matrix effects. After extraction and centrifugation, the pH of the supernatant was adjusted to 3.0 and partitioned twice against di-ethyl-ether (Panreac). The organic layers were combined and evaporated in a centrifuge vacuum evaporator (Jouan, Saint-Herblain, France). The dry residue was thereafter resuspended in a water:methanol (9 : 1) solution, filtered, and injected into an HPLC system (Alliance 2695, Waters Corp., Milford, USA). Hormones were then separated in a reversed-phase Kromasil 100 C18 column (100×2.1 mm 5-μm particle size) using methanol and ultrapure water both supplemented with glacial acetic acid to a concentration of 0.05%. The mass spectrometer, a triple quadrupole (Quattro LC, Micromass Ltd., Manchester, UK), was operated in negative ionization electrospray mode and plant hormones were detected according to their specific transitions using a multiresidue mass spectrometric method [28].

2.7. Statistical Analyses. Data mean comparisons and regression analyses were performed with STATGRAPHICS PLUS v.5.1 software (Statistical Graphics Corporation, Herndon, VA). One-way ANOVA and comparisons between means were made following the LSD test at $P \leq 0.05$.

3. Results

To study the effect of osmotic stress in citrus, Carrizo citrange genotype, a commercial rootstock sensitive to salinity and water stress, was selected. Adjustment of PEG concentration was performed to achieve medium osmotic potentials of $-0,75$ MPa$/-1,50$ MPa and $-0,50$ MPa$/-0,75$ MPa to provide moderate- and severe-osmotic stress conditions in whole plants and micropropagated shoots, respectively. These final concentrations were fixed after comparing the effect of six different PEG concentrations in the two kind of plant material in preliminary experiments (data not shown).

3.1. Visible Symptoms of Plant Damage. From the beginning of the experiments, whole plants and micropropagated shoots cultured under osmotic stress conditions showed plant tissue damage. The percentage of affected plants and shoots was proportional to the severity of the imposed stress (Figure 1). Regardless the plant material, the percentage of affected individuals increased with time, being higher than 70% in all cases after 30 days of treatment.

A significant length reduction in plants and micropropagated shoots was observed in response to the osmotic stress

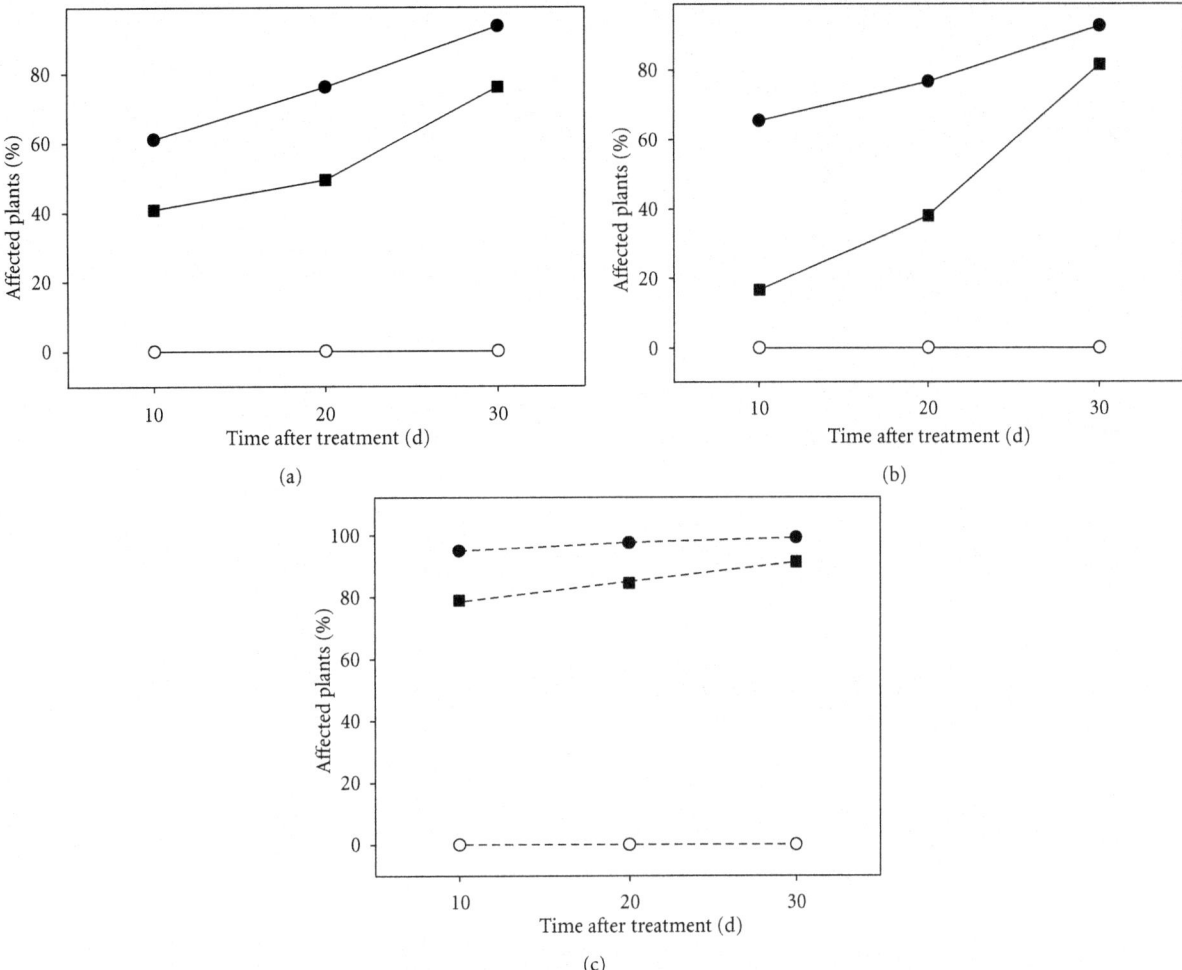

FIGURE 1: Effect of osmotic stress on plant performance in Carrizo citrange cultured *in vitro*. (a): intact plants affected in the aerial part; (b): intact plants affected in the root tissue; (c): micropropagated shoots affected. Control plants (open circle), plants cultured under moderate (filled square) or severe (filled circle) osmotic stress conditions.

conditions from the first day of measurement (Figure 2). In intact plants, this reduction occurred in both organs: root and shoot (Figures 2(a) and 2(b)). Plants under severe stress did not elongate neither shoot or roots throughout the experimental period. Micropropagated shoots were also very sensitive to stress and shoots did not virtually grow under any of the two stress conditions assayed.

3.2. Malondialdehyde Concentration. In this work, MDA concentration was used as a marker of oxidative stress. MDA contents were measured in shoot and root tissues of *in vitro* cultured intact plants and in micropropagated shoots, cultured without root system, after 10, 20, and 30 days of treatment (Figure 3). In shoot tissue, either from intact plants or micropropagated shoots, MDA levels significantly increased with the imposition of osmotic stress (ranging from 2,0- to 3,0-fold increase 10 days after the onset of experiments, in shoots of entire plant under moderate stress and microshoots under severe stress, respectively, in comparison to controls). At day 30, increased levels of MDA were found only in shoots from intact plants cultured under severe stress conditions.

On the contrary, in the case of root tissue (Figure 3(b)), higher levels of MDA were measured in control plants 10 and 30 days after the onset of the experiment. At day 20, a significant increase in MDA levels was observed in roots of plants subjected to moderate stress (1,7-fold increase with respect control plants).

3.3. Antioxidant Enzyme Activity. Antioxidant enzyme activities in intact plants and micropropagated shoots are shown in Figures 4, 5, and 6.

APX activity increased, both in the aerial part and in the roots of plants subjected to moderate or severe stress. The differences were statistically significant between treatments from the first sampling in the case of the aerial part (2.0- and 2.2- fold increase with respect control plants in moderate and severe stress, resp). This pattern was observed throughout all the experimental period (Figure 4(a)). In the root tissue, there was no difference between stressed and control plants at the beginning of the experiment. However, ten days later, APX activity was significantly higher in roots of plants cultured under stress conditions. At day 30, APX activity drastically increased in roots of plants cultured on PEG150,

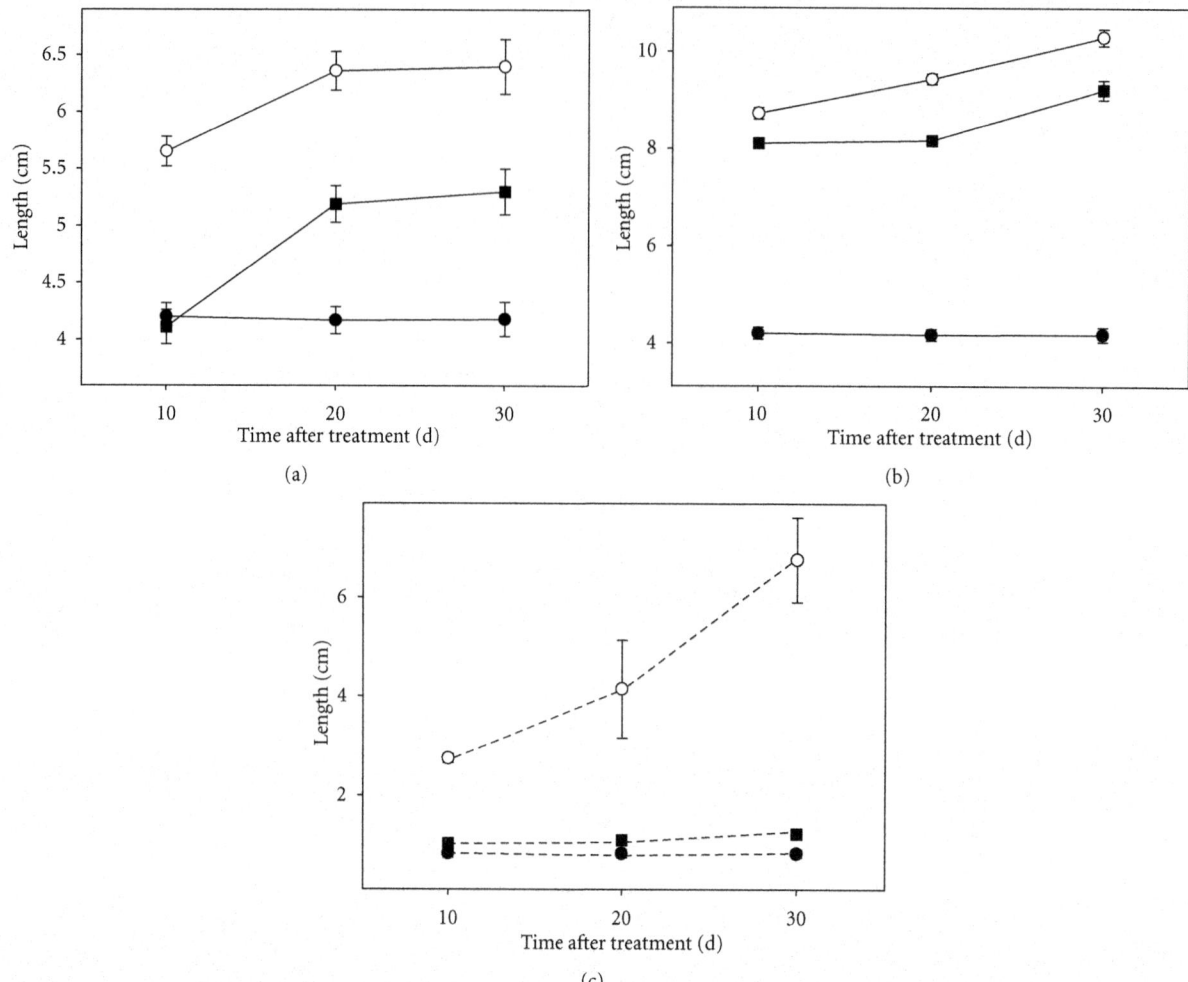

FIGURE 2: Effect of osmotic stress on organ length in Carrizo citrange cultured *in vitro*. (a): aerial part of intact plants; (b): root tissue of intact plants; (c): micropropagated shoots. Control plants (open circle), plants cultured under moderate (filled square) or severe (filled circle) osmotic stress conditions.

reaching values 3.5 times higher than controls. Roots of plants cultured in PEG075 exhibited a moderate increase in APX activity but still significantly higher than that observed in roots of control plants (Figure 4(b)).

When subjected to osmotic stress (both conditions), lower levels of CAT activity were measured in all assayed plant material, being significantly lower than values recorded in controls, except in the case of roots of plants cultured 30 days under moderate stress conditions, that exhibited values similar to controls (Figure 5).

Enzymatic activity was determined in micropropagated shoots 30 days after the onset of the experiment (Figure 6), APX activity was higher in plants subjected to severe stress treatment. No differences were detected between shoots cultured in moderate stress and control conditions. On the contrary, in the case of CAT, control shoots exhibited CAT activities higher than those determined in stressed ones.

3.4. Proline Content. As shown in Figure 7, as a result of the imposed stress, there was a significant accumulation of proline from the first day of measurement, both in intact plants (in shoot and root tissues) and in micropropagated shoots. Proline content in the aerial part of stressed plants increased to very high levels (that varied from 4.5-fold to 6.6-fold increase). The pattern of proline accumulation in root tissue was similar to that described for shoots, with high levels in stressed plants ranging from 1.9- to 3.5-fold increase with respect to controls.

Proline contents in micropropagated shoots cultured under stress conditions were significantly higher than those determined in controls throughout the experimental period.

In all assayed plant material, proline levels were proportional to the stress severity (Figure 7).

3.5. Hormone Content. Throughout the experimental period, ABA levels were higher in shoots than in roots (Figures 8(a) and 8(b)). Plants under stress showed foliar ABA concentrations higher than those observed in control plants. After 10 and 20 days of stress treatment, leaves of plants subjected to moderate stress had higher concentrations of ABA, (4.0-fold increase with respect to controls, in both cases), than those subjected to severe stress

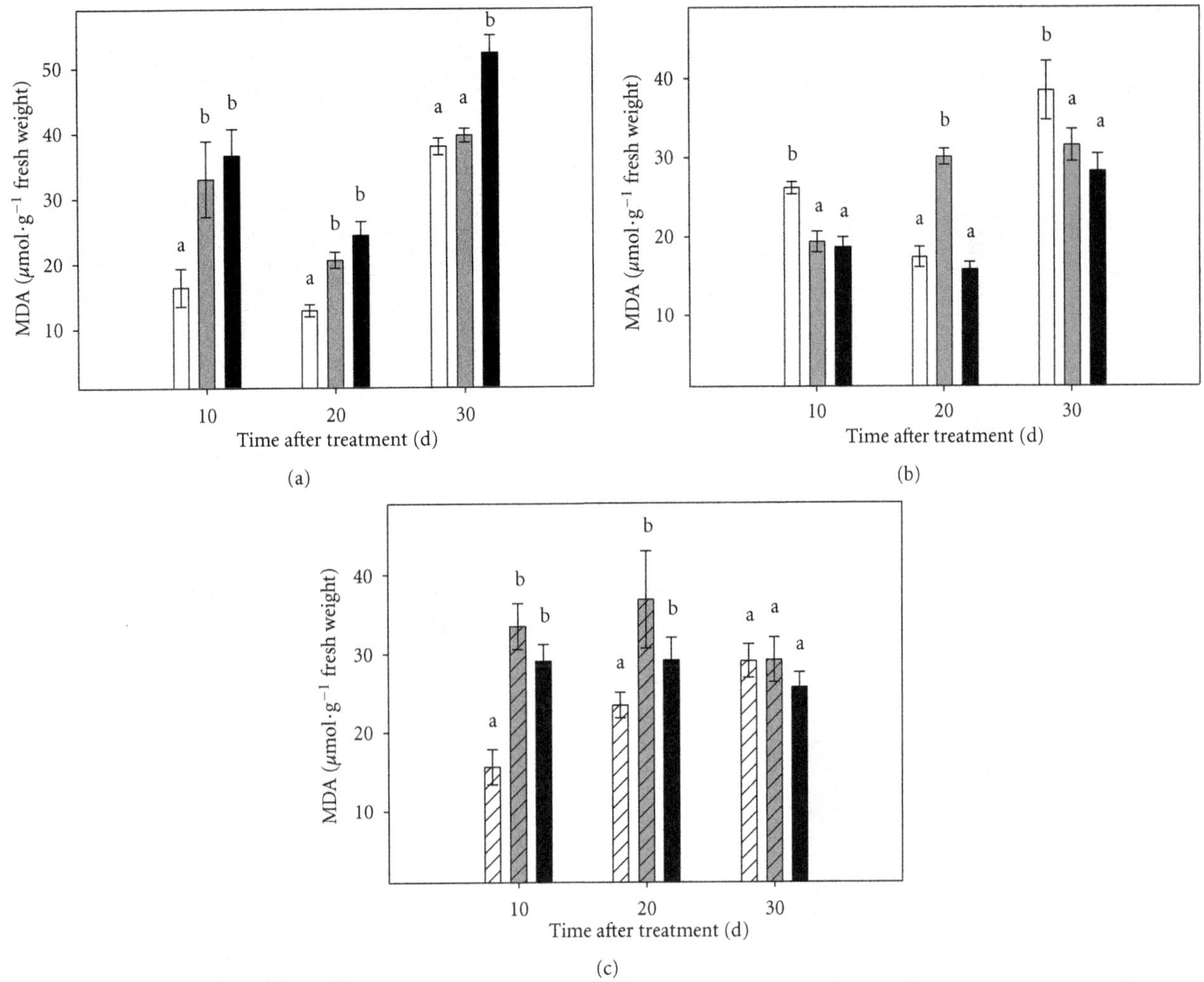

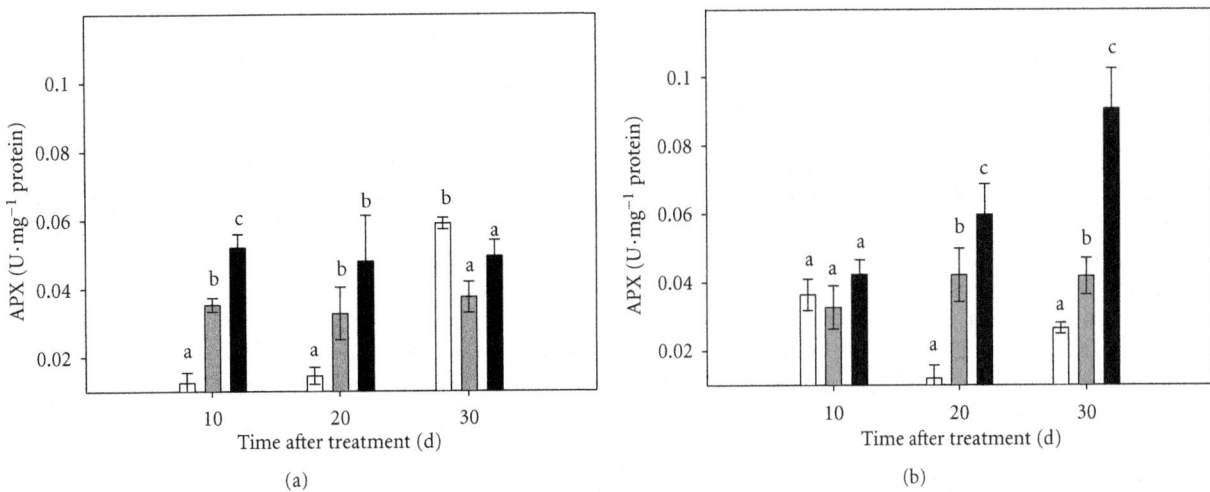

FIGURE 3: Effect of osmotic stress on MDA content in Carrizo citrange cultured *in vitro*. (a): aerial part of intact plants; (b): root tissue of intact plants; (c): micropropagated shoots. Control plants (white bar), plants cultured under moderate (grey bar) or severe (black bar) osmotic stress conditions. Each point corresponds to the average ± standard error of six independent determinations. Different letters denote statistical significance at $P < 0.05$.

FIGURE 4: Effect of osmotic stress on APX activity in Carrizo citrange plants cultured *in vitro*. (a): aerial part of intact plants; (b): root tissue of intact plants. Control plants (white bar), plants cultured under moderate (grey bar) or severe (black bar) osmotic stress conditions. Each point corresponds to the average ± standard error of six independent determinations. Different letters denote statistical significance at $P < 0.05$.

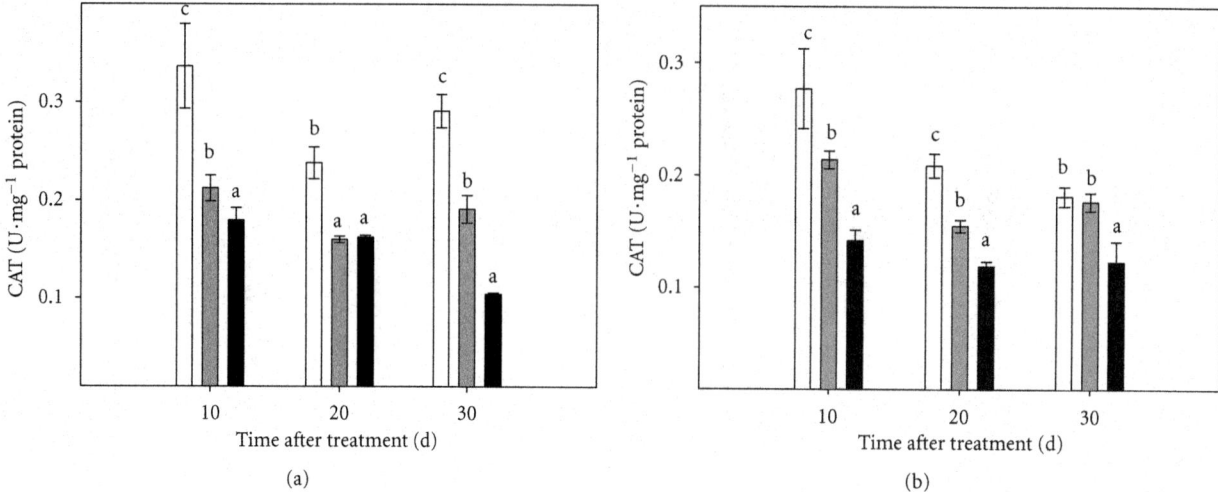

FIGURE 5: Effect of osmotic stress on CAT activity in Carrizo citrange plants cultured *in vitro*. (a): aerial part of intact plants; (b): root tissue of intact plants. Control plants (white bar), plants cultured under moderate (grey bar) or severe (black bar) osmotic stress conditions. Each point corresponds to the average ± standard error of six independent determinations. Different letters denote statistical significance at $P <$ 0.05.

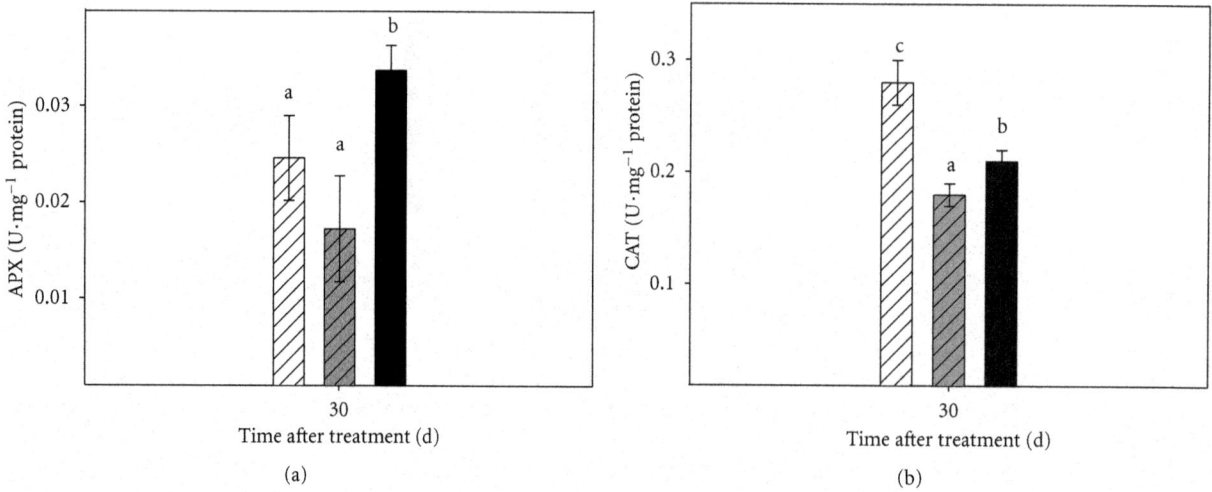

FIGURE 6: Effect of osmotic stress on enzymatic activity in micropropagated Carrizo citrange shoots. (a): APX activity; (b): CAT activity. Control plants (white bar), plants cultured under moderate (grey bar) or severe (black bar) osmotic stress conditions Each point corresponds to the average ± standard error of six independent determinations. Different letters denote statistical significance at $P < 0.05$.

(3.3- and 3.2-fold increase). This situation was reversed at day 30, when plants grown in the PEG150 medium showed ABA concentrations higher than those grown in the PEG075 medium (5.2-fold versus 3.5-fold increase with respect to controls). ABA concentration in roots of plants subjected to severe stress treatment was higher than in controls throughout the experimental period (Figure 8(b)).

As described for entire plants, ABA content was higher in micropropagated shoots subjected to osmotic stress than in controls (Figure 8(c)). The highest values were observed, throughout the experimental period in microshoots cultured under severe stress conditions.

The content of SA was significantly higher in the aerial part of plants cultured under stress conditions from day 20 of experiment (Figure 9(a)). At day 30, differences

increased drastically, reaching values 2.5 and 2.9 times higher than controls in plants under severe and moderate stress respectively. In root tissue, plants cultured under severe stress conditions exhibited, throughout the experimental period, lower concentrations of SA than those cultured under moderate stress or in control conditions, except at day 20 (Figure 9(b)). No differences were found in SA levels in plants cultured in control and PEG75 media.

The osmotic stress caused an increase in endogenous SA levels in micropropagated shoots throughout the experimental period (Figure 9(c)). At the end of the experimental period, there was a marked increase in the levels of this hormone in shoots cultured under stress (5.6- and 5.9-fold increase in shoots subjected to moderate and severe stress, resp.).

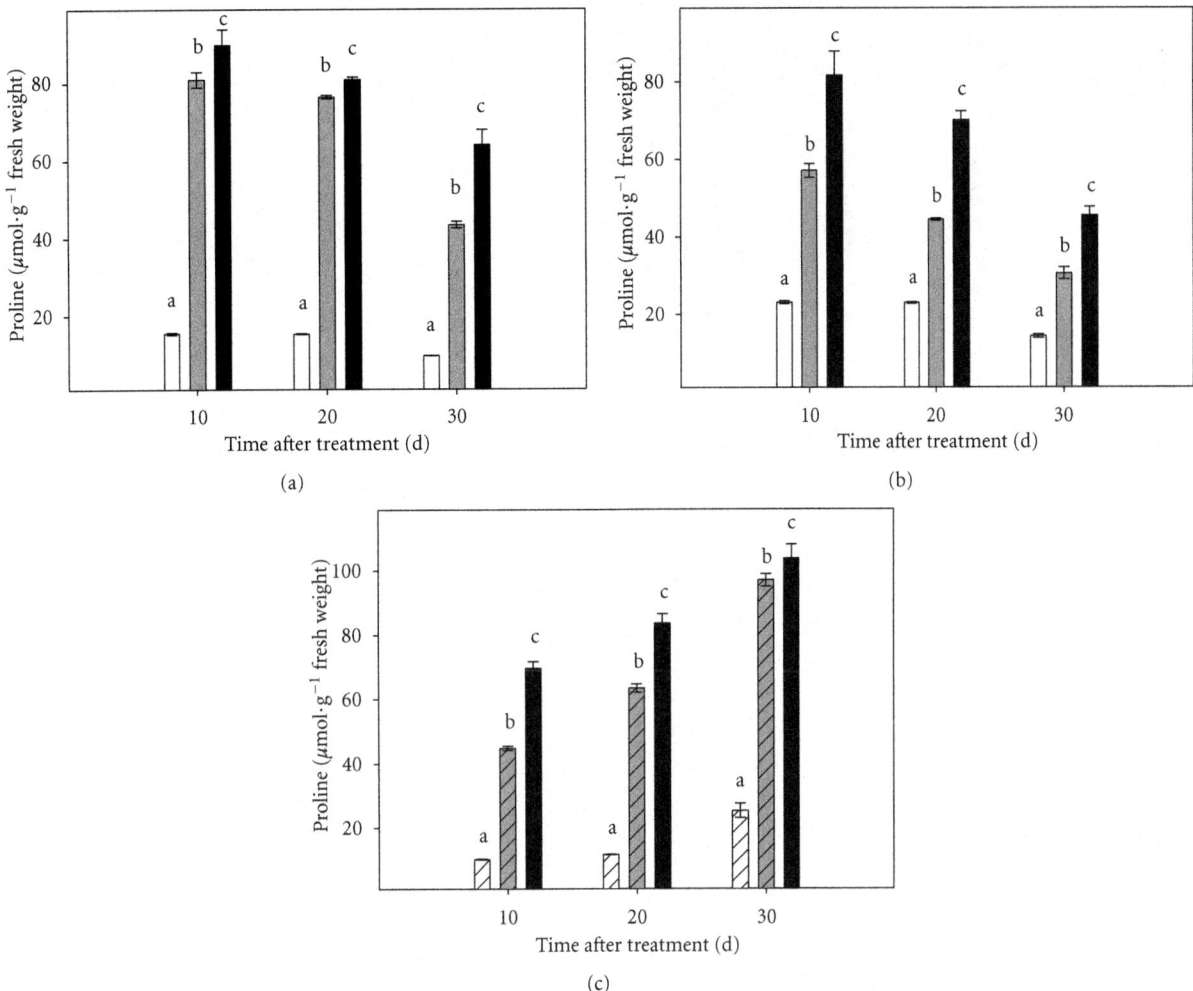

FIGURE 7: Effect of osmotic stress on proline content in Carrizo citrange plants cultured *in vitro*. (a): aerial part of intact plants; (b): root tissue of intact plants; (c): micropropagated shoots. Control plants (white bar), plants cultured under moderate (grey bar) or severe (black bar) osmotic stress conditions. Each point corresponds to the average ± standard error of six independent determinations. Different letters denote statistical significance at $P < 0.05$.

Ten days after the onset of the experiment, control plants showed JA contents, both in leaf and root tissues, significantly higher than those measured in plants cultured under moderate or severe stress conditions (Figure 10). From this point, JA concentrations erratically varied throughout the experimental period both in roots and shoots.

The endogenous JA levels in stressed micropropagated shoots were also significantly lower than in controls at day 10. After this, the values recorded for this hormone were similar between control and stressed plants (Figure 10(c)).

4. Discussion

Nowadays, *in vitro* culture of plant tissues, allow us to reproduce in laboratory conditions new possibilities for the study of processes that cannot be addressed in plants grown in greenhouses or field conditions [30]. In the field, plants are exposed to variable biological and environmental conditions that can make some basic studies very difficult. *In vitro* plant tissue culture has proved to be a valuable tool for the study of

the biochemical processes involved in the response of citrus to a singular abiotic stress condition without the interference of any other factor [9].

In a similar manner to what was observed in previous experiments conducted *ex vitro* [6], *ex vitro* culturing of plants or shoots under osmotic stress conditions caused yellowing and browning of leaves, being the rate of deleterious symptom onset proportional to the severity of the imposed stress. The system also allowed the study of oxidative damage induced by osmotic stress and it was noted that its incidence was higher in the aerial part of plants, which is consistent with previous studies made in plants grown in greenhouses under conditions of continuous flooding of the substrate [17]. In both systems, *in vitro* and *ex vitro*, aerial tissues accumulated higher amounts of MDA, an indirect marker of oxidative damage. When shoots were cultured without the root system, a similar pattern of accumulation was observed, a high increase in MDA content was detected in stressed tissues during the first 20 d of treatment. These values contrast with previous data obtained

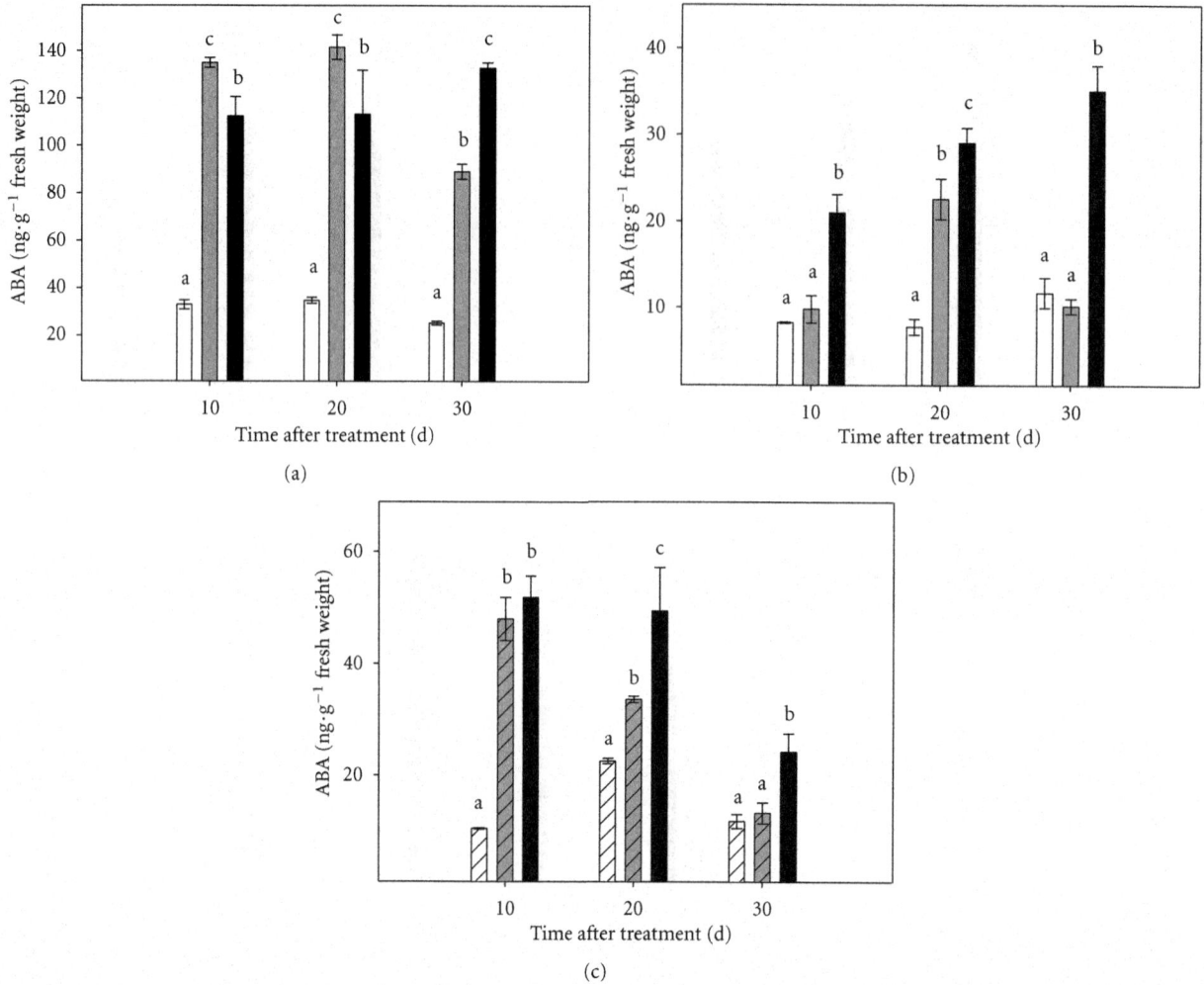

FIGURE 8: Effect of osmotic stress on ABA content in Carrizo citrange plants cultured *in vitro*. (a): aerial part of intact plants; (b): root tissue of intact plants; (c): micropropagated shoots. Control plants (white bar), plants cultured under moderate (grey bar) or severe (black bar) osmotic stress conditions. Each point corresponds to the average ± standard error of six independent determinations. Different letters denote statistical significance at $P < 0.05$.

in a similar system where CC microshoots were cultured under salt stress conditions. In that case, MDA content was similar in stressed and control plants throughout the entire experimental period [9]. Under saline stress conditions, *in vitro* cultured CC shoots accumulated high levels of Cl^- ions and exhibited the characteristic necrosis in leaf tissue. This, together with the absence of MDA accumulation, allowed us to conclude that there was no correlation between foliar damage and oxidative stress in the case of that specific abiotic stress.

It has been proposed that the lack of antioxidant response in *ex vitro* growing plants of the salt tolerant citrus rootstock Cleopatra mandarin (CM) is due to a restricted chloride uptake due to its particular anatomical and physiological characteristics [31]. In our previous work [9] it was proved that under salt stress conditions, *in vitro* microshoots of both genotypes, salt tolerant and sensitive (CM and CC respectively), exhibited the same pattern of Cl- intoxication and, as a consequence, the same foliar damage. In both genotypes oxidative damage in leaf tissue was no detected, as

no MDA accumulation was recorded throughout the stress treatment. On the contrary, as described in the present work, when CC microshoots were subjected to osmotic stress, the occurrence of leaf tissue damage was concomitant with the increase in oxidative damage. These results reinforce the idea that different abiotic stresses cause different physiological responses in citrus plants.

Proline concentration increased in all studied plant material when plants or micropropagated shoots were cultured in media supplemented with PEG-6000. It has been reported that, in *ex vitro* conditions, proline levels increased in citrus leaves in response to salt stress conditions [2, 5], soil flooding [17] and drought [20]. High proline levels have been correlated with tolerance to different abiotic stresses, such as drought, salinity and high temperatures [20, 32]. In transgenic "Swingle" citrumelo plants over-expressing P5CSF129A gene, which codes for the key-enzyme for proline synthesis, it seems that the high proline content mitigates the effect of ROS, by directly scavenging free radicals and by activating antioxidant systems [18].

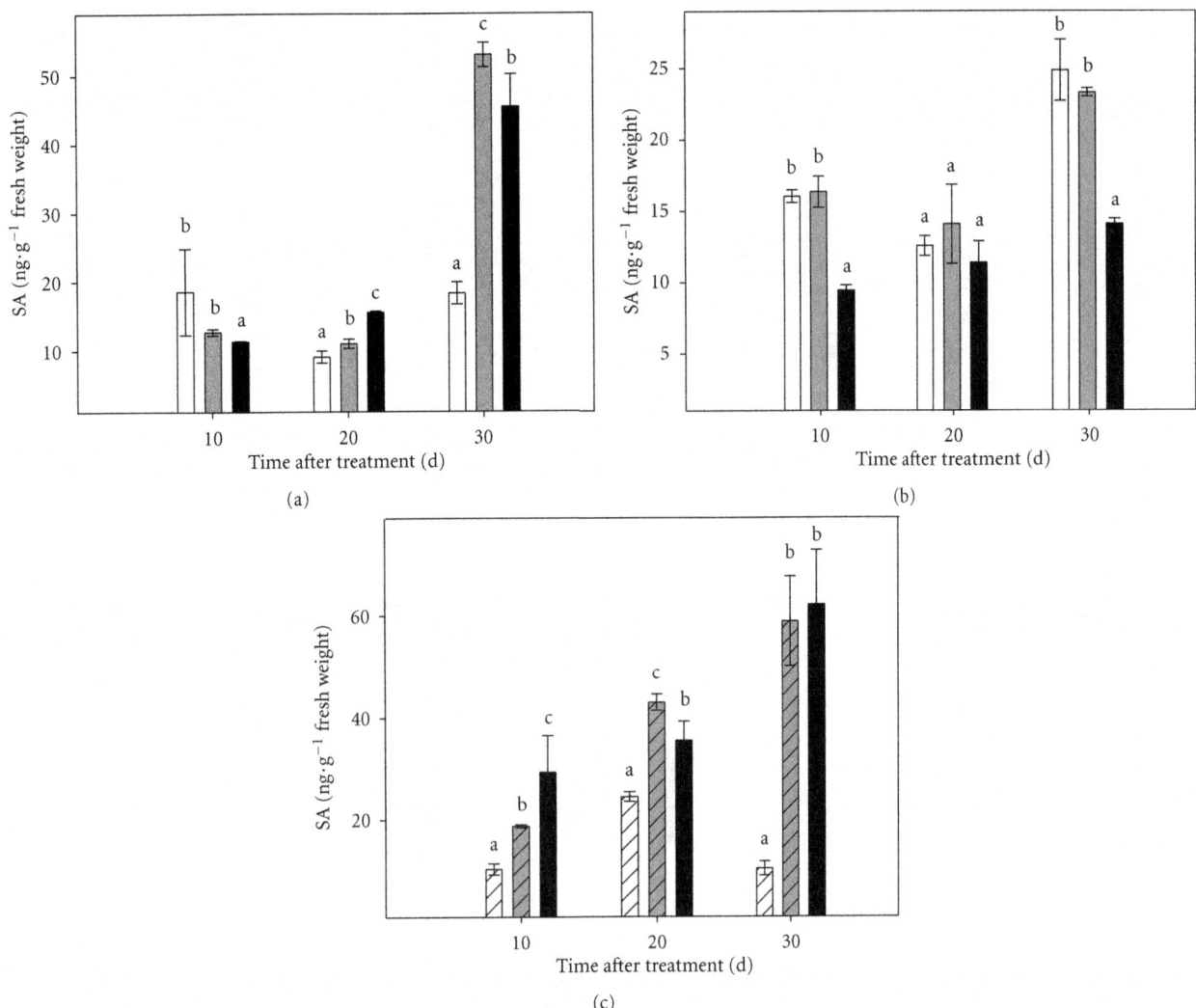

FIGURE 9: Effect of osmotic stress on SA content in Carrizo citrange plants cultured *in vitro*. (a): aerial part of intact plants; (b): root tissue of intact plants; (c): micropropagated shoots. Control plants (white bar), plants cultured under moderate (grey bar) or severe (black bar) osmotic stress conditions. Each point corresponds to the average ± standard error of six independent determinations. Different letters denote statistical significance at $P < 0.05$.

However, the putative protective effect of proline is not very effective in the conditions described in this work because plants subjected to increasing concentrations of PEG-6000 accumulated significant amounts of proline in parallel with the increase of oxidative damage. These observations are consistent with those shown *ex vitro* in citrus grown under conditions of continuous flooding of the substrate [17], where the highest concentrations of this amino acid were found in the most damaged genotypes. Therefore, despite the controversy of the effectiveness of the accumulation of endogenous proline in wild-type plants under stress as a protective mechanism, it seems that the increase of this metabolite in citrus leaves could be considered as a good stress marker.

Data presented here also demonstrate that root tissue is not necessary for osmotic stress to cause a high increase in proline content as high levels of this amino acid were detected in micropropagated shoots cultured in media supplemented with PEG. Different response was observed under salt stress conditions and in that experimental system no proline accumulation was detected in stressed microshoots cultured without roots [9]. This adds new evidence confirming that plant is differently affected by the osmotic and toxic components of salt stress.

It is known that the activities of several enzymes are affected by stress conditions; as a general pattern SOD, CAT and APX activities increase in response to different abiotic stresses, such as water deficit [33] or flooding [17]. Under osmotic stress conditions, plants grown *in vitro* are able to activate, their antioxidant machinery. As it could be expected, APX activity increased, not only in roots and leaves of intact plants but also in micropropagated shoots subjected to stress conditions. On the contrary, CAT activity decreased both, in plants and in micropropagated shoots grown in medium supplemented with PEG-6000. Similar trend was observed in CC plants cultured *ex vitro* under stress conditions in response to flooding of the substrate [34] and in citrumelo plants, when transgenic plants

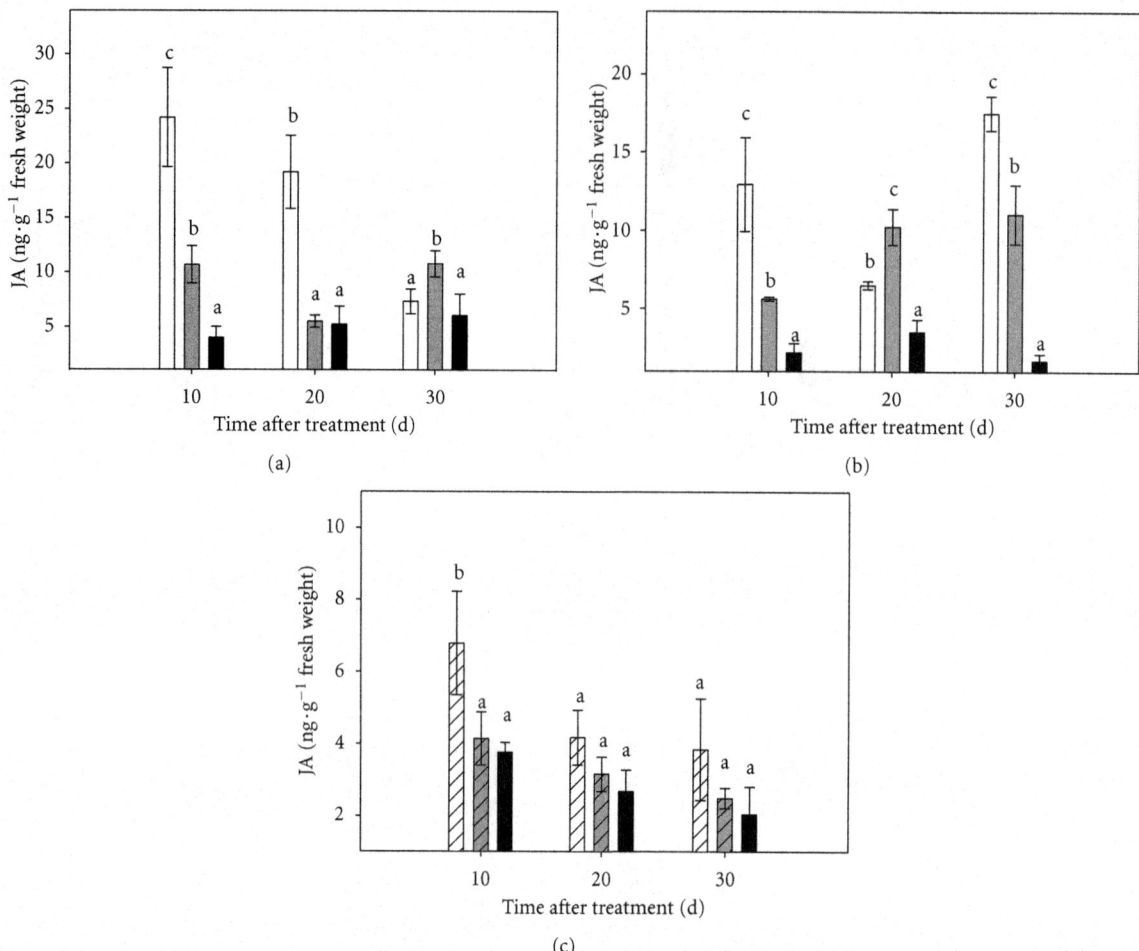

FIGURE 10: Effect of osmotic stress on JA content in Carrizo citrange plants cultured *in vitro*. (a): aerial part of intact plants; (b): root tissue of intact plants; (c): micropropagated shoots. Control plants (white bar), plants cultured under moderate (grey bar) or severe (black bar) osmotic stress conditions. Each point corresponds to the average ± standard error of six independent determinations. Different letters denote statistical significance at $P < 0.05$.

over-expressing the *P5CSF129A* gene were cultured under osmotic stress conditions [18]. The increase in APX and the reduction in CAT activities as a consequence of the osmotic stress treatment in *in vitro* cultured CC plants, concomitant with the high increase detected in proline contents may suggest that CAT is not participating in the protection against oxidative stress. Data also show that in *in vitro* conditions, plants retain their antioxidant machinery intact even when micropropagated shoots are cultured without root system.

Osmotic stress caused a significant accumulation of ABA in both, intact plants and micropropagated shoots. This agrees with the data obtained plants grown in greenhouse and field [5, 29, 35] under conditions of water stress, high salinity and flooding of the substrate and reinforces the role of ABA as a mediator of plant responses. It is important to point out that citrus leaves accumulate the same amount of ABA under osmotic stress conditions even when cultured without root system. On the contrary, no accumulation of ABA was observed when CC micropropagated shoots were cultured *in vitro*, under salt stress conditions [9]. On the view of these results it can be suggested that the pattern of

signaling is dependent on the type of adverse condition that plant has to cope with.

The hormone SA is considered to be involved in plant resistance to several plant pathogens [36] and also appears to have an important role in plant response to oxidative stress [37]. The increased SA content in the aerial part of the plant at the end of the experimental period support this hypothesis, as it is concomitant with the occurrence of severe oxidative damage in this tissue.

The decrease in JA levels, observed throughout the experimental period in both leaf and root tissue of intact plants and in micropropagated shoots, is compatible with previous data which showed that JA acts as mediator between the perception of the stress and the induction of physiological responses and, therefore, its action should be early [29, de Ollas et al., 2012, unpublished data]. Although it was not the main point of this work, it is possible that the first data collection point was carried out too late to find a possible transient increase in JA concentration. However, the data presented here show a decreased levels of JA in stressed plants, similar to that described previously in citrus plants cultured *ex vitro* under abiotic stress conditions [29].

It has been reported that, in plants grown in field, the responses to water and salt stress are essentially identical [38]. When cultured *in vitro* without roots, CC shoots growing under osmotic stress conditions shown a significant increase of MDA, proline, and ABA levels. On the contrary, no increase in the endogenous concentrations of these compounds was observed when cultured under saline stress conditions [9]. These results evidence that the mechanisms of response of CC shoots to saline or osmotic stress are different. There is an organ-dependent response to stress: in salt stress conditions it is necessary the presence of the roots for the signaling of stress to the aerial part. On the contrary, the presence of this organ is not necessary to modulate the response of shoots to osmotic stress caused by PEG6000. Therefore, the extrapolation of the information obtained studying the effects of an abiotic stress factor to that obtained when considering other stress factors should be avoided. On the other hand, it is important to note that the organ involved in the stress perception/signaling depends on the type of adverse condition to which plants/shoots are subjected.

Acknowledgments

This work was supported by the Spanish Ministerio de Ciencia e Innovación and Universitat Jaume I/Fundació Bancaixa through Grant nos. AGL2010-22195-C03-01/AGRand P11B2009-01, respectively. Hormone determinations were performed at the central facilities (Servei Central d'Instrumentació Científica, SCIC) of Universitat Jaume I.

References

[1] D. S. Veselov, A. R. Mustafina, I. B. Sabirjanova et al., "Effect of PEG-treatment on the leaf growth response and auxin content in shoots of wheat seedlings," *Plant Growth Regulation*, vol. 38, no. 2, pp. 191–194, 2002.

[2] V. Arbona, V. Flors, J. Jacas, P. García-Agustín, and A. Gómez-Cadenas, "Enzymatic and non-enzymatic antioxidant responses of Carrizo citrange, a salt-sensitive citrus rootstock, to different levels of salinity," *Plant and Cell Physiology*, vol. 44, no. 4, pp. 388–394, 2003.

[3] M. F. López-Climent, V. Arbona, R. M. Pérez-Clemente, and A. Gómez-Cadenas, "Relationship between salt tolerance and photosynthetic machinery performance in citrus," *Environmental and Experimental Botany*, vol. 62, no. 2, pp. 176–184, 2008.

[4] J. G. Pérez-Pérez, J. P. Syvertsen, P. Botía, and F. García-Sánchez, "Leaf water relations and net gas exchange responses of salinized Carrizo citrange seedlings during drought stress and recovery," *Annals of Botany*, vol. 100, no. 2, pp. 335–345, 2007.

[5] A. Gómez-Cadenas, F. R. Tadeo, E. Primo-Millo, and M. Talon, "Involvement of abscisic acid and ethylene in the responses of citrus seedlings to salt shock," *Physiologia Plantarum*, vol. 103, no. 4, pp. 475–484, 1998.

[6] V. Arbona, A. J. Marco, D. J. Iglesias, M. F. López-Climent, M. Talon, and A. Gómez-Cadenas, "Carbohydrate depletion in roots and leaves of salt-stressed potted *Citrus clementina* L.," *Plant Growth Regulation*, vol. 46, no. 2, pp. 153–160, 2005.

[7] J. Gopal and K. Iwama, "*In vitro* screening of potato against water-stress mediated through sorbitol and polyethylene glycol," *Plant Cell Reports*, vol. 26, no. 5, pp. 693–700, 2007.

[8] Y. J. Zhang, Y. Q. Qian, X. J. Mu, Q. G. Cai, Y. L. Zhou, and X. P. Wei, "Plant regeneration from in vitro-cultured seedling leaf protoplasts of *Actinidia eriantha* Benth," *Plant Cell Reports*, vol. 17, no. 10, pp. 819–821, 1998.

[9] A. Montoliu, M. F. López-Climent, V. Arbona, R. M. Pérez-Clemente, and A. Gómez-Cadenas, "A novel in vitro tissue culture approach to study salt stress responses in citrus," *Plant Growth Regulation*, vol. 59, no. 2, pp. 179–187, 2009.

[10] M. Almansouri, J. M. Kinet, and S. Lutts, "Effect of salt and osmotic stresses on germination in durum wheat (*Triticum durum* Desf.)," *Plant and Soil*, vol. 231, no. 2, pp. 243–254, 2001.

[11] N. Carpita, D. Sabularse, D. Montezinos, and D. P. Delmer, "Determination of the pore size of cell walls of living plant cells," *Science*, vol. 205, no. 4411, pp. 1144–1147, 1979.

[12] P. E. Verslues, E. S. Ober, and R. E. Sharp, "Root growth and oxygen relations at low water potentials. Impact of oxygen availability in polyethylene glycol solutions," *Plant Physiology*, vol. 116, no. 4, pp. 1403–1412, 1998.

[13] H. Medrano, J. M. Escalona, J. Bota, J. Gulías, and J. Flexas, "Regulation of photosynthesis of C3 plants in response to progressive drought: stomatal conductance as a reference parameter," *Annals of Botany*, vol. 89, pp. 895–905, 2002.

[14] K. Apel and H. Hirt, "Reactive oxygen species: metabolism, oxidative stress, and signal transduction," *Annual Review of Plant Biology*, vol. 55, pp. 373–399, 2004.

[15] B. Halliwell and J. M. C. Gutteridge, *Free Radicals in Biology and Medicine*, Clarendon, Oxford, UK, 1989.

[16] J. G. Scandalios, "Oxidative stress: molecular perception and transduction of signals triggering antioxidant gene defenses," *Brazilian Journal of Medical and Biological Research*, vol. 38, no. 7, pp. 995–1014, 2005.

[17] V. Arbona, Z. Hossain, M. F. López-Climent, R. M. Pérez-Clemente, and A. Gómez-Cadenas, "Antioxidant enzymatic activity is linked to waterlogging stress tolerance in citrus," *Physiologia Plantarum*, vol. 132, no. 4, pp. 452–466, 2008.

[18] M. K. F. de Campos, K. de Carvalho, F. S. de Souza et al., "Drought tolerance and antioxidant enzymatic activity in transgenic "Swingle" citrumelo plants over-accumulating proline," *Environmental and Experimental Botany*, vol. 72, no. 2, pp. 242–250, 2011.

[19] R. Grene, "Oxidative stress and acclimation mechanisms in plants," in *The Arabidopsis Book*, C. R. Somerville and E. M. Myerowitz, Eds., American Society of Plant Biologists, Rockville, Md, USA, 2002, http://www.aspb.org/publications/Arabidopsis/.

[20] H. B. C. Molinari, C. J. Marur, J. C. B. Filho et al., "Osmotic adjustment in transgenic citrus rootstock Carrizo citrange (*Citrus sinensis* Osb. x *Poncirus trifoliata* L. Raf.) overproducing proline," *Plant Science*, vol. 167, no. 6, pp. 1375–1381, 2004.

[21] H. B. C. Molinari, C. J. Marur, E. Daros et al., "Evaluation of the stress-inducible production of proline in transgenic sugarcane (*Saccharum* spp.): osmotic adjustment, chlorophyll fluorescence and oxidative stress," *Physiologia Plantarum*, vol. 130, no. 2, pp. 218–229, 2007.

[22] S. D. McNeil, M. L. Nuccio, M. J. Ziemak, and A. D. Hanson, "Enhanced synthesis of choline and glycine betaine in transgenic tobacco plants that overexpress phosphoethanolamine N-methyltransferase," *Proceedings of the National Academy of*

Sciences of the United States of America, vol. 98, no. 17, pp. 10001–10005, 2001.

[23] M. A. Hoque, E. Okuma, M. N. A. Banu, Y. Nakamura, Y. Shimoishi, and Y. Murata, "Exogenous proline mitigates the detrimental effects of salt stress more than exogenous betaine by increasing antioxidant enzyme activities," *Journal of Plant Physiology*, vol. 164, no. 5, pp. 553–561, 2007.

[24] T. Murashige and D. P. H. Tucker, "Growth factors requirements of citrus tissue," in *Proceedings of the 1st International Citrus Symposium*, vol. 3, pp. 1155–1161, 1969.

[25] B. E. Michel and M. R. Kaufmann, "The osmotic potential of polyethylene glycol 6000," *Plant Physiology*, vol. 51, no. 5, pp. 914–916, 1973.

[26] T. Murashige and F. Skoog, "A revised medium for rapid growth and bioassays with tobacco cultures," *Physiol Plant*, vol. 15, no. 5, pp. 473–497, 1962.

[27] D. M. Hodges, J. M. DeLong, C. F. Forney, and R. K. Prange, "Improving the thiobarbituric acid-reactive-substances assay for estimating lipid peroxidation in plant tissues containing anthocyanin and other interfering compounds," *Planta*, vol. 207, no. 4, pp. 604–611, 1999.

[28] A. Durgbanshi, V. Arbona, O. Pozo, O. Miersch, J. V. Sancho, and A. Gómez-Cadenas, "Simultaneous determination of multiple phytohormones in plant extracts by liquid chromatography-electrospray tandem mass spectrometry," *Journal of Agricultural and Food Chemistry*, vol. 53, no. 22, pp. 8437–8442, 2005.

[29] V. Arbona and A. Gómez-Cadenas, "Hormonal modulation of citrus responses to flooding," *Journal of Plant Growth Regulation*, vol. 27, no. 3, pp. 241–250, 2008.

[30] R. M. Pérez Clemente, A. Montoliu, P. López, M. F. López Climent, V. Arbona, and A. Gómez-Cadenas, "*In vitro* tissue culture approaches for the study of salt stress in citrus," in *Biosaline Agriculture and High Saline Tolerance*, Birkhänser, Basel, Switzerland, 2008.

[31] J. L. Moya, A. Gómez-Cadenas, E. Primo-Millo, and M. Talon, "Chloride absorption in salt-sensitive Carrizo citrange and salt-tolerant *Cleopatra mandarin* citrus rootstocks is linked to water use," *Journal of Experimental Botany*, vol. 54, no. 383, pp. 825–833, 2003.

[32] E. W. Hamilton and S. A. Heckathorn, "Mitochondrial adaptations to NaCl. Complex I is protected by anti-oxidants and small heat shock proteins, whereas Complex II is protected by proline and betaine," *Plant Physiology*, vol. 126, no. 3, pp. 1266–1274, 2001.

[33] S. Jung, "Variation in antioxidant metabolism of young and mature leaves of *Arabidopsis thaliana* subjected to drought," *Plant Science*, vol. 166, no. 2, pp. 459–466, 2004.

[34] Z. Hossain, M. F. López-Climent, V. Arbona, R. M. Pérez-Clemente, and A. Gómez-Cadenas, "Modulation of the antioxidant system in citrus under waterlogging and subsequent drainage," *Journal of Plant Physiology*, vol. 166, no. 13, pp. 1391–1404, 2009.

[35] A. Gómez-Cadenas, F. R. Tadeo, M. Talon, and E. Primo-Millo, "Leaf abscission induced by ethylene in water-stressed intact seedlings of *Cleopatra mandarin* requires previous abscisic acid accumulation in roots," *Plant Physiology*, vol. 112, no. 1, pp. 401–408, 1996.

[36] M. E. Alvarez, "Salicylic acid in the machinery of hypersensitive cell death and disease resistance," *Plant Molecular Biology*, vol. 44, no. 3, pp. 429–442, 2000.

[37] Q. Hayata, S. Hayata, M. Irfana, and A. Ahmadb, "Effect of exogenous salicylic acid under changing environment," *Review of Environmental and Experimental Botany*, vol. 68, no. 1, pp. 14–25, 2010.

[38] R. Munns, "Comparative physiology of salt and water stress," *Plant, Cell and Environment*, vol. 25, no. 2, pp. 239–250, 2002.

16

Botanicals to Control Soft Rot Bacteria of Potato

**M. M. Rahman,[1] A. A. Khan,[1] M. E. Ali,[2] I. H. Mian,[1]
A. M. Akanda,[1] and S. B. Abd Hamid[2]**

[1] Department of Plant Pathology, Bangabadhu Sheikh Mujibur Rahman Agricultural University,
  Gazipur 1706, Bangladesh
[2] Center for Research in Nanotechnology and Catalysis, University of Malaya,
  50603 Kuala Lumpuor, Malaysia

Correspondence should be addressed to M. E. Ali, eaqubali@gmail.com

Academic Editors: A. Ferrante and R. Pohjanvirta

Extracts from eleven different plant species such as jute (*Corchorus capsularis* L.), cheerota (*Swertia chiraita* Ham.), chatim (*Alstonia scholaris* L.), mander (*Erythrina variegata*), bael (*Aegle marmelos* L.), marigold (*Tagetes erecta*), onion (*Allium cepa*), garlic (*Allium sativum* L.), neem (*Azadiracta indica*), lime (*Citrus aurantifolia*), and turmeric (*Curcuma longa* L.) were tested for antibacterial activity against potato soft rot bacteria, *E. carotovora* subsp. *carotovora (Ecc)* P-138, under *in vitro* and storage conditions. Previously, *Ecc* P-138 was identified as the most aggressive soft rot bacterium in Bangladeshi potatoes. Of the 11 different plant extracts, only extracts from dried jute leaves and cheerota significantly inhibited growth of *Ecc* P-138 *in vitro*. Finally, both plant extracts were tested to control the soft rot disease of potato tuber under storage conditions. In a 22-week storage condition, the treated potatoes were significantly more protected against the soft rot infection than those of untreated samples in terms of infection rate and weight loss. The jute leaf extracts showed more pronounced inhibitory effects on *Ecc* 138 growth both in *in vitro* and storage experiments.

1. Introduction

Indiscriminate use of chemical pesticides to control various pests and pathogenic microorganisms of crops plants is causing health hazard both in terrestrial and aquatic lives through their residual toxicity [1, 2]. Considering the adverse and alarming effects of synthetic pesticides on environment and natural habitats, this study was undertaken to find out an alternative and nontoxic biological control agents [3] to control the soft rot bacterial pathogens in Bangladeshi potatoes. Green plants are a huge reservoir of various effective chemotherapeutics and could serve as an environmentally friendly natural alternative to the toxic chemical pesticides [4].

During the recent decades, many herbal extracts have been extensively tested and a myriad of reports have been documented outlining the uses of plant extracts to control the animal and plant diseases [5–7]. A good number of reports outlined the antimicrobial effects of some medicinal plants for plant disease control [7]. Some plant extracts were documented as effective inhibitors of phytopathogenic bacteria [5, 6]. Antimicrobial activities of several plant extracts against bacterial soft rot of potatoes were evaluated and a quite satisfactory result was obtained [8, 9]. The liquid extract of hemp flowers and essential oils were tested against *Ecc*, the causal bacterium of potato soft rot, and satisfactory results were documented [8]. However, no attempts have been made to identify and characterize the antibacterial plant extracts to control the soft rot bacterial pathogens of potatoes in Bangladesh. In this paper, we investigated the anti-Ecc P-138 activity of 11 different plant extracts and documented antibacterial activity in jute leaf and cheerota extracts in *in vitro* and storage experiments.

2. Materials and Methods

2.1. Selection of Plants and Preparation of Extracts. A total of 11 plants, namely, jute (*Corchorus capsularis* L.), cheerota (*Swertia chirata* Ham.), chatim (*Alstonia scholaris* L.), mander (*Erythrina variegata*), bael (*Aegle marmelos* L.), marigold

TABLE 1: List of plants tested to control bacterial soft rot pathogens of potato.

Name of plants		Family	Parts used
English/Bangla	Scientific name		
Jute	*Corchorus capsularis* L.	Tiliaceae	Dry leaves
Cheerota	*Swertia chirata* Ham.	Gentianaceae	Whole plant
Devils tree	*Alstonias cholaris* L.	Apocynaceae	Bark
Coral tree	*Erythrina indica*	Leguminosae	Bark
Bael	*Aegle marmelos* L.	Rutaceae	Young fruits and leaves
Marigold	*Tagetes serecta*	Compositae	Leaves and roots
Onion	*Allium cepa*	Lilliaceae	Bulbs and leaves
Garlic	*Allium sativum* L.	Lilliaceae	Bulbs and leaves
Neem	*Azadirachta indica*	Meliaceae	Leaves
Lime	*Citrus aurantifolia*	Rutaceae	Leaves
Turmeric	*Curcuma longa* L.	Zingiberaceae	Rhizome

(*Tagete serecta*), onion (*Allium cepa*), garlic (*Allium sativum* L.), neem (*Azadirachta indica*), lime (*Citrus aurantifolia*), and turmeric (*Curcuma longa* L.) were tested in this investigation (Table 1). Dried jute leaves, whole plant of cheerota, bark of chatim and mandar were used for the preparation of extracts at the ratio of 1 : 10 (w/v) in water. Plant parts were soaked or submerged in distilled water for 20–24 h. Water was chosen as an extraction media because of its low cost, easier availability, and biocompatibility. The water extracts were collected by passing through double-layered muslin cloth at least two times. To prepare extracts of other plants, different plant parts like leaves, roots, bulbs, and rhizomes were crushed in a mortar and pestle. The crushed materials were mixed with distilled water at 1 : 1 (w/v) and blended in an electrical blender. They were filtered through double layered muslin cloth at least two times. The extracts were poured into conical flasks and used as stock solution. Mouth of each flask was closed with aluminum foil and preserved in a refrigerator at 4°C for future uses.

2.2. Bioassay of Plant Extracts against Soft Rot Bacteria.
Antibacterial activity of each plant extracts (Table 1) was tested against *Ecc* P-138, the most virulent soft rot bacterial strain of Bangladeshi potatoes, through the growth inhibition test *in vitro* [8, 10]. *Ecc* P-138 (10^8 cfu/ml) was inoculated on autoclaved YPDA media at 28°C for 24 h to obtain pure culture of *Ecc* P-138. A fresh YPDA medium was then amended with 30, 50, 75 and 90% plant extracts and was autoclaved. The medium was poured into petri dishes at the rate of 20 ml/dish. After solidification, the amended medium was spot inoculated with the pure culture of *Ecc* P-138. The spot inoculated plates were incubated in an incubator at 30°C for 14 days and the bacterial growth was recorded to determine the inhibitory effects of the plant extracts. The plates were arranged in the incubator following complete randomized design with three replications (petri dishes).

2.3. Effect of Jute Leaf and Cheerota Extracts on Soft Rot Disease of Storage Potatoes.
Based on encouraging results of the bioassay, extracts of dry jute leaf and cheerota plant were selected to evaluate their efficacy to protect potatoes against soft rot bacteria under storage conditions. Two extracts were prepared in distilled water at a ratio of 1 : 10 (v/v) following the procedures as described above. Seven hundred grams of potato tubers were treated with each of the plant extracts. The potato tuber bulbs were submerged in the extracts for 30 minutes and air-dried at room temperature. Inocula of the soft rot bacteria, *Ecc* P-138 were prepared at a concentration of 10^8 cfu mL^{-1} following the same procedures as described under *in vitro* test. Plant extract treated potato tubers were inoculated with the inocula of *Ecc* P-138. For inoculation, the inoculum suspensions were sprayed over the tubers uniformly using an automizer. Inoculated tubers were air-dried and stored at room temperature. An uninoculated control was maintained for each variety. Visual observations were made after 2, 6, 10, 14, 18, and 22 weeks of inoculation and data on the number of soft rot infected tubers was recorded and loss in weight due to soft rot in storage was computed and expressed in percentage (w/w) using the formula [11] given below:

$$\text{Infection \%} = \frac{\text{No. of infected tubers}}{\text{Total no. of tubers}} \times 100,$$

$$\text{Loss of weight \%} = \frac{\text{Initial weight} - \text{weight after discarding the infected sample}}{\text{Initial weight}} \times 100.$$

(1)

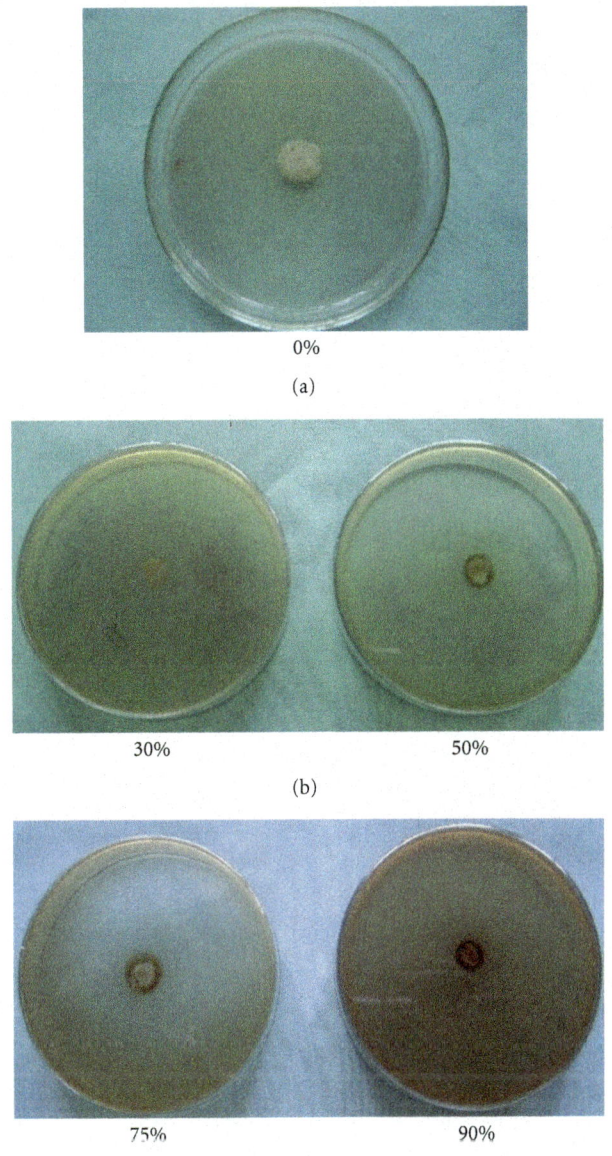

0%

(a)

30%   50%

(b)

75%   90%

(c)

FIGURE 1: Antibacterial activity of jute leaf extract against *Ecc* P-138 at different concentrations of the extract in YPDA medium.

Percentage of disease reduction (PDR) was calculated following formula shown below [12]:

$$PDR = \frac{Ack - Atr}{Ack} \times 100, \qquad (2)$$

where, Ack is disease severity/loss (by weight) in control and Atr is disease severity/loss (by weight) in treatment.

3. Results

3.1. Antibacterial Assay of Plant Extracts In Vitro. Out of 11 different plant extracts, only extracts of dry jute leaf and cheerota suppressed the growth of the soft rot bacteria, *Ecc* P-138, in 50–90% extracts containing YPDA medium (Table 2). This was confirmed with the visual appearance

of inhibition zones around the soft rot bacterium *Ecc* P-138 (Figures 1 and 2). Higher antibacterial activity of the extracts was observed at higher concentration. This was reflected by the higher thickness of the inhibition zones around the soft rot bacterial strain. The jute leaf extract demonstrated more inhibition than that of the cheerota against potato soft rot *Ecc* P-138 in triplicate experiments. On the basis of *in vitro* test, jute leaf and cheerota plant extracts were selected for treatment of potato tubers against soft rot disease under storage. Other nine plant extracts did not show antibacterial activity (Table 3).

3.2. Effect of Jute Leaf and Cheerota Extracts on Potato Tubers under Storage. The rates of soft rot infection and tuber

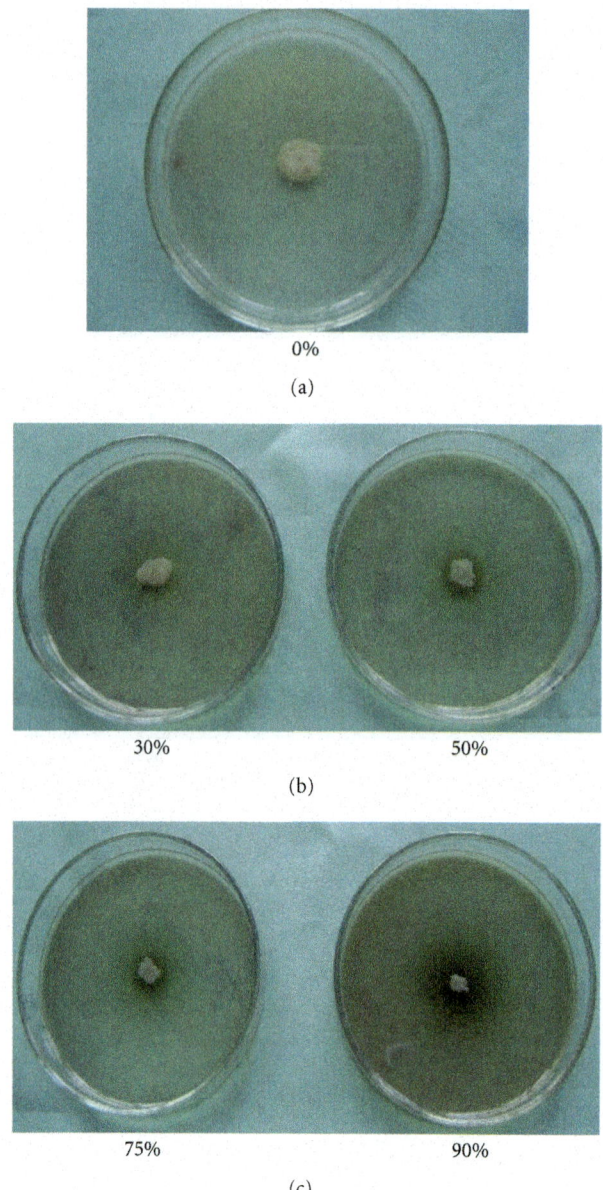

FIGURE 2: Antibacterial activity of cheerota extract against *Ecc* P-138 at different concentrations of extract in YPDA medium.

damage in treated and untreated potatoes are demonstrated in Figures 3 and 4, respectively. In case of untreated tubers, the infection rate was much higher and 100% potato tubers were damaged within 14 weeks of storage. On the other hand, the infection rate and damage were significantly lowered (20–50%) in treated samples (Figures 3 and 4). Approximately ~40–70% of treated samples survived even in 22 weeks of storage. The maximum infection rate and damage were observed in the cardinal and diamante varieties and the lowest infection rate and damage were found in granola varieties. The cardinal and diamante varieties of potatoes also demonstrated the highest incidence of disease and the granola varieties showed lowest incidence of that in the jute leaf and cheerota-plant-extract-treated samples after 22 weeks of storage (Figure 5).

4. Discussions

The use of herbal extracts to control plant diseases is an environment friendly approach and an effective alternative to toxic chemical pesticides. Krebs and Jaggir [8] investigated a water extract of hemp flowers and essential oils against *E. carotovora* causal agent of potato soft rot. The extracts were tested *in vitro* using a pure bacterial culture and *in vivo* on potatoes latently infected with the pathogen. The protective effect was most pronounced with the extract of hemp flowers. Since very few botanical extracts have been documented to control the soft rot pathogens and many works reported the antibacterial action of herbal extracts against a good number of bacterial pathogens [7, 13, 14], the identification of effective plant extracts against the potato

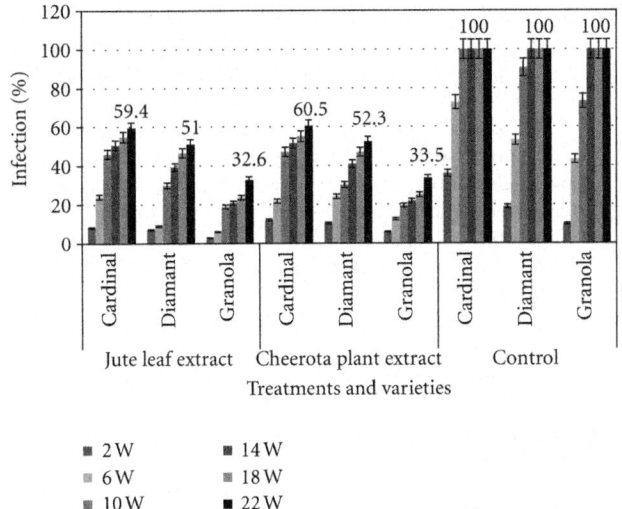

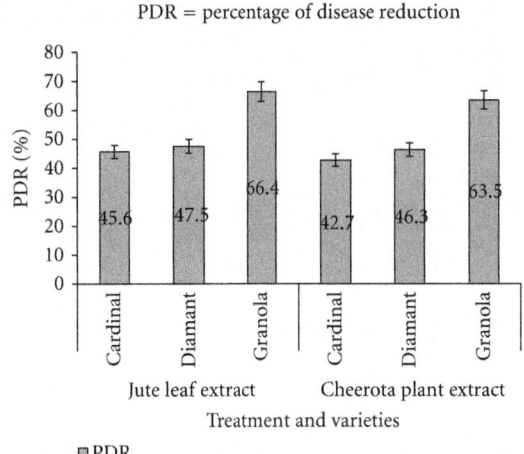

PDR = percentage of disease reduction

FIGURE 3: Effect of tuber treatment with jute leaf extract and cheerota plant extract on soft rot disease incidence of potato in storage.

FIGURE 5: Effect of two botanical extracts on percentage of disease reduction (PDR) of potato after 22 weeks of inoculation in storage condition.

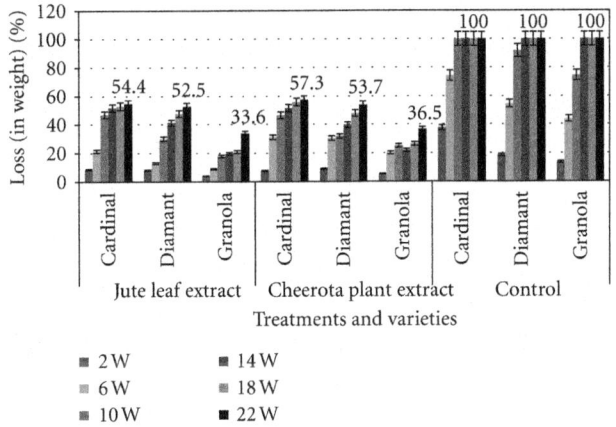

FIGURE 4: Effect of tuber treatment with jute leaf extract and cheerota plant extract on percentage of loss in weight of potato in storage condition.

TABLE 2: *In vitro* evaluation of antibacterial activity of dried jute leaf extract and cheerota plant extract against potato soft rot bacteria (*E. carotovora* subsp. *Carotovora* P-138).

Concentration of plant extract (%)	Jute leaf extract *E. carotovora* subsp. *carotovora* (P-138)
30	−
50	−
75	++
90	++++
Control*	− −
	Cheerota plant extract
30	−
50	−
75	++
90	+++
Control*	− −

−/−−: did not show antibacterial activity; ++: medium antibacterial activity; +++/++++: strong antibacterial activity; −−: good growth of soft rot bacteria; *: plant extract was not added in medium.

soft rot bacterium, *Ecc* P-138, was undertaken to control the potato soft rot diseases in Bangladesh.

In this study, the effects of plant extracts and their efficiencies against bacterial soft rot of potatoes were evaluated *in vitro* and in storage conditions, and quite satisfactory results were obtained with jute leaf and cheerota extracts. The inhibitory activity of plant extracts was most likely due to antimicrobial components present in plant extracts. However, the exact chemical compounds and their controlling mechanism to the soft rot bacteria need to be elucidated. Since jute is the first economic plant in Bangladesh, the jute leaves are readily available without any cost. Additionally, the decomposition of jute leaves further increases the soil fertility. Thus the use of jute leaf extracts should not have any phytotoxic effects on other plants. However, its effects

on epiphytic beneficial microorganisms need to be addressed before making any recommendation.

5. Conclusion

Jute leaf (dried) (*Corchorus capsularis* L.) and cheerota plant (*Swertia chirata* Ham.) extracts showed antibacterial activity against soft rot bacteria *Ecc* P-138 *in vitro* and effectively reduced the bacterial soft rot disease of different potato varieties in storage conditions. Since jute leaves are readily available without any cost in Bangladesh, the application of dry jute leaf extracts is a viable alternative to toxic chemical pesticides to control the soft rot diseases in Bangladeshi potatoes.

TABLE 3: *In vitro* evaluation of antibacterial activity of nineplant extracts against soft rot bacteria of potato.

Treatments with plant extracts (low-to-high concentration)	Antibacterial activity against soft rot bacteria at different time of intervals
	E. carotovora subsp. *carotovora* (P-138)
Aegle marmelos L.	−
Alstonia scholaris L.	−
Erythrina variegata	−
Tagetes erecta	−
Allium cepa	−
Allium sativum L.	−
Azadirachta indica	−
Citrus aurantifolia	−
Curcuma longa L.	−
Control*	− −

−/−−: no antibacterial activity showed; −−: good growth of soft rot bacteria; *: no treatment with plant extracts.

Acknowledgments

The authors acknowledge the Ministry of Science and Information and Communication Technology of Bangladesh for providing (NSICT) funds for this research and the Universiti Malaya for providing publication cost.

References

[1] E. M. Ambridge and T. H. Haines, "Some aspects of pesticide use and human safety in South East Asia," in *Proceedings of 11th International Congress of Plant Protection*, E. D. Magallona, Ed., pp. 219–224, Manila Philippines, October 1987.

[2] Anonymous, "Pesticides incidents up for 1996/1997 compared with previous year," *International Journal of Pest Control*, vol. 40, pp. 1–8, 1998.

[3] M. J. Rice, M. Legg, and K. A. Powell, "Natural products in agriculture—a view from the industry," *Pesticide Science*, vol. 52, no. 2, pp. 184–188, 1998.

[4] K. Hostettmann and J. L. Wolfender, "The search for biologically active secondary metabolites," *Pesticide Science*, vol. 51, no. 4, pp. 471–482, 1997.

[5] C. Leksomboon, N. Thaveechai, and W. Kositratana, "Effect of Thai medicinal plant extracts on growth of phytopathogenic bacteria," in *Proceeding of the 36th Kasetsart University Annual Conference, Plant Section*, Kasetsart University, Bangkok, Thailand, February 1998.

[6] C. Leksomboon, N. Thaveechai,, W. Kositratana, and Y. Paisooksantivatana, "Antiphytobacterial activity of medicinal plant extracts," *Science*, vol. 54, pp. 91–97, 2000.

[7] E. U. Opara and F. T. Obani, "Performance of some plant extracts and pesticides in the control of bacterial spot diseases of solanum," *Agricultural Journal*, vol. 5, no. 2, pp. 45–49, 2010.

[8] H. Krebs and W. Jaggir, "Effect of plant extracts against soft rot of potatoes: *Erwiniacarotovora* Flora and Fauna n Industrial Crops," *Agrarforschung*, vol. 6, no. 1, pp. 17–20, 1999.

[9] B. S. Bdliya and H. U. Haruna, "Efficacy of solar heat in the control of bacterial soft of potato tubers caused by *Erwiniacarotovora* subsp. *Carotovora*," *Journal of Plant Protection Research*, vol. 47, no. 1, pp. 11–17, 2007.

[10] H. H. Long, N. Furuya, D. Kurose, M. Takeshita, and Y. Takanami, "Isolation of endophytic bacteria from Solanum sp. and their antibacterial activity against plant pathogenic bacteria," *Journal of the Faculty of Agriculture, Kyushu University*, vol. 48, no. 1-2, pp. 21–28, 2003.

[11] H. Abd-El-Khair and H. E. H. Karima, "Application of some bactericides and bioagents for controlling the soft rot disease in potato," *Research Journal of Agricultural and Biological Sciences*, vol. 3, no. 5, pp. 463–473, 2007.

[12] A. A. Hajhamed, W. M. A. E. Sayed, A. A. E. Yazied, and N. Y. A. E. Ghaffar, "Suppression of bacterial soft rot disease of potato," *Egyptian Journal Phytopathology*, vol. 35, no. 2, pp. 69–80, 2007.

[13] M. A. Khan, A. Rashid, and A. C. Riaz, "Biological control of bacterial blight of cotton using some plant extracts," *Pakistan Journal of Agricultural Sciences*, vol. 37, pp. 3–4, 2000.

[14] V. Sasitorn, "Potential of some Thai herbal extracts for inhibiting growth of *Ralstonia solanacearum*, the causal agent of bacterial wilt of tomato," *Kamphaengsaen Academic Journal*, vol. 1, no. 2, pp. 70–76, 2003.

17

Growth and Anatomical Parameters of Adventitious Roots Formed on Mung Bean Hypocotyls Are Correlated with Galactoglucomannan Oligosaccharides Structure

K. Kollárová,[1] I. Zelko,[1] M. Henselová,[2] P. Capek,[1] and D. Lišková[1]

[1] Institute of Chemistry, Slovak Academy of Sciences, Dúbravská cesta 9, 845 38 Bratislava, Slovakia
[2] Department of Plant Physiology, Faculty of Natural Sciences, Comenius University, Mlynská dolina B-2, 842 15 Bratislava, Slovakia

Correspondence should be addressed to K. Kollárová, karin.kollarova@savba.sk

Academic Editors: R. K. Kohli and E. Olmos

The effect of galactoglucomannan oligosaccharides (GGMOs) compared with chemically modified oligosaccharides, GGMOs-g (with reduced number of D-galactose side chains) and GGMOs-r (with reduced reducing ends) on mung bean (*Vigna radiata* (L.) Wilczek) adventitious roots formation, elongation, and anatomical structure have been studied. All types of oligosaccharides influenced adventitious root formation in the same way: stimulation in the absence of exogenous auxin and inhibition in the presence of exogenous auxin. Both reactions are probably related with the presence/content of endogenous auxin in plant cuttings. However, the adventitious root length was inhibited by GGMOs both in the absence as well as in the presence of auxin (IBA or NAA), while GGMOs-g inhibition was significantly weaker compared with GGMOs. GGMOs-r were without significant difference on both processes, compared with GGMOs. GGMOs affected not only the adventitious root length but also their anatomy in dependence on the combination with certain type of auxin. The oligosaccharides influenced cortical cells division, which was reflected in the cortex area and in the root diameter. All processes followed were dependent on oligosaccharides chemical structure. The results suggest also that GGM-derived oligosaccharides may play an important role in adventitious roots elongation but not in their formation.

1. Introduction

The research in the recent years has focused on ascertainment of various plant growth regulators or some chemical products influence on growth and development of agricultural plants [1, 2]. Mung bean is in tropical countries a common and widely cultivated nutritious legume crop with antioxidant activity [3], seedlings of which have been used as a model to examine adventitious root formation [4, 5]. Adventitious root formation is important for the vegetative propagation of plants and their growth. Various plant growth regulators have been tested for rooting of mung bean hypocotyl cuttings [5–7]. Besides growth regulators, oligosaccharides isolated from plant cell walls are the most important factors acting in plant growth and development [8]. Xylooligosaccharides stimulate, for example, the rooting

of birch and black pine shoots [9] and induce callus formation and somatic embryogenesis in explants of common mallow (*Malva silvestris* L.) and cotton [10]. Oligogalacturonides support root elongation growth of lettuce [11] and were shown to promote cytokinin-induced vegetative shoot formation from tobacco leaf explants [12]. Trisaccharide fragment of xyloglucan stimulated callus growth and increased the number of embryos in suspension culture of cotton [13]. Hepta- and octa-saccharide (linear oligomers composed of glucose and mannose) isolated from the water extract of the rhizomes of *Paris polyphylla* var. *yunnanensis* stimulated shoot formation of *P. polyphylla* var. *yunnanensis* and root hairs growth of *Panax japonicus* var. *major* [14]. A pentasaccharide synthesized by *Paris polyphylla* var. *yunnanensis* showed a significant stimulus on tobacco seedling growth [15].

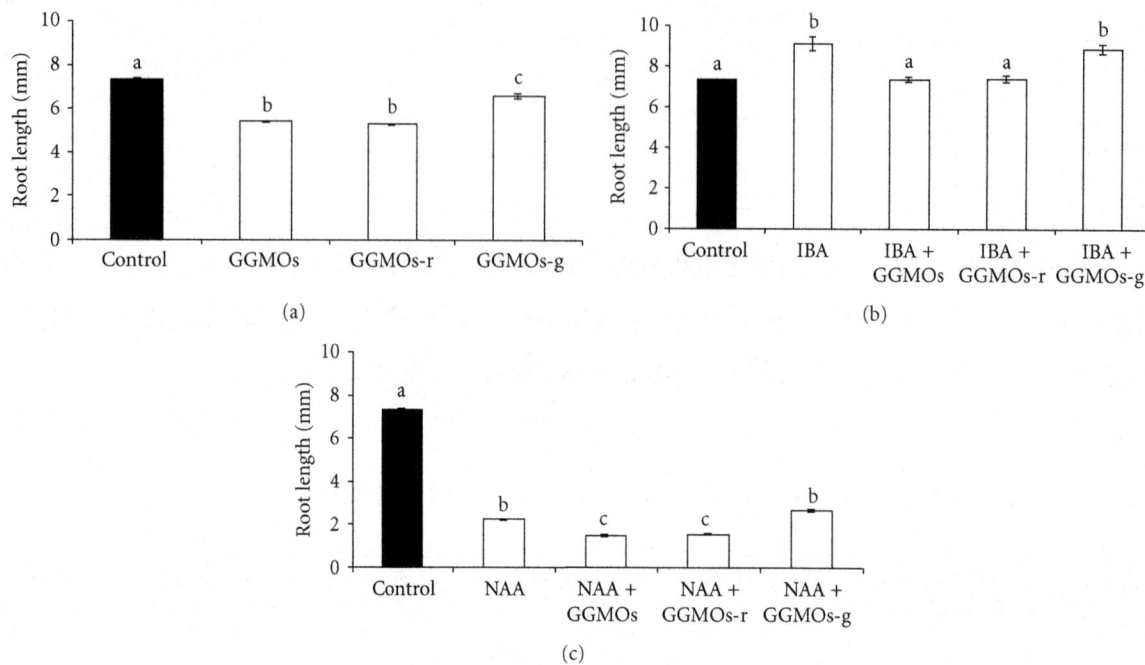

FIGURE 1: Effect of GGMOs, GGMOs-g, and GGMOs-r alone, and in combination with IBA or NAA on mung bean adventitious root elongation. Control—without any plant growth regulators. Different letters above bars indicate significant differences at $P < 0.05$ according to LSD test.

Galactoglucomannan oligosaccharides (GGMOs) derived from plant cell walls galactoglucomannan influence growth, developmental processes, and defence reactions in plant cells [16–18]. GGMOs showed inhibition effect on elongation growth of pea and spruce stem segments induced by auxins and gibberellin at very low concentrations [19, 20] and their inhibitory effect depended on their chemical structure [20, 21]. GGMOs also inhibited adventitious root formation and elongation of mung bean hypocotyl cuttings in the presence of auxins [22]. Morphology and anatomy of *in vitro* cultivated *Karwinskia humboldtiana* root culture was examined, and the results have shown a dependency on GGMOs concentration and interaction with certain type of auxin [23]. However, the effect of chemically modified forms of GGMOs on adventitious root formation and elongation in plant cuttings has not been studied yet. Therefore, the aim of our work was to compare the effect of GGMOs and their modified forms GGMOs-r (with reduced reducing ends) and GGMOs-g (with reduced number of D-galactose side chains) alone, or in combination with auxins (IBA or NAA) on mung bean adventitious roots formation, elongation, and their anatomy.

2. Materials and Methods

2.1. Preparation of Galactoglucomannan Oligosaccharides (GGMOs).
GGMOs with d.p. 4–8 were obtained from spruce galactoglucomannan by partial acid hydrolysis as described previously [24]. Galactoglucomannan consists of a backbone of $(1 \rightarrow 4)$-linked β-D-mannopyranosyl and β-D-glucopyranosyl residues distributed at random, having single stubs of $(1 \rightarrow 6)$-linked α-D-galactopyranosyl residues attached to both mannosyl and glucosyl residues, with slightly preferred substitution of mannosyl residues. GGMOs consist of galactose (4.5%), glucose (21.1%), and mannose (70.4%). Galactoglucomannan oligomers (d.p. 4–8) were composed of tetramers (46%), pentamers (28%), hexamers (12%), heptamers (9%), and octamers (5%). Their number-average molecular mass (M_n) was calculated to be 827.

2.2. Preparation of Partly Degalactosylated Galactoglucomannan Oligosaccharides (GGMOs-g).
GGMOs-g, with reduced number of D-galactose units to about 50%, were prepared by treatment of GGMOs with purified α-galactosidase (EC 3.2.1.22) from coffee beans (Sigma Aldrich, St. Louis, MO, USA) as described previously [25]. Monosaccharide analysis of GGMOs-g by glucose revealed the presence of galactose (2.4%), glucose (21.6%), and mannose (72.0%) residues. Not complete splitting of side chains (deglycosylation only to 47%) is a phenomenon, which may occur by exoenzymes digestion, in this case by the cleavage of α-linked galactose residues with α-galactosidase. The most plausible causes for this state are inhibition of the reaction by the end product, or steric properties of the molecule. The structural features of the individual oligomers in this mixture did not change in comparison with GGMOs.

2.3. Preparation of Modified Oligosaccharides (GGMOs-r).
Modified oligosaccharides GGMOs-r (with reduced reducing ends) were prepared by method described previously [26]. GGMOs of d.p. 4–8 were dissolved in distilled water and treated with 2 M solution of NaBH₄. Excess of reagent

Growth and Anatomical Parameters of Adventitious Roots Formed on Mung Bean Hypocotyls Are Correlated with
Galactoglucomannan Oligosaccharides Structure

143

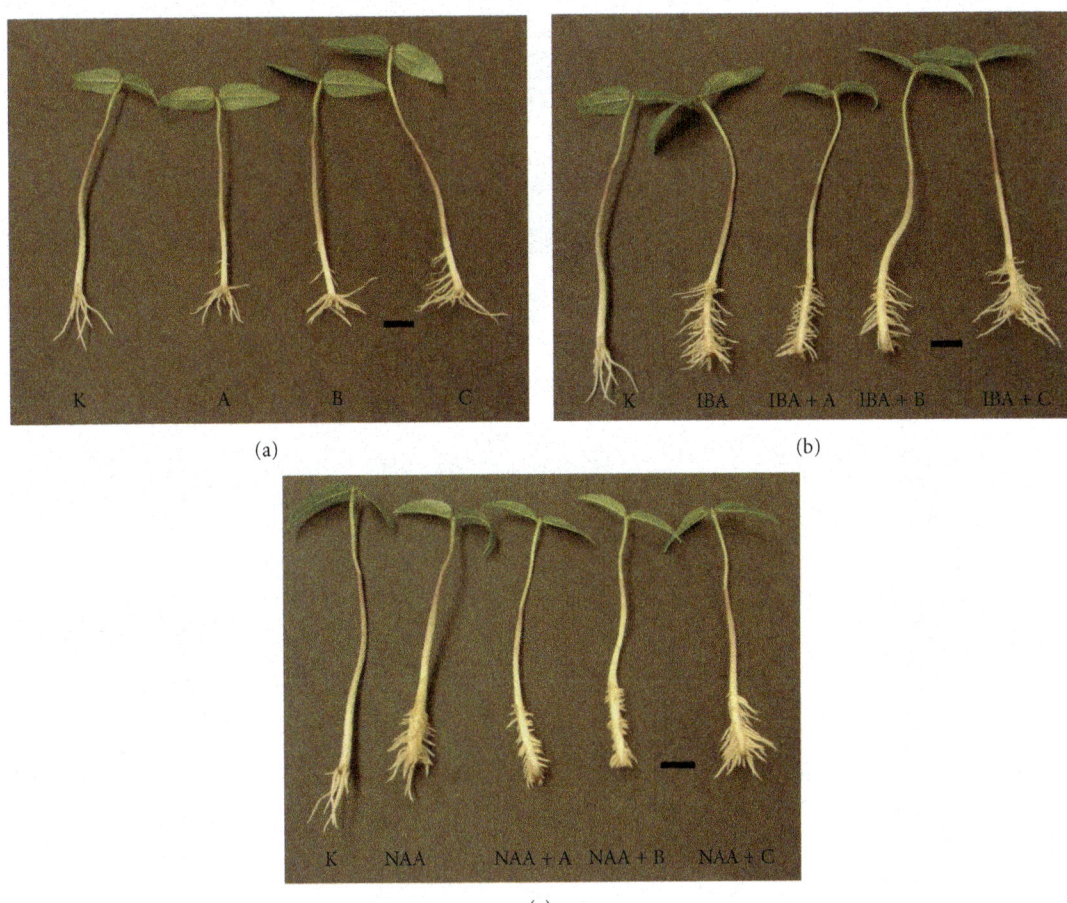

FIGURE 2: Effect of GGMOs, GGMOs-r, and GGMOs-g alone, and in combination with IBA or NAA on the rooting and adventitious root elongation of mung bean hypocotyl cuttings. K: control, without any plant growth regulators, A: GGMOs, B: GGMOs-r, C: GGMOs-g, bar = 1 cm.

was destroyed by Dowex (H^+), filtered, concentrated to dryness, and the boric acid was removed by codistillation with methanol. Modified oligosaccharides were dissolved in distilled water and freeze-dried. The mutual ratio of single oligomers in GGMOs-r was the same as in nonmodified GGMOs.

2.4. Plant Material and Growth Conditions. Seeds of mung bean (*Vigna radiata* (L.) Wilczek var. Emmerald) (Breeding Station Co., Horná Streda, Slovakia) were soaked in water for 3 hours and sown on cellulose wadding. The seeds were kept in the thermostat for 72 hours at $27 \pm 1°C$, 80% relative humidity in the dark. Uniform seedlings with 6-7 cm long hypocotyls were cut 5 cm below the cotyledons and roots were removed. For precise dosing, the bases of hypocotyl cuttings were immersed for 24 h in test solutions according to effective and simple method for promoting adventitious root formation [6, 7]. The following treatments were used: IBA and NAA in 10^{-4} M concentration either alone or in combination with GGMOs and/or with their modified form (10^{-8} M). IBA, NAA, and GGMOs in their most effective concentrations tested previously were applied [22]. For control variant, distilled water was used. After the treatment

with test solutions, cuttings were grown in the substrate (wet sand + peat in the ratio 3 : 1). This substrate is suitable for easy extraction of roots, which is needed for structural studies. Cultivation conditions were the following: $27 \pm 1°C$, 60–70% relative humidity, 12 h photoperiod, irradiance of $180 \mu mol\, m^{-2}\, s^{-1}$, and daily watering to maintain constant water saturation of the substrate at cca 75%. Number and length of adventitious roots and their anatomy were determined after six days of growth.

2.5. Microscopy. For light microscopy, root segments (3.5–4 mm from the apex in the case of roots treated in GGMOs/GGMOs-g alone or in combination with IBA, and 1 mm from the apex in the case of roots treated in NAA or GGMOs/GGMOs-g in combination with NAA) were fixed in 5% glutaraldehyde and postfixed in 0.5% osmium tetroxide, both in 0.1% sodium cacodylate buffer (pH 7.2). The samples were dehydrated in ethanol and propylene oxide, embedded in Spurr medium, and cut with glass knives using Tesla BS 490 ultramicrotome. Semithin sections made at the distance 3 mm from the root apex were stained with toluidine blue and basic fuchsin [27]. Microscopic samples were recorded with digital camera Sony Exwave

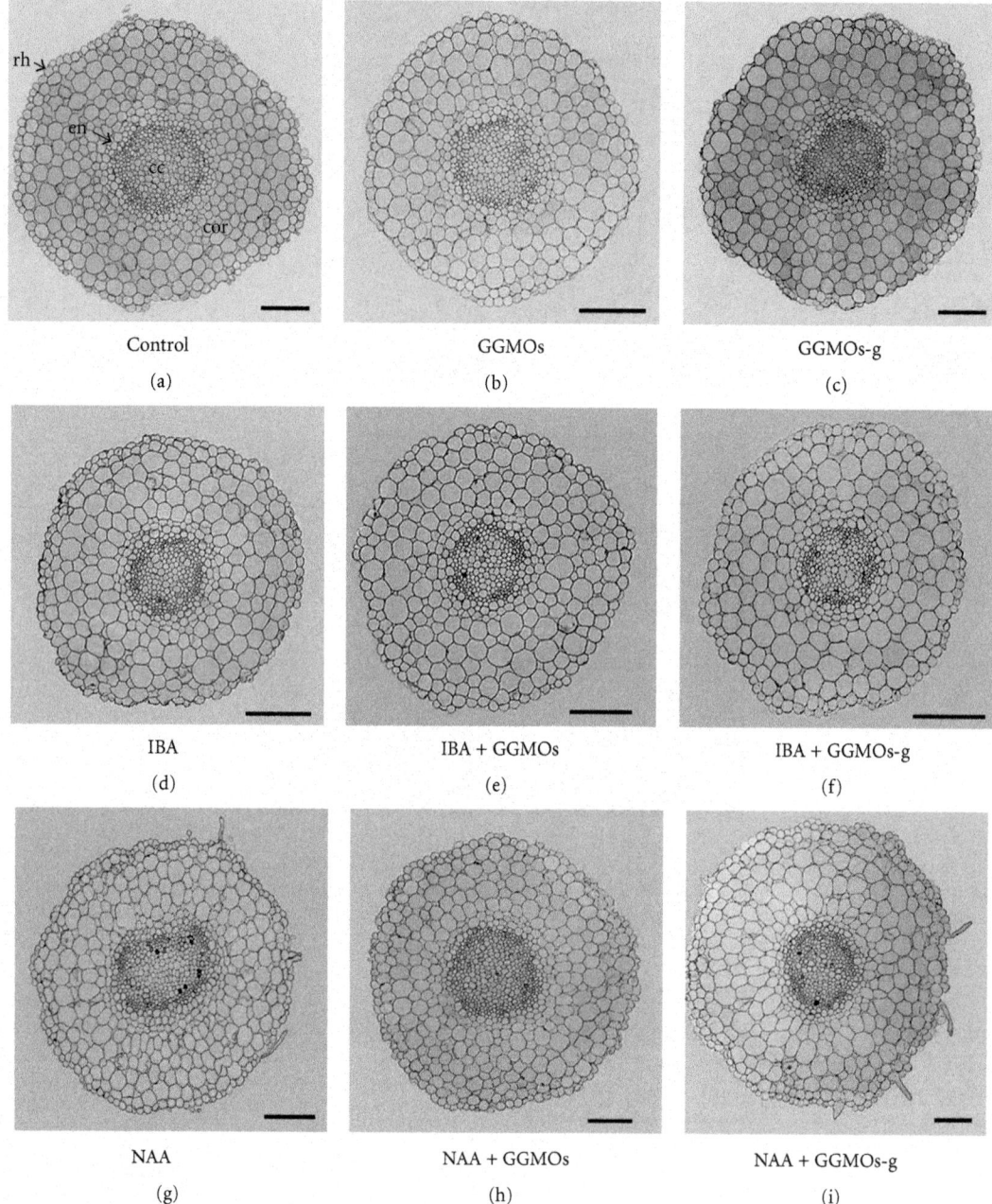

FIGURE 3: Cross sections of mung bean adventitious roots. Control—without any plant growth regulators, rh—rhizodermis, cor—cortex, en—endodermis, cc—central cylinder, bar = 100 μm.

HAD. Adventitious root morphometric parameters were determined by Lucia image analysis system (Lucia 4.8, 1991–2002 Laboratory Imaging, Prague, Czech Republic). Diameter of roots (μm), area of rhizodermis, cortex, endodermis, central cylinder (μm²), and number of primary cortical cells were measured on the root cross sections.

2.6. Statistical Analysis. The values represent the means of three separate experiments with 15 samples per treatment. The data were evaluated by analysis of variance (ANOVA),

and comparisons between the mean values were made by least significant difference (LSD) test at $P < 0.05$, and standard error (SE) was calculated.

3. Results and Discussion

3.1. Root Formation and Elongation. GGMOs stimulated adventitious roots formation in the absence of auxins, though their effect was weaker compared with IBA and NAA (Table 1). On the contrary, in the presence of exogenous

TABLE 1: Effect of GGMOs, GGMOs-g, and GGMOs-r alone, and in combination with IBA or NAA on mung bean adventitious root formation. Control—without any plant growth regulators. Means followed by the same letters are not significantly different at $P < 0.05$ according to LSD test.

	Adventitious roots number
Control	10.26 ± 0.04 a
GGMOs	13.82 ± 1.20 b
GGMOs-r	13.93 ± 0.65 b
GGMOs-g	13.41 ± 0.05 b
IBA	55.71 ± 0.96 c
IBA + GGMOs	49.23 ± 0.54 d
IBA + GGMOs-r	46.00 ± 1.80 d
IBA + GGMOs-g	48.13 ± 1.97 d
NAA	50.20 ± 1.37 d
NAA + GGMOs	40.62 ± 2.49 e
NAA + GGMOs-r	40.40 ± 2.21 e
NAA + GGMOs-g	40.31 ± 3.05 e

auxin GGMOs inhibited adventitious roots formation. All forms of oligosaccharides influenced adventitious roots formation in the same range, no significant differences were determined. Effect of GGMOs on adventitious root formation was independent on their chemical structure.

On the other hand, GGMOs and GGMOs-r inhibited root elongation in the absence, as well as in the presence of IBA or NAA, while GGMOs-g inhibition was significantly weaker compared with GGMOs (Figure 1). Moreover, GGMOs-g + IBA and IBA stimulated adventitious root elongation compared with the control (Figures 1 and 2). The impact of GGMOs and their modified forms on adventitious root elongation in the presence of IBA or NAA may be connected with the distinct action of these auxins in the rooting process [28, 29], as well as with their interaction with oligosaccharides used [30]. In addition to this, both reactions (formation and elongation of roots) are probably related with the presence or content of endogenous auxin in such plant cuttings. The reducing ends of GGMOs did not influence their action in root elongation growth similarly as reducing ends of glucan and chitin oligosaccharides did not affect their biological activity [31, 32], while Spiro et al. [33] observed that the modification at the reducing end of oligogalacturonides influenced in different ways their biological activity in morphogenic bioassays. It seems that the inhibitory effect of GGMOs on root elongation could be related to the presence of galactosyl side chains likewise their inhibitory effect in pea stem segments [20]. Similarly, the stimulating or inhibiting effects of oligogalacturonides on root formation in thin-layer explants of buckwheat were dependent on the monosaccharide content [34]. The biological activity of xyloglucan oligosaccharides in plant growth and development was dependent also on their chemical structure [35, 36]. It is evident that the GGMOs chemical structure influences their action in elongation growth of

aboveground plant parts [20] and of roots but has no effect on root formation.

3.2. *Root Anatomy.* Differences in structural aspects of adventitious roots were compared from samples cultured in the presence of auxins, GGMOs, GGMOs-g, and under the coaction of auxins with GGMOs or GGMOs-g. The impact of GGMOs-r on adventitious root structure is not shown because GGMOs-r did not influence the root elongation compared with GGMOs. After the GGMOs treatment, it has been ascertained that the diameter of roots, cortex area and central cylinder, and the number of cortical cells decreased in comparison with the control, though in GGMOs-g treated roots these parameters were higher compared to GGMOs treatment (Table 2, Figure 3). From results obtained, it can be supposed that GGMOs inhibit not only adventitious root elongation but also the enlargement of root diameter connected with the inhibition of cortical cells division.

The effect of GGMOs in the presence of both types of auxin on adventitious root anatomy was significantly different in comparison with the previous experiment. GGMOs + IBA increased the diameter of roots, cortex area and central cylinder, and the number of cortical cells in comparison with IBA-treated roots (Table 2, Figure 3). The diameter of roots, cortex area and central cylinder, and the number of cortical cells was significantly lower in the presence of GGMOs-g + IBA compared with GGMOs + IBA. Adventitious roots treated with GGMOs + NAA were larger in diameter, cortex area and central cylinder, and number of cortical cells compared to NAA. Treatment of GGMOs-g + NAA increased the diameter of roots, cortex area, and the number of cortical cells in comparison with GGMOs + NAA. From results obtained, it can be supposed that GGMOs influence cortical cell division likewise in the case of GGMOs action in zinnia xylogenic cultures [37]. The impact of GGMOs on cortical cell division in the presence of different types of auxin is dependent on the chemical structure of oligosaccharides, but probably also on different mechanisms of action of certain type of auxin [38–40].

From our results, it can be concluded that galactose side chains can notably modify the biological activity of GGMOs in elongation of adventitious roots, but not in their formation. The anatomy of adventitious roots affected by GGM-derived oligosaccharides of different chemical structure and combination with certain auxin was then reflected in the root diameter resulting from variations mainly in the cell number and the dimension of cortex area.

Abbreviations

d.p.:	Degree of polymerisation
GGM:	Galactoglucomannan
GGMOs:	Galactoglucomannan oligosaccharides
GGMOs-g:	GGMOs with reduced number of D-galactose side chains
GGMOs-r:	GGMOs with reduced reducing ends
IBA:	Indole-3-butyric acid
NAA:	1-naphthaleneacetic acid.

TABLE 2: Effect of GGMOs, GGMOs-g alone, or in combination with IBA or NAA on the root diameter, area of rhizodermis, cortex, endodermis, central cylinder, and number of cortical cells of mung bean adventitious roots. Particular tissues were measured on transversal root sections. Means in each column followed by the same letters are not significantly different at $P < 0.05$ according to LSD test.

	Root diameter (μm)	Area of rhizodermis $(1000\,\mu m^2)$	Area of cortex $(1000\,\mu m^2)$	Area of endodermis $(1000\,\mu m^2)$	Area of central cylinder $(1000\,\mu m^2)$	Number of cortical cells
Control	643.1 ± 7.8 a	23.84 ± 1.06 a	259.51 ± 4.40 a	5.44 ± 0.92 a	25.54 ± 2.66 a	411.5 ± 14.5 a
GGMOs	518.2 ± 14.2 b	21.57 ± 3.47 a	153.31 ± 3.65 b	4.71 ± 0.52 ab	18.19 ± 0.59 b	244.0 ± 2.3 b
GGMOs-g	644.4 ± 9.1 a	22.88 ± 2.37 a	249.01 ± 4.11 ac	6.27 ± 0.62 a	23.65 ± 1.15 a	319.3 ± 15.3 c
IBA	422.4 ± 1.4 c	15.01 ± 0.27 b	106.89 ± 4.20 d	2.91 ± 0.10 b	8.21 ± 0.17 d	186.0 ± 6.0 d
IBA + GGMOs	469.9 ± 7.7 d	15.40 ± 1.60 b	135.19 ± 3.71 b	3.42 ± 0.10 b	12.18 ± 1.14 c	247.3 ± 6.8 b
IBA + GGMOs-g	426.5 ± 0.9 c	13.73 ± 0.92 b	108.40 ± 1.64 d	3.20 ± 0.08 b	10.95 ± 0.42 ad	197.7 ± 6.3 d
NAA	596.4 ± 13.2 e	23.48 ± 1.60 a	224.44 ± 7.38 c	8.94 ± 0.25 c	31.87 ± 1.77 e	294.3 ± 5.8 c
NAA + GGMOs	655.4 ± 20.2 a	28.84 ± 1.17 c	307.59 ± 1.65 e	9.07 ± 1.24 c	37.26 ± 0.80 f	357.0 ± 15.2 e
NAA + GGMOs-g	761.5 ± 34.4 f	32.09 ± 1.14 c	426.84 ± 2.49 f	8.61 ± 0.77 c	38.89 ± 0.07 f	398.3 ± 22.3 a

Acknowledgments

The authors are grateful to J. Kohanová, Department of Plant Physiology, Faculty of Natural Sciences, Comenius University in Bratislava for technical assistance. This work was supported by the Grant of Slovak Grant Agency for Science (no. 2/0046/10). Besides, this contribution is the result of the project implementation: Centre of Excellence for White-Green Biotechnology, ITMS 26220120054, supported by the Research & Development Operational Programme funded by the ERDF.

References

[1] P. C. Endler, W. Matzer, C. Reich et al., "Seasonal variation of the effect of extremely diluted agitated gibberellic acid (10e-30) on wheat stalk growth: a multiresearcher study," *The Scientific World Journal*, vol. 11, pp. 1667–1678, 2011.

[2] M. İsfendiyaroğlu and E. Özeker, "Rooting of *Olea europaea* "Domat" cuttings by auxin and salicylic acid treatments," *Pakistan Journal of Botany*, vol. 40, no. 3, pp. 1135–1141, 2008.

[3] F. Anwar, S. Latif, R. Przybylski, B. Sultana, and M. Ashraf, "Chemical composition and antioxidant activity of seeds of different cultivars of mungbean," *Journal of Food Science*, vol. 72, no. 7, pp. S503–S510, 2007.

[4] G. J. de Klerk and J. Hanecakova, "Ethylene and rooting of mung bean cuttings. The role of auxin induced ethylene synthesis and phase-dependent effects," *Plant Growth Regulation*, vol. 56, no. 2, pp. 203–209, 2008.

[5] R. Pan and X. Tian, "Comparative effect of IBA, BSAA and 5,6-Cl2-IAA-Me on the rooting of hypocotyl in mung bean," *Plant Growth Regulation*, vol. 27, no. 2, pp. 91–98, 1999.

[6] R. Pan and Z. Zhao, "Synergistic effects of plant growth retardants and IBA on the formation of adventitious roots in hypocotyl cuttings of mung bean," *Plant Growth Regulation*, vol. 14, no. 1, pp. 15–19, 1994.

[7] R. C. Pan and H. Gui, "Physiological basis of the synergistic effects of IBA and triadimefon on rooting of mung bean hypocotyls," *Plant Growth Regulation*, vol. 22, no. 1, pp. 7–11, 1997.

[8] M. John, H. Röhring, J. Schmidt, R. Walden, and J. Schell, "Cell signalling by oligosaccharides," *Trends in Plant Science*, vol. 2, no. 3, pp. 111–115, 1997.

[9] K. Ishii, I. Kinoshita, H. Shigenaga et al., "The effects of oligosaccharides on tissue culture of white birch and black pine," *Transactions of the Japanese Forestry Society*, vol. 103, pp. 485–486, 1992.

[10] P. Katapodis, A. Kavarnou, S. Kintzios et al., "Production of acidic xylo-oligosaccharides by a family 10 endoxylanase from *Thermoascus aurantiacus* and use as plant growth regulators," *Biotechnology Letters*, vol. 24, no. 17, pp. 1413–1416, 2002.

[11] K. I. Iwasaki and Y. Matsubara, "Purification of pectate oligosaccharides showing root-growth-promoting activity in lettuce using ultrafiltration and nanofiltration membranes," *Journal of Bioscience and Bioengineering*, vol. 89, no. 5, pp. 495–497, 2000.

[12] G. Falasca, F. Capitani, F. Della Rovere et al., "Oligogalacturonides enhance cytokinin-induced vegetative shoot formation in tobacco explants, inhibit polyamine biosynthetic gene expression, and promote long-term remobilisation of cell calcium," *Planta*, vol. 227, no. 4, pp. 835–852, 2008.

[13] V. Y. Rakitin, Y. I. Dolgikh, E. Y. Shaikina, and V. V. Kuznetsov, "Oligosaccharide inhibits ethylene synthesis and stimulates somatic embryogenesis in a cotton cell culture," *Russian Journal of Plant Physiology*, vol. 48, no. 5, pp. 628–632, 2001.

[14] L. Zhou, C. Yang, J. Li, S. Wang, and J. Wu, "Heptasaccharide and octasaccharide isolated from *Paris polyphylla* var. *yunnanensis* and their plant growth-regulatory activity," *Plant Science*, vol. 165, no. 3, pp. 571–575, 2003.

[15] H. Liu, J. Yang, Y. Du, X. Bai, and Y. Du, "Synthesis of four oligosaccharides derived from *Paris polyphylla* var. *yunnanensis* and their tobacco (*Nicotiana tabacum* L.) growth-regulatory activity," *Plant Growth Regulation*, vol. 60, no. 1, pp. 69–75, 2010.

[16] D. Kákoniová, E. Hlinková, D. Lišková, and K. Kollárová, "Oligosaccharides induce changes in protein patterns of regenerating spruce protoplasts," *Central European Journal of Biology*, vol. 5, no. 3, pp. 353–363, 2010.

[17] D. Lišková, O. Auxtová, D. Kákoniová, M. Kubačková, Š. Karácsonyi, and L. Bilisics, "Biological activity of galacto-glucomannan-derived oligosaccharides," *Planta*, vol. 196, no. 3, pp. 425–429, 1995.

[18] L. Slováková, D. Lišková, P. Capek, M. Kubačková, D. Kákoniová, and Š. Karácsonyi, "Defence responses against TNV infection induced by galactoglucomannan-derived oligosaccharides in cucumber cells," *European Journal of Plant Pathology*, vol. 106, no. 6, pp. 543–553, 2000.

Growth and Anatomical Parameters of Adventitious Roots Formed on Mung Bean Hypocotyls Are Correlated with
Galactoglucomannan Oligosaccharides Structure

147

[19] O. Auxtová, D. Lišková, D. Kákoniová, M. Kubačková, S. Karácsonyi, and L. Bilisics, "Effect of galactoglucomannan-derived oligosaccharides on elongation growth of pea and spruce stem segments stimulated by auxin," *Planta*, vol. 196, no. 3, pp. 420–424, 1995.

[20] K. Kollárová, D. Lišková, and P. Capek, "Further biological characteristics of galactoglucomannan oligosaccharides," *Biologia Plantarum*, vol. 50, no. 2, pp. 232–238, 2006.

[21] K. Kollárová, L. Slováková, E. Kollerová, and D. Lišková, "Galactoglucomannan oligosaccharides inhibition of elongation growth is in pea epicotyls coupled with peroxidase activity," *Biologia*, vol. 64, no. 5, pp. 919–922, 2009.

[22] K. Kollárová, M. Henselová, and D. Lišková, "Effect of auxins and plant oligosaccharides on root formation and elongation growth of mung bean hypocotyls," *Plant Growth Regulation*, vol. 46, no. 1, pp. 1–9, 2005.

[23] K. Kollárová, D. Lišková, and A. Lux, "Influence of galactoglucomannan oligosaccharides on root culture of *Karwinskia humboldtiana*," *Plant Cell, Tissue and Organ Culture*, vol. 91, no. 1, pp. 9–19, 2007.

[24] P. Capek, M. Kubačková, J. Alföldi, L. Bilisics, D. Lišková, and D. Kákoniová, "Galactoglucomannan from the secondary cell wall of *Picea abies* L. Karst," *Carbohydrate Research*, vol. 329, no. 3, pp. 635–645, 2000.

[25] L. Bilisics, J. Vojtaššák, P. Capek, K. Kollárová, and D. Lišková, "Changes in glycosidase activities during galactoglucomannan oligosaccharide inhibition of auxin induced growth," *Phytochemistry*, vol. 65, no. 13, pp. 1903–1909, 2004.

[26] L. Bilisics and M. Kubačková, "Biosynthesis of water-soluble metabolites of UDP-D-galactose containing D-galactose by an enzymic preparation isolated from tissue culture of poplar (*Populus alba* L., var. *pyramidalis*)," *Collection of Czechoslovak Chemical Communications*, vol. 54, pp. 819–833, 1989.

[27] A. Lux, "A rapid method for staining semithin sections of plant material," *Biologia*, vol. 36, pp. 753–757, 1981.

[28] S. Gantait, N. Mandal, and P. K. Das, "Impact of auxins and activated charcoal on *in vitro* rooting of *Dendrobium chrysotoxum* Lindl. cv. Golden Boy," *Journal of Tropical Agriculture*, vol. 47, pp. 84–86, 2009.

[29] A. Husen and M. Pal, "Metabolic changes during adventitious root primordium development in *Tectona grandis* Linn. f. (teak) cuttings as affected by age of donor plants and auxin (IBA and NAA) treatment," *New Forests*, vol. 33, no. 3, pp. 309–323, 2007.

[30] G. H. Mata, B. Sepúlveda, A. Richards, and E. Soriano, "The architecture of *Phaseolus vulgaris* root is altered when a defense response is elicited by an oligogalacturonide," *Brazilian Journal of Plant Physiology*, vol. 18, no. 2, pp. 351–355, 2006.

[31] J. J. Cheong, W. Birberg, P. Fügedi et al., "Structure-activity relationships of oligo-β-glucoside elicitors of phytoalexin accumulation in soybean," *Plant Cell*, vol. 3, no. 2, pp. 127–136, 1991.

[32] K. Baureithel, G. Felix, and T. Boller, "Specific, high affinity binding of chitin fragments to tomato cells and membranes. Competitive inhibition of binding by derivatives of chitooligosaccharides and a Nod factor of Rhizobium," *The Journal of Biological Chemistry*, vol. 269, no. 27, pp. 17931–17938, 1994.

[33] M. D. Spiro, B. L. Ridley, S. Eberhard et al., "Biological activity of reducing-end-derivatized oligogalacturonides in tobacco tissue cultures," *Plant Physiology*, vol. 116, no. 4, pp. 1289–1298, 1998.

[34] O. A. Zabotina, N. N. Ibragimova, A. I. Zabotin, O. I. Trofimova, and A. P. Sitnikov, "Biologically active oligosaccharides from pectins of *Pisum sativum* L. seedlings affecting root generation," *Biochemistry*, vol. 67, no. 1, pp. 227–232, 2002.

[35] B. Baldan, A. Bertoldo, L. Navazio, and P. Mariani, "Oligogalacturonide-induced changes in the developmental pattern of *Daucus carota* L. somatic embryos," *Plant Science*, vol. 165, no. 2, pp. 337–348, 2003.

[36] C. Dunand, C. Gautier, G. Chambat, and Y. Liénart, "Characterization of the binding of α-L-Fuc (1 → 2)-β-D-Gal (1 →), a xyloglucan signal, in blackberry protoplasts," *Plant Science*, vol. 151, no. 2, pp. 183–192, 2000.

[37] A. Beňová-Kákošová, C. Digonnet, F. Goubet et al., "Galactoglucomannans increase cell population density and alter the protoxylem/metaxylem tracheary element ratio in xylogenic cultures of Zinnia," *Plant Physiology*, vol. 142, no. 2, pp. 696–709, 2006.

[38] B. Singh, H. D. Cheek, and C. H. Haigler, "A synthetic auxin (NAA) suppresses secondary wall cellulose synthesis and enhances elongation in cultured cotton fiber," *Plant Cell Reports*, vol. 28, no. 7, pp. 1023–1032, 2009.

[39] M. D. Spiro, J. F. Bowers, and D. J. Cosgrove, "A comparison of oligogalacturonide- and auxin-induced extracellular alkalinization and growth responses in roots of intact cucumber seedlings," *Plant Physiology*, vol. 130, no. 2, pp. 895–903, 2002.

[40] S. Tereso, C. M. Miguel, M. Mascarenhas et al., "Improved *in vitro* rooting of *Prunus dulcis* Mill. cultivars," *Biologia Plantarum*, vol. 52, no. 3, pp. 437–444, 2008.

Antimicrobial Activity and Phytochemical Screening of *Buchenavia tetraphylla* (Aubl.) R. A. Howard (Combretaceae: Combretoideae)

Ygor Lucena Cabral de Oliveira,[1] Luís Cláudio Nascimento da Silva,[2] Alexandre Gomes da Silva,[1] Alexandre José Macedo,[3] Janete Magali de Araújo,[4] Maria Tereza dos Santos Correia,[2] and Márcia Vanusa da Silva[1]

[1] *Laboratório de Produtos Naturais, Departamento de Bioquímica, Centro de Ciências Biológicas, Universidade Federal de Pernambuco, Avenida Professor Moraes Rêgo, s/n Cidade Universitária, 50670-420 Recife, PE, Brazil*
[2] *Laboratório de Glicoproteínas, Departamento de Bioquímica, Centro de Ciências Biológicas, Universidade Federal de Pernambuco, Avenida Professor Moraes Rêgo, s/n Cidade Universitária, 50670-420 Recife, PE, Brazil*
[3] *Faculdade de Farmácia e Centro de Biotecnologia, Universidade Federal do Rio Grande do Sul, Avenida Ipiranga, CEP 90610-000 Porto Alegre, RS, Brazil*
[4] *Laboratório de Genética de Microrganismos, Departamento de Antibióticos, Centro de Ciências Biológicas, Universidade Federal de Pernambuco, Avenida Professor Moraes Rêgo, s/n Cidade Universitária, 50670-420 Recife, PE, Brazil*

Correspondence should be addressed to Márcia Vanusa da Silva, marciavanusa@yahoo.com.br

Academic Editors: W. Hunziker and D. M. Lloyd

This study evaluated the antimicrobial and hemolytic activities and phytochemical constituents of hydroalcoholic extract and its fractions from *Buchenavia tetraphylla* leaves. Cyclohexane (BTCF), ethyl acetate (BTEF), and n-butanol-soluble (BTSBF) and non-soluble (BTNBF) fractions were obtained from a liquid-liquid partition of hydroalcoholic extract (BTHE) from *B. tetraphylla* leaves. The hemolytic activity of active fractions was checked. The BTHE inhibited the growth of *Micrococcus luteus* (MIC: 0.10 mg/mL), *Pseudomonas aeruginosa* (MIC: 0.20 mg/mL), *Mycobacterium smegmatis* (MIC: 0.39 mg/mL), *Proteus vulgaris*, and *Staphylococcus aureus* (MIC: 0.78 mg/mL for both). The more active fractions were BTCF and BTSBF. BTCF showed better potential to inhibit *M. luteus* (0.10 mg/mL), *P. aeruginosa* (0.20 mg/mL), *S. enteritidis* (0.39 mg/mL), and *S. aureus* (1.56 mg/mL). BTBSF showed the best results for *M. luteus* (0.10 mg/mL), *M. smegmatis*, *B. subtilis* (0.39 mg/mL for both), and *P. vulgaris* (0.10 mg/mL). The HC50 were greater than observed MIC: 20.30, 4.70 and 2.53 mg/mL, respectively, to BTBF, BTHE and BTCF, which. The phytochemical analysis detected the presence of flavanoids, triterpene, carbohydrate, and tannin. Our work showed for the first time the broad-spread antimicrobial activity of *B. tetraphylla*, which has nonhemolytic action, creating a new perspective on the interesting association of traditional and scientific knowledge.

1. Introduction

Traditional medicine is used by a large proportion of the semiarid Brazilian population as the major health need of humans and animals [1, 2]. Caatinga medicinal plants have become the focus of intense study recently in terms of conservation and as to whether their traditional uses are supported by actual pharmacological effects or merely folklore [3, 4]. With the increasing acceptance of herbal medicine as an alternative form of health care, the screening of Caatinga medicinal plants for bioactive compounds is important and has been confirmed by the traditional uses [5–9].

In this context, many species of the Combretaceae are used medicinally in several continents in the world. In northeastern Brazil, Agra et al. [10] listed many more traditional medicinal uses of the Combretaceae, which include anthelminthic, treatment of acute enteritis, colitis, constipation, dental caries, diuretic, inflammations in general, malaria, tuberculosis, and cancer, among others.

Buchenavia is a genus of Combretaceae family comprising about 25 species distributed on Central America (Cuba, Trinidad, Panama, and West Indies), Venezuela, Colombia, Guyana, Brazil, Peru, and Bolivia. In the Amazon region, there is the highest concentration of species (20), six occur in the southeast and one reaches the southern Brazil (Santa Catarina). *Buchenavia tetraphylla* (Aubl.) R. A. Howard (Combretaceae: Combretoideae) is a neotropical species with distribution from Cuba Island (Central America) to Rio de Janeiro state, southern Brazil (South America) [11]. In Brazil this plant is known as "tanimbuca" and it is related as an ethnomedicinal plant by traditional communities in the region northeast of Brazil, including indigenous groups [4, 12]. An anti-HIV alkaloid was previously isolated from the leaves of this plant but its cytotoxicity led to a lower therapeutic index [13].

In this work we performed a phytochemical screening of *B. tetraphylla* leaves, examined the antimicrobial activity of hydroalcoholic crude extract and its fractions, and checked the hemolytic effect of more active samples.

2. Materials and Methods

2.1. Plant Collection and Plant Storage. Leaves of *B. tetraphylla* were collected *in Parque Nacional do Catimbau*, Pernambuco, Brazil, northeastern Brazil, in September 2010. Botanical identification was made by staff of the herbarium of Instituto de Pesquisa Agronômica de Pernambuco (IPA), Brazil, and voucher specimens were deposited in the herbarium (IPA 84.104). Leaves were dried at room temperature. The dried plants were milled to a fine powder in a Macsalab mill (Model 200 LAB), Eriez, Bramley, and stored at room temperature in closed containers in the dark until used.

2.2. Preparation of the Crude Hydroalcoholic Extract. B. tetraphylla leaves were dried at room temperature for 7 days, ground into a fine powder and used for extraction. The powder (20 g) was mixed with 50 mL ethanol : water (7 : 3) and submitted to agitation for 15 hours. Then the extracts were filtered and the powder residue was mixed again with 50 mL ethanol-water and the entire extraction process was repeated. The supernatants collected were mixed in a round bottom flask and concentrated at 45°C. The residue was dissolved in DMSO (dimethyl sulfoxide) and kept at −20°C until use.

2.3. Phytochemical Analysis. The phytochemical tests to detect the presence of tannins, flavonoids, anthocyanins, saponins, coumarins, quinones, anthraquinones, reducers compounds, and alkaloids were performed according to the method described by Kokate [14] and Harborne [15].

2.4. Fractionation of the Hydroalcoholic Extract. The hydroalcoholic extract was dissolved in water, producing a solution that was submitted to liquid-liquid partitions successively with cyclohexane, ethyl acetate, and n-butanol. The solutions

produced were dried in anhydrous Na_2SO_4 and submitted to filtration under reduced pressure. Thereafter, the solvents were evaporated under reduced pressure in a rotary evaporator oven at 60°C, producing hexane, ethyl acetate, n-butanol soluble, and n-butanol nonsoluble phases. The residues obtained were kept at −20°C for future use.

2.5. Microbial Strains. The antimicrobial activity of *B. tetraphylla* leaves extract and its fractions were tested against the following microorganisms: *Staphylococcus aureus* (UFPEDA02), *Mycobacterium smegmatis* (UFPEDA71), *Bacillus subtilis* (UFPEDA82), *Micrococcus luteus* (UFPEDA-100), *Enterococcus faecalis* (UFPEDA138), *Escherichia coli* (UFPEDA 224), *Klebsiella pneumoniae* (UFPEDA 396), *Salmonella enteritidis* (UFPEDA 414), *Pseudomonas aeruginosa* (UFPEDA416), *Proteus vulgaris* (UFPEDA740), *Candida krusei* (UFPEDA1002), *Candida albicans* (UFPEDA1007), and *Aspergillus niger* (UFPEDA2003). All strains were provided by Departamento de Antibióticos, Universidade Federal de Pernambuco (UFPEDA) (Table 1) and maintained in Nutrient Agar (NA) and stored at 4°C.

2.6. Determination of Antibacterial Activity Using the Disc Diffusion Method. The antibacterial activity of the extracts was determined by the disc diffusion method [16]. Briefly, bacterial strains were grown on Mueller-Hinton Agar (MHA) medium at 37°C for 18 hours, suspended in distillated water (approximately 1.5×10^8 CFU/mL). An aliquot of $100\,\mu L$ of bacterial suspension was immediately inoculated in petri dishes containing MHA medium. Sterile paper discs containing $2000\,\mu g$ of extracts were added to the culture plates and the samples were incubated at 37°C for an additional 18 hours. After incubation, the diameter of the zone of growth inhibition was examined. Antibiotics and DMSO were used as the negative control.

2.7. Minimum Inhibitory Concentration and the Minimum Bactericidal Concentration. Minimum inhibitory concentration (MIC) was determined by the microdilution method [17]. A twofold serial dilution of the extract/fractions was prepared in Mueller Hinton Broth (MHB) and $100\,\mu L$ (approximately 1.5×10^8 CFU/mL) of bacteria suspension was added. The samples were incubated for 24 h at 37°C. Resazurin solution (0.01%) was used as an indicator by color change visualization: any color changes from purple to pink were recorded as bacterial growth. The lowest concentration at which no color change occurred was taken as the MIC. Afterwards, cultures were seeded in MHA medium and incubated for 24 h at 37°C to determine the minimum bactericidal concentration (MBC) which corresponds to the minimum concentration of extract/fractions that eliminated the bacteria.

2.8. In Vitro Hemolytic Assay. Blood (5–10 mL) was obtained from healthy nonsmoking volunteers by venipuncture, after a written informed consent was obtained. Human erythrocytes from citrated blood were immediately isolated by centrifugation at 1500 rpm for 10 min at 4°C. After removal of plasma and buffy coat, the erythrocytes were washed three times with phosphate-buffered saline (PBS; pH 7.4) and then resuspended using the same buffer and a 1% erythrocyte suspension was prepared. The hemolytic activity of the crude extract was tested under *in vitro* conditions. Each tube received 1.1 mL of erythrocyte suspension and 0.4 mL of extract of various concentrations (50–500 μg/mL) were added. The negative control was only solvent and the positive control received 0.4 mL of Quillaja saponin (0.0025%). After 60-min incubation at room temperature, cells were centrifuged and the supernatant was used to measure the absorbance of the liberated hemoglobin at 540 nm. The average value was calculated from triplicate assays. The hemolytic activity was expressed in relation to ascorbic acid and calculated by the following formula [18]:

$$\text{hemolytic activity } (\%) = \frac{(A_s - A_b)}{(A_c - A_b)} \times 100, \qquad (1)$$

where A_c was the absorbance of the control (blank, without extract), A_s was the absorbance in the presence of the extract, and A_c was the absorbance of saponin solution.

2.9. Statistical Analysis. Each experiment was performed in triplicate and results are expressed as the mean ± SD (standard deviation). Statistical analysis was performed by Student's *t*-test. Differences were considered significant at $P < 0.05$.

3. Results and Discussion

The results from the present study showed that at least one of BTHE and its fractions displayed antimicrobial activities against all the pathogens tested, except for *A. niger* (Table 1). However, the inhibition varied according to extract/fractions and microorganism tested. In addition, the extract and its fractions exhibited broad spectrum of activity.

The BTHE showed inhibition diameter zones (IDZs) ranging from 0 to 27 mm, with the highest IDZs observed against *M. smegmatis* and *M. luteus* (27 mm), followed by *S. aureus* (21.70 mm), *P. aeruginosa* (18.5 mm), *P. vulgaris* (18 mm), *S. enteritidis* (12 mm), and *B. subtilis* (9.7 mm). The IDZs against two yeasts were in a partial way with IDZs of 17.7 and 22 mm to *C. krusei* and *C. albicans*. This extract did not show activity against *E. faecalis*, *E. coli*, *K. pneumaniae*, and *A. niger* in disc paper assay.

In this context, BTHE inhibited strongly the growth of *M. luteus* (MIC: 0.10 mg/mL), *P. aeruginosa* (MIC: 0.20 mg/mL), *M. smegmatis* (MIC: 0.39 mg/mL), *P. vulgaris*, and *S. aureus* (MIC: 0.78 mg/mL for both). Antimicrobial substances are considered as bacteriostatic agents when the ratio MBC/MIC > 4 and bactericidal agents when the ratio MBC/MIC ≤ 4 [19]. Thus, BHTE was a bacteriostatic agent for these pathogens. For the other pathogens the MIC values were greater than 1 mg/mL.

In relation to antimicrobial activities of fractions, the best results were found in cyclohexane (BTCF) and n-butanol soluble fractions (BTBSF), followed by n-butanol non-soluble (BTNBF) and ethyl acetate fractions (BTEF).

The most active fraction was BTCF which showed better potential (MIC < 1 mg/mL) to inhibit the growth of *M. luteus* (MIC: 0.10 mg/mL), *P. aeruginosa* (MIC: 0.20 mg/mL), *S. enteritidis* (MIC: 0.39 mg/mL), and *S. aureus* (MIC: 1.56 mg/mL). The BTBSF showed the best results for *M. luteus* (0.10 mg/mL), *M. smegmatis*, *B. subtilis* (0.39 mg/mL for both), and *P. vulgaris* (0.10 mg/mL) (Table 1).

The results of phytochemical screening of *B. tetraphylla* leaves showed the presence of flavanoids (luteolin), proanthocyanidin, leucoanthocianidin, triterpene, Carbohydrate, and Gallic Tannin. Our results revealed that all of the fractions showed antimicrobial activity suggesting that all solvents are able to solubilize at least one kind of active compounds.

Flavonoids are ubiquitous in photosynthesizing cells and therefore occur widely in plant kingdom [20]. The antibacterial activity of flavonoids has been documented in several earlier studies [21, 22]. Flavonoids have multiple cellular targets and may act as nucleic acid synthesis, cytoplasmic membrane function, or energy metabolism inhibitor. Also, flavonoids are bacteriostatic compounds which induce the formation of bacterial aggregates thereby reducing the number of viable colonies [23]. The presence of luteolin, which has a hydroxyl group at the 3′ position, was detected, being known as a powerful antimicrobial agent [21]. The proantocyanidin have showed ability to protect the urinary tract infections and antioxidant activity [24].

Terpenoids are the largest and the most diverse class of plant compounds and they have numerous functional roles in metabolism and in ecological interactions [25]. These products are soluble in nonpolar solvent and have been showed a lot of biotechnologic activity and were found in *B. tetraphylla* leaves. Terpenes are active against bacteria and fungi [26].

Another compound detected in our study was the gallic tannin, which belongs to the tannins class, a group of polymeric phenolic substances capable of tanning leather or precipitating gelatin from solution, a property known as astringency, commonly found in higher herbaceous and woody plants. Many human physiological activities, such as stimulation of phagocytic cells, host-mediated tumor activity, and a wide range of anti-infective actions, have been assigned to tannin. Their mode of antimicrobial action, as described in the section on quinones, may be related to their ability to inactivate microbial adhesins, enzymes, cell envelope transport proteins, and so forth [27].

Cellular toxicity of the extract and most active fractions (BTBSF, BTCF) was also evaluated using human erythrocytes as a test system. These extract and fractions showed HC50 (the concentration needed for 50% of hemolysis) of 20.30, 4.70, and 2.53 mg/mL, respectively, to BTBF, BTHE, and BTCF. It is important to note that these concentrations are much lower than MIC values.

In conclusion, our work showed that *B. tetraphylla* leaves have antimicrobial activity in a broad-spread way. This plant was able to inhibit strongly the growth of *S. aureus*, *B. subtilis*, *S. enteritidis*, *M. smegmatis*, *M. luteus*, and *P. aeruginosa*. The active extract and fractions did not show hemolytic activity at MIC values, advocating thereby their safety

TABLE 1: Antimicrobial activity of *Buchenavia tetraphylla* leaves.

Microorganism[1]	BTHE				BTCF				BTEF				BTBSF				BTBNF			
	IDZ[2]	MIC[3]	MBC[3]	MBC/MIC	IDZ	MIC	MBC	MBC/MIC	IDZ	MIC	MBC	MBC/MIC	IDZ	MIC	MBC	MBC/MIC	IDZ	MIC	MBC	MBC/MIC
02	21.70	0.78	3.13	4	24.00	1.56	3.13	2	27.70	3.13	12.50	4	24.70	3.13	6.25	2	26.00	6.25	12.50	2
71	27.00	0.39	6.25	16	27.00	0.78	3.13	4	29.00	0.78	6.25	8	26.00	0.39	3.13	8	30.00	0.78	12.50	16
82	28.70	0.39	0.78	2	29.00	0.20	0.78	4	33.00	0.78	1.56	2	28.00	0.39	0.78	2	29.00	0.78	6.25	8
100	27.00	0.10	0.39	4	30.00	0.10	0.20	2	34.00	0.20	1.56	8	29.00	0.10	0.39	4	31.00	0.20	1.56	8
138	0.00	25.00	>25	>1	0.00	25.00	>25	>1	0.00	25.00	>25	>1	0.00	12.50	25.00	2	0.00	25.00	25.00	1
224	0.00	6.25	>25	>4	0.00	3.13	25.00	8	0.00	6.25	12.50	2	19.70	3.13	25.00	8	0.00	12.50	25.00	2
396	0.00	12.50	25.0	2	18.00	6.25	12.50	2	0.00	12.50	25.00	2	19.00	12.50	12.50	1	19.30	12.50	12.50	1
414	12.00	0.78	12.5	16	13.00	0.39	0.78	2	15.00	0.78	1.56	2	13.00	0.78	0.78	1	14.00	1.56	3.13	2
416	18.50	0.20	1.56	8	21.00	0.20	0.78	4	24.00	0.20	3.13	16	22.00	0.20	1.56	8	22.70	0.78	3.13	4
740	18.00	0.78	1.56	2	20.00	0.39	1.56	4	23.00	0.39	3.13	8	21.00	0.10	0.78	8	25.00	0.78	6.25	8
1002	17.7	6.25	25.0	4	19.00	3.13	12.50	4	21.00	3.13	25.00	8	20.00	6.25	12.50	2	21.00	6.25	12.50	2
1007	22	12.50	>25	>2	25.00	12.50	25.00	2	28.00	12.50	25.00	2	22.00	12.50	25.00	2	24.00	12.50	25.00	2
2003	0.00	25.00	>25	>1	0.00	25.00	>25	>1	0.00	25.00	>25	>1	0.00	25.00	>25	>1	0.00	25.00	>25	>1

[1]*Staphylococcus aureus* (UFPEDA 02), *Mycobacterium smegmatis* (UFPEDA 71), *Bacillus subtilis* (UFPEDA 82), *Micrococcus luteus* (UFPEDA 100), *Enterococcus faecalis* (UFPEDA 138), *Escherichia coli* (UFPEDA 224), *Klebsiella pneumoniae* (UFPEDA 396), *Salmonella enteritidis* (UFPEDA 414), *Pseudomonas aeruginosa* (UFPEDA 416), *Proteus vulgaris* (UFPEDA 740), *Candida krusei* (UFPEDA 1002), *Candida albicans* (UFPEDA 1007), and *Candida albicans* (UFPEDA 1007).

[2]IDZ is expressed in mm.

[3]MIC and MMC are expressed in mg/mL.

in therapeutic use. To the best of our knowledge this is the first paper about antimicrobial activity of *B. tetraphylla*. The isolation and chemical characterization of these extracts are being performed by our group and represent a sustainable possibility to the utilization of the natural resources from Caatinga.

Conflict of Interests

The authors declare that there is no conflict of interests.

Acknowledgments

The authors express their gratitude to the Conselho Nacional de Desenvolvimento Científico e Tecnológico (CNPq), to the Coordenação de Aperfeiçoamento de Pessoal de Nível Superior (CAPES), and to the Fundação de Amparo à Ciência e Tecnologia do Estado de Pernambuco (FACEPE) for research Grants. Scott V. Heald, North American teacher at CIEC, bilingual school, is acknowledged for the review of English.

References

[1] M. I. G. Silva, C. T. V. de Melo, L. F. Vasconcelos, A. M. R. de Carvalho, and F. C. F. Sousa, "Bioactivity and potential therapeutic benefits of some medicinal plants from the Caatinga (semi-arid) vegetation of Northeast Brazil: a review of the literature," *Revista Brasileira de Farmacognosia*, vol. 22, pp. 193–207, 2012.

[2] U. P. de Albuquerque, P. M. de Medeiros, A. L. S. de Almeida et al., "Medicinal plants of the caatinga (semi-arid) vegetation of NE Brazil: a quantitative approach," *Journal of Ethnopharmacology*, vol. 114, no. 3, pp. 325–354, 2007.

[3] U. P. De Albuquerque and L. D. H. C. Andrade, "Uso de recursos vegetais da caatinga: o caso do agreste do Estado de Pernambuco (Nordeste do Brasil)," *Interciencia*, vol. 27, no. 7, pp. 336–346, 2002.

[4] M. F. Agra, P. F. França, and J. M. Barbosa-Filho, "Synopsis of the plants known as medicinal and poisonous in Northeast of Brazil," *Revista Brasileira de Farmacognosia*, vol. 17, pp. 114–140, 2007.

[5] C. F. C. B. R. Almeida, D. L. V. Cabral, C. C. B. R. Almeida, E. L. C. Amorim, J. M. Araújo, and U. P. Albuquerque, "Comparative study of the antimicrobial activity of native and exotic plants from the Caatinga and Atlantic Forest selected through an ethnobotanical survey," *Pharmaceutical Biology*, vol. 50, pp. 201–220, 2012.

[6] L. C. N. Da Silva, C. A. Da Silva, R. M. De Souza, A. J. Macedo, M. V. Da Silva, and M. T. S. Correia, "Comparative analysis of the antioxidant and DNA protection capacities of *Anadenanthera colubrina*, *Libidibia ferrea* and *Pityrocarpa moniliformis* fruits," *Food and Chemical Toxicology*, vol. 49, no. 9, pp. 2222–2228, 2011.

[7] D. S. Trentin, R. B. Giordani, K. R. Zimmer et al., "Potential of medicinal plants from the Brazilian semi-arid region (Caatinga) against *Staphylococcus epidermidis* planktonic and biofilm lifestyles," *Journal of Ethnopharmacology*, vol. 137, pp. 327–335, 2011.

[8] L. C. N. Silva, J. M. Sandes, M. M. Paiva et al., "Anti-*Staphylococcus aureus* action of three Caatinga fruits evaluated by electron microscopy," *Natural Product Research*. In press.

[9] A. P. Frasson, O. dos Santos, M. Duarte et al., "First report of anti-*Trichomonas vaginalis* activity of the medicinal plant *Polygala decumbens* from the Brazilian semi-arid region, Caatinga," *Parasitology Research*, vol. 110, pp. 2581–2587, 2012.

[10] M. F. Agra, G. S. Baracho, K. N. Silva, I. J. L. D. Basílio, and V. P. M. Coelho, "Medicinal and poisonous diversity of the flora of "Cariri Paraibano", Brazil," *Journal of Ethnopharmacology*, vol. 111, pp. 383–395, 2007.

[11] P. L. Weaver, "*Buchenavia tetraphylla* (Vahl.) Eichler: Granadillo," Tech. Rep. SO-ITF-SM-43, Department of Agriculture, Southern Forest Experiment Station, New Orleans, 1991.

[12] M. D. F. Agra, K. N. Silva, I. J. L. D. Basílio, P. F. De Freitas, and J. M. Barbosa-Filho, "Survey of medicinal plants used in the region Northeast of Brazil," *Brazilian Journal of Pharmacognosy*, vol. 18, no. 3, pp. 472–508, 2008.

[13] J. A. Beutler, J. H. Cardellina, J. B. McMahon, M. R. Boyd, and G. M. Cragg, "Anti-HIV and cytotoxic alkaloids from *Buchenavia Capitata*," *Journal of Natural Products*, vol. 55, no. 2, pp. 207–213, 1992.

[14] C. K. Kokate, *Practical Pharmacognosy*, Vallabh Prakashan, New Delhi, India, 1994.

[15] J. B. Harborne, *Phytochemical Methods*, Chapman & Hall, London, UK, 1998.

[16] A. W. Bauer, W. M. Kirby, J. C. Sherris, and M. Turck, "Antibiotic susceptibility testing by a standardized single disk method," *American Journal of Clinical Pathology*, vol. 45, no. 4, pp. 493–496, 1966.

[17] Clinical and Laboratory Standards Institute, *Performance Standards for Antimicrobial Susceptibility Testing*, M100-S21, Clinical and Laboratory Standards Institute, Wayne, Pa, USA, 21st edition, 2011.

[18] S. M. Hassan, A. U. Haq, J. A. Byrd, M. A. Berhow, A. L. Cartwright, and C. A. Bailey, "Haemolytic and antimicrobial activities of saponin-rich extracts from guar meal," *Food Chemistry*, vol. 119, no. 2, pp. 600–605, 2010.

[19] D. Gatsing, J. A. Mbah, I. H. Garba, P. Tane, P. Djemgou, and B. F. Nji-Nkah, "An antisalmonellal agent from the leaves of *Glossocalyx brevipes* Benth (Monimiaceae)," *Pakistan Journal of Biological Sciences*, vol. 9, no. 1, pp. 84–87, 2006.

[20] J. A. Manthey, N. Guthrie, and K. Grohmann, "Biological properties of citrus flavonoids pertaining to cancer and inflammation," *Current Medicinal Chemistry*, vol. 8, no. 2, pp. 135–153, 2001.

[21] S. Süzgeç-Selçuka and A. S. Birteksözb, "Flavonoids of *Helichrysum chasmolycicum* and its antioxidant and antimicrobial activities," *South African Journal of Botany*, vol. 77, pp. 170–174, 2011.

[22] H. Liu, Y. Mou, J. Zhao et al., "Flavonoids from *Halostachys caspica* and their antimicrobial and antioxidant activities," *Molecules*, vol. 15, no. 11, pp. 7933–7945, 2010.

[23] T. P. T. Cushnie and A. J. Lamb, "Antimicrobial activity of flavonoids," *International Journal of Antimicrobial Agents*, vol. 26, pp. 343–356, 2005.

[24] M. A. S. Marles, H. Ray, and M. Y. Gruber, "New perspectives on proanthocyanidin biochemistry and molecular regulation," *Phytochemistry*, vol. 64, no. 2, pp. 367–383, 2003.

[25] S. C. Trapp and R. B. Croteau, "Genomic organization of plant terpene synthases and molecular evolutionary implications," *Genetics*, vol. 158, no. 2, pp. 811–832, 2001.

[26] M. P. Popova, I. B. Chinou, I. N. Marekov, and V. S. Bankova, "Terpenes with antimicrobial activity from *Cretan* propolis," *Phytochemistry*, vol. 70, no. 10, pp. 1262–1271, 2009.

[27] M. M. Cowan, "Plant products as antimicrobial agents," *Clinical Microbiology Reviews*, vol. 12, no. 4, pp. 564–582, 1999.

19

Nemesia Root Hair Response to Paper Pulp Substrate for Micropropagation

Pascal Labrousse,[1,2] David Delmail,[3] Raphaël Decou,[2] Michel Carlué,[2] Sabine Lhernould,[2] and Pierre Krausz[2]

[1] *Groupement de Recherche Eau, Sol, Environnement (GRESE EA4330), Laboratoire de Botanique et Cryptogamie, Faculté de Pharmacie, Université de Limoges, 2 rue du Docteur Marcland, 87025 Limoges Cedex, France*
[2] *Laboratoire de Chimie des Substances Naturelles (LCSN EA 1069), Faculté des Sciences et Techniques, Université de Limoges, 123 Avenue Albert Thomas, 87060 Limoges Cedex, France*
[3] *Laboratoire de Pharmacognosie et de Mycologie, UMR CNRS 6226 SCR, Université de Rennes 1, Equipe PNSCM, 2 Avenue du Professeur Léon Bernard, 35043 Rennes Cedex, France*

Correspondence should be addressed to Pascal Labrousse, pascal.labrousse@unilim.fr

Academic Editors: H. P. Bais and C. Gehring

Agar substrates for *in vitro* culture are well adapted to plant micropropagation, but not to plant rooting and acclimatization. Conversely, paper-pulp-based substrates appear as potentially well adapted for *in vitro* culture and functional root production. To reinforce this hypothesis, this study compares *in vitro* development of nemesia on several substrates. Strong differences between nemesia roots growing in agar or in paper-pulp substrates were evidenced through scanning electron microscopy. Roots developed in agar have shorter hairs, larger rhizodermal cells, and less organized root caps than those growing on paper pulp. In conclusion, it should be noted that in this study, *in vitro* microporous substrates such as paper pulp lead to the production of similar root hairs to those found in greenhouse peat substrates. Consequently, if agar could be used for micropropagation, rooting, and plant acclimatization, enhancement could be achieved if rooting stage was performed on micro-porous substrates such as paper pulp.

1. Introduction

Micropropagation is a powerful biotechnology for plant multiplication [1, 2], but plant losses during acclimatization in greenhouse reduced, for some species, the asset of *in vitro* culture multiplication. *In vitro* rooting induction can be mediated by adding plant growth regulators or hormone-like substances to the culture medium [3]. However, the survival rate of these plants during acclimatization is low [3, 4]. In fact, greenhouse culture conditions like hygrometry, CO_2 levels, and nutrient bioavailability in culture medium are drastically different from those used for *in vitro* micropropagation. Most of the time, *in vitro* culture medium is composed of macro- and micronutrients, vitamins, carbohydrates, and eventually plant growth regulators gelified by polysaccharidic substances like agar. So, root formation *in vitro* could be drastically different from in classical greenhouse

substrates. Gonçalves et al. [3] suggested that the lower survival rate during plant acclimatization is due to nonfunctionality of the *in vitro* developed rooting system. Root hairs constitute the major plant/substrate interface as they represent as much as 70% of the plant root surface [5, 6]. So, it could be assumed that root-hair nonfunctionality can drastically reduce water and mineral nutrient uptake, thus representing a limiting key step to acclimatization in peat substrate.

As first proposed by Afreen-Zobayed et al. [4] for sweet potato, paper pulp could be a potentially suitable substrate for *in vitro* culture and functional root-hair production. In order to clarify this assumption, this study compares *in vitro* development of an ornamental plant, *Nemesia denticulata* (Scrophulariaceae), on several substrates like agar and paper pulp. Moreover, enhancement of nemesia acclimatization through the use of paper-pulp substrate was evaluated.

2. Experimental Procedures

2.1. Preparation of Paper-Pulp Miniplugs. Paper pulp (a mixture of wood fibers from deciduous trees) was kindly provided by L. Harvengt from AFOCEL (http://www.fcba.fr/). Paper pulp was rehydrated in boiling water (200 g dry mass·L^{-1}) for 30 min and then vigorously mixed during 30 min in order to eliminate remaining aggregates. After supplemental water draining, paper pulp was manually pressed in plug molds (16 × 15 mm, Ø × H) and dried at 50°C for 24 h.

2.2. Plant Culture and Acclimatization. *Nemesia denticulata* (Scrophulariaceae) plants were cultivated on Murashige and Skoog's (MS) modified by Van der Salm et al. [7] medium supplemented with 20 g·L^{-1} sucrose and 7 g·L^{-1} agar HP-696 (Kalys). The pH was adjusted to 5.8 before autoclaving at 121°C (106 kPa) for 20 min. Cultures were maintained at 22 ± 2°C under fluorescent lights (20 μmol·m^{-2}·s^{-1} of PAR light (photosynthetically active radiation), photoperiod 16 h/24 h) (Grolux 36W). After 3 subcultures, plants were placed on 4 different rooting substrates: agar 7 g·L^{-1} (A), paper pulp prepared as miniplugs (PP), sorbarod (S) (cellulose plugs from Baumgartner Papiers), and peat (fertil miniplug) (P) as control. All substrates were supplemented with 5 mL of liquid half-strength MS Van der Salm medium. After 25 days of *in vitro* culture, the plants were divided into 2 batches: 24 plants per treatment were harvested and 24 other plants per treatment were then transferred to greenhouse for 21 days of acclimatization under fog (cycle of 3 min per hour, 4 times per day during 7 days). Root and shoot fresh and dry masses were measured.

2.3. Scanning Electron Microscopy. For scanning electron microscopy, 2-cm-long root tips from the apex were dehydrated in an ethanol-graded series (10 min at 10°, 10 min at 30°, 10 min at 50°, 10 min at 70°, 10 min at 90°, and three 15 min times at 100°). After critical point drying with CO_2 (FL9496 critical point dryer, Balzers Union), samples were mounted on stubs and coated with 17 nm of gold/paladium (SCD050 sputter coater, Baltec). Root observations were realized using a Philips XL30 scanning electron microscope at 10 kV. Root hair length was measured using Visilog Viewer 6.820 (Noesis).

2.4. Data Analysis. The data were analyzed using R.2.9.2 software. For all further statistical tests, the null hypothesis was the data normality or homogeneity, and the alpha level was set at 0.05 (data are nonnormal or heterogeneous when P value <0.05).

Normality of the measurement data matrix for culture was tested with multivariate Shapiro-Francia test [8] which indicates that the results are not normally distributed ($P = 2.649e − 05$). Thus, only nonparametric tests will be used to process the matrix.

The experiment was laid down in a randomized complete block design. Thus, for each treatment, experiments were carried out with 24 plants and repeated in duplicate. As the data distribution was not normal, the nonparametric ANOVA Friedman test [9] was used to check if duplicates were homogenous, and no difference between duplicates was evidenced. The Friedman test [9] adapted to plant data [10] and the nonlinear principal component analysis [11] were used for medium comparisons.

3. Results

3.1. Scanning Electron Microscopy. Major differences in root-hair morphology and length between plants growing on agar and on paper-pulp substrate (Figures 1(a) and 1(b), resp., Figure 4) were evidenced whereas roots from paper pulp and sorbarod were quite similar (Figures 1(b) and 1(d), Figure 4). It should be noted that root hairs were drastically shortened on agar in comparison with those obtained on paper-pulp substrate (Figures 1(a1), 1(a3), and 1(b1), resp., Figure 4). Moreover, root apex (epidermal cells and cap) strongly differed between the two treatments. Root cap in agar was less organized and epidermal cells were inflated and ovoid (Figure 1(a2)), whereas roots from plants cultivated on paper-pulp substrate (Figure 1(b)) presented a quiet similar morphology to roots from control plants (roots growing in greenhouse on peat substrate) which exhibited long root hairs (Figure 1(c1), Figure 4) and highly organized root cap (Figure 1(c2)).

3.2. Plant Biomass and Water Content. Plant biomass and water content were determined before and after 21 days of acclimatization (Figures 2(a) and 2(b) and Table 1). During the *in vitro* culture phase, no significant differences were observed between the paper pulp and agar ($P = 1.000$) and between agar and sorbarod ($P = 0.317$) even if sorbarod appeared as the best substrate for this stage in terms of biomass production (Figures 2(a) and 3). All the substrates appeared more potent for this micropropagation phase than peat ($P = 0.046$). For the acclimatization stage (Figure 2(b)), sorbarod differed from agar ($P = 4.678e − 3$) and peat ($P = 0.034$) but not from paper pulp ($P = 0.479$) which differed from peat ($P = 4.678e − 3$). Nonlinear principal component analysis (Figure 5) evidenced that paper-pulp-based substrates were the best for *in vitro* culture and acclimatization phase of nemesia. It should be noted that the two paper-pulp based substrates were very similar as the PC2 axis contributes only to 0.78% of the discrimination. Contrariwise, PC1 axis, contributing to 99.16% of the discrimination, clearly segregates these two media from peat and agar. Root fresh and dry masses of *in vitro* plants contributed, respectively, to 13.02% and 11.30% of the discrimination along the PC1 axis (Table 2). Moreover, shoot fresh mass and root dry mass of acclimatized plant contributed to 11.33% and 10.43%, respectively. Along the PC2 axis, the main discriminant was the shoot fresh mass of the acclimatized plants (11.56%). It could be noted that root dry mass of *in vitro* cultured and acclimatized plants contributed to 6.91% and 8.10% of the discrimination, respectively. For the plant water content, no significant difference could be observed between all substrates during the acclimatization phase. A slight increase in water content was evidenced in roots of *in vitro* nemesia from agar and peat but not in shoots.

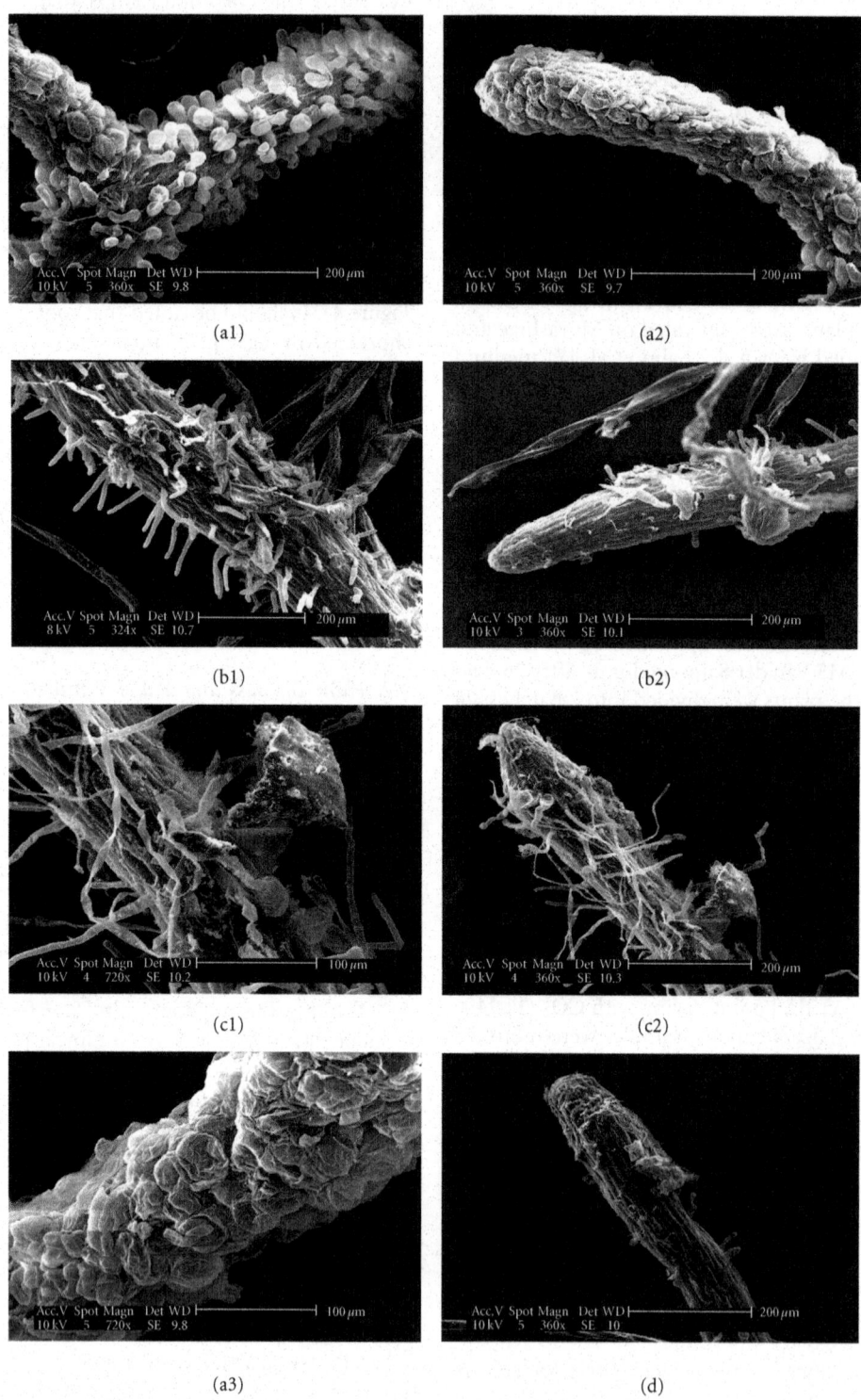

FIGURE 1: Nemesia roots cultivated on agar (a), paper pulp (b), and control plant (c) from greenhouse. (a1) Root hair of nemesia plant cultivated on agar medium. Note the strongly modified cellular morphology even at higher magnification in (a3). (a2) Root cap of nemesia cultivated in agar medium. No root cap is clearly identifiable. (b1) Root hair structure in nemesia root cultivated on paper pulp. (b2) Root cap of nemesia cultivated on paper pulp. Note the quiet similar structure to the control plant from greenhouse (c2). (c1) Root hair structure of the nemesia root cultivated in peat in greenhouse conditions. (c2) Root cap of nemesia cultivated in greenhouse conditions peat. (d) Root cap of nemesia cultivated in sorbarod.

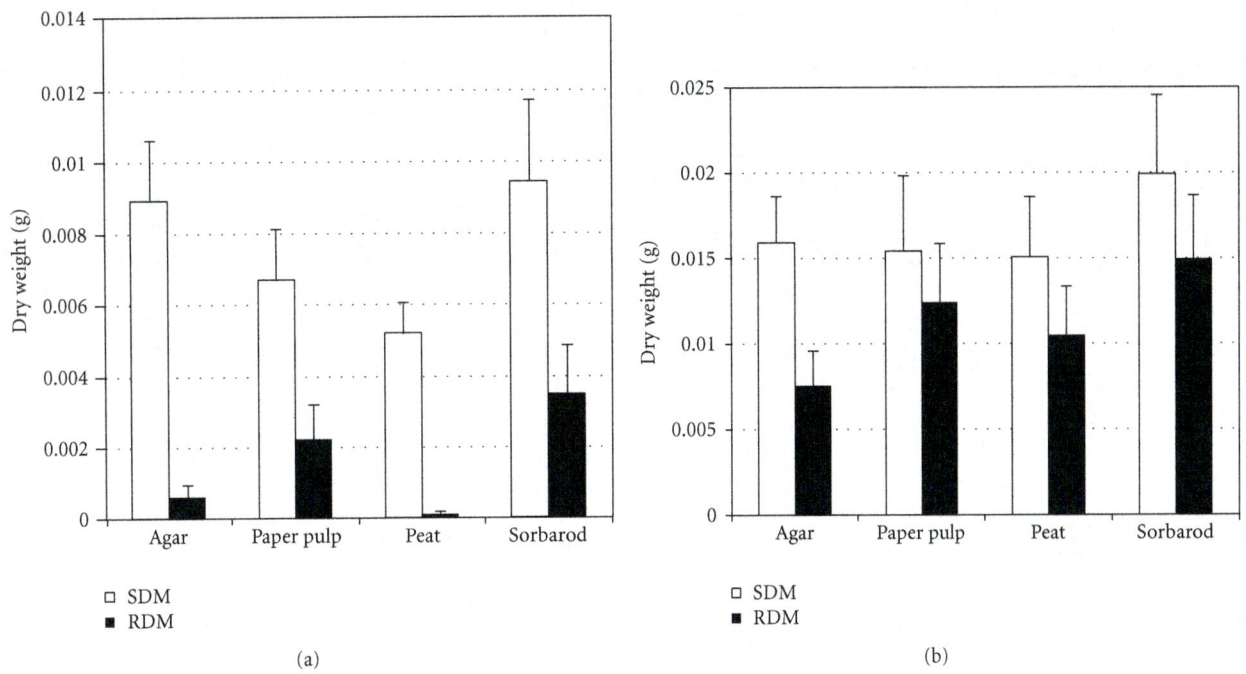

FIGURE 2: Dry mass of nemesia after 25 days of *in vitro* culture (a) and after 21 days of acclimatization (b). SFM: shoot fresh mass, RFM: root fresh mass, SDM: shoot dry mass, and RDM: root dry mass. Data are mean ± se, *n* = 24.

TABLE 1: Nemesia water content (%) after 25 days of *in vitro* culture and after 21 days of acclimatization in greenhouse conditions.

Water content (%)	*In vitro* culture		Acclimatization	
	Shoot	Root	Shoot	Root
Agar	90.6 ± 1.41	95.36 ± 2.40	89.45 ± 0.84	90.81 ± 1.38
Paper pulp	92.06 ± 0.61	88.90 ± 1.90	88.97 ± 0.75	90.74 ± 1.22
Peat	91.36 ± 2.15	98.54 ± 0.24	87.94 ± 0.79	90.05 ± 0.99
Sorbarod	91.95 ± 0.65	92.36 ± 2.46	88.22 ± 0.76	89.18 ± 0.82

FIGURE 3: Aspect of nemesia plants on the different substrates during *in vitro* culture phase. S: sorbarod; P: peat; PP: paper pulp; A: agar.

4. Discussion

Roots of nemesia cultivated on agar medium have similar phenotype to hairless root mutants [6]. Absences of root hair and poor growth are attributed by several authors to hypoxia in agar medium [12, 13]. In addition, Bidel et al. [14] reported that root meristems emerging from the agar gel thereafter progressed quicker than meristems remaining in the gel. These authors hypothesized the presence of several limiting factors for root growth in agar medium in addition of O_2 depletion: progressive dehydration, acidification, and mineral depletion around the older root segments may also have reduced the meristem growth. Moreover, actively tip-growing root hair cells are characterized by a polarized apex rich in Golgi vesicles and mitochondria [15] suggesting important ATP needs for root-hair growth. High amounts of ATP in root hair imply a good O_2 pressure in the substrate [14]. The diffusion of O_2 in agar medium is lower than those found in conventional substrates. In fact, substrates other than agar, including sorbarod [16–19], foam [20–22], vermiculite [23], a vermiculite/gelrite mixture [24], peat [25], rockwool [26], coir [27], and a paper-pulp/vermiculite mixture [4], have been used to prevent low O_2 pressure and poor rooting in agar medium. Decreased O_2 level in a medium could be directly associated with a decrease in root-hair length and to

TABLE 2: Contribution percentage of each variable to discrimination between the four substrates along PC1 and PC2 axis in the nonlinear principal component analysis. RDM Ac: root dry mass of acclimatized nemesia; RDM C: root dry mass of *in vitro* cultured nemesia; RFM Ac: root fresh mass of acclimatized nemesia; RFM C: root fresh mass of *in vitro* cultured nemesia; SDM Ac: shoot dry mass of acclimatized nemesia; SDM C: shoot dry mass of *in vitro* cultured nemesia; SFM Ac: shoot fresh mass of acclimatized nemesia; SFM C: Shoot fresh mass of *in vitro* culture nemesia; TFM Ac: Total fresh mass of acclimatized nemesia; TFM C: total fresh mass of *in vitro* cultured nemesia; TDM Ac: total dry mass of acclimatized nemesia; TDM C: total dry mass of *in vitro* cultured nemesia.

Variable	Contribution along PC1 axis	Contribution along PC2 axis
SFM C	0.1%	8.23%
SFM Ac	11.33%	11.56%
SDM C	2.20%	4.99%
SDM Ac	1.93%	4.70%
RFM C	13.02%	5.19%
RFM Ac	7.83%	0.05%
RDM C	11.29%	6.91%
RDM Ac	10.43%	8.10%
TFM C	0.11%	6.57%
TFM Ac	9.65%	0.04%
TDM C	2.28%	4.39%
TDM Ac	1.38%	5.11%

a complementary extent with a decrease in root respiration [14]. This could result from the direct effect of redox state on gene expression as Sánchez-Fernández et al. [28] demonstrated that the redox state of cellular thiols plays a key role in root-hair growth (for an update see [29, 30]). Consequences are a decrease in water, nutrient uptake, and biomass production. In a controlled and confined environment like a culture tube, plants growing on agar medium do not suffer from this poorly functional rooting system and absorb water and nutrients directly through epidermal and/or rhizodermal cells. But in greenhouse environment, atmospheric water amount is limited, and roots must assume the water and nutrient supply. Even under fog, more than 21 days of culture were necessary for tending towards a complete restoration of physiological processes. Then, the nonfunctionality of the *in vitro* rooting system developed in agar has no consequence on *in vitro* plants but has deleterious effects on plant acclimatization in greenhouse (for a review, see Hazarika [31]).

On the other hand, cheap alternatives to agar for micropropagation are currently under research from low-cost gelling agent to vegetables fibers or vegetables byproducts like Isabgol [32–35], sugarcane bagasse [36, 37], plant gums [38–40], plant fibers [41–43], starch [44, 45], or other systems devoided of agar [46–52]; for a review, see Gangopadhyay et al. [53]. In this way, paper pulp could be evaluated alone or in association with compounds leading to enhance the porosity of the substrate. Similarly, Barrett-Lennard and Dracup [12] demonstrated that plant growth was increased even in po-

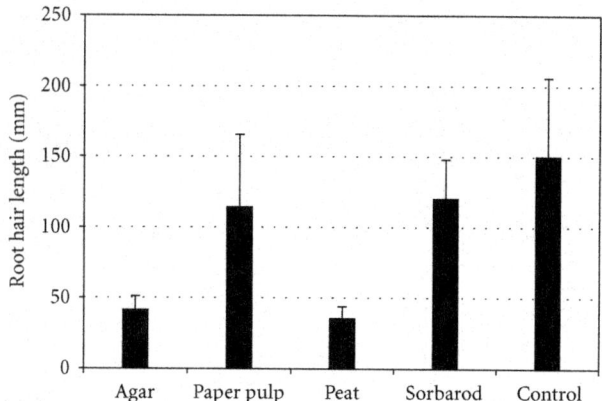

FIGURE 4: Root hair length in the different substrates *in vitro*. Control: root hair of the nemesia cultivated in peat in greenhouse conditions. Data are mean ± s.e., *n* = 30.

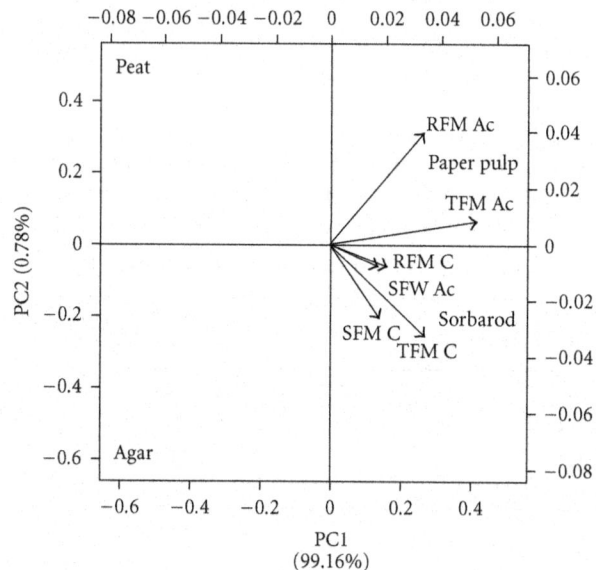

FIGURE 5: Biplot of nonlinear principal component analysis of *in vitro* cultured and acclimatized nemesia in the four substrates. RFM Ac: root fresh mass of acclimatized nemesia; RFM C: root fresh mass of *in vitro* cultured nemesia; SFM Ac: shoot fresh mass of acclimatized nemesia; SFM C: shoot fresh mass of *in vitro* culture nemesia; TFM Ac: total fresh mass of acclimatized nemesia; TFM C: total fresh mass of *in vitro* cultured nemesia. N.B.: dry masses are not visible due to high clustering near the origin.

rous agar-gelled media. Cellulose plugs like sorbarod constitute a good alternative for agar-gelled media but in the sorbarod system, plant roots pass through the pore of the plug, and only few ramifications were produced. Moreover, roots grown on filter paper matrix were often problematic to take out without injury. Paper-pulp plugs with enhanced porous structure could combine the advantages of sorbarod with a well-ramified rooting system like these obtained in paper-pulp experiments.

In that sense paper, pulp appears as a good alternative to agar for rooting *in vitro* cultured plants before acclimatization even if a best aeration of paper-pulp miniplugs should be achieved in order to enhance the rooting-system development.

Abbreviations

MS: Murashige and Skoog medium.

Acknowledgments

This study was supported by a grant of Conseil Régional du Limousin and European Social Fund (ESF). Thanks were due to Pierre Carles for his help with the scanning electron microscope. Also, the authors sincerely thank Dr. Guy Costa and Dr. John Martins for critical reading of the paper.

References

[1] D. Delmail, P. Labrousse, P. Hourdin, L. Larcher, C. Moesch, and M. Botineau, "Physiological, anatomical and phenotypical effects of a cadmium stress in different-aged chlorophyllian organs of *Myriophyllum alterniflorum* DC (Haloragaceae)," *Environmental and Experimental Botany*, vol. 72, no. 2, pp. 174–181, 2011.

[2] D. Delmail, P. Labrousse, P. Hourdin, L. Larcher, C. Moesch, and M. Botineau, "Differential responses of *Myriophyllum alterniflorum* DC (Haloragaceae) organs to copper: physiological and developmental approaches," *Hydrobiologia*, vol. 664, no. 1, pp. 95–105, 2011.

[3] J. C. Gonçalves, G. Diogo, and S. Amâncio, "*In vitro* propagation of chestnut (*Castanea sativa* x *C. crenata*): effects of rooting treatments on plant survival, peroxidase activity and anatomical changes during adventitious root formation," *Scientia Horticulturae*, vol. 72, no. 3-4, pp. 265–275, 1998.

[4] F. Afreen-Zobayed, S. M. A. Zobayed, C. Kubota, T. Kozai, and O. Hasegawa, "A combination of vermiculite and paper pulp supporting material for the photoautotrophic micropropagation of sweet potato," *Plant Science*, vol. 157, no. 2, pp. 225–231, 2000.

[5] R. L. Peterson and M. J. Farquhar, "Root hairs: specialized tubular cells extending root surfaces," *Botanical Review*, vol. 62, no. 1, pp. 1–40, 1996.

[6] J. S. Parker, A. C. Cavell, L. Dolan, K. Roberts, and C. S. Grierson, "Genetic interactions during root hair morphogenesis in Arabidopsis," *Plant Cell*, vol. 12, no. 10, pp. 1961–1974, 2000.

[7] T. P. M. Van der Salm, C. J. G. Van der Toorn, C. H. Hanisch ten Cate, L. A. M. Dubois, D. P. De Vries, and H. J. M. Dons, "Importance of the iron chelate formula for micropropagation of *Rosa hybrida* L. Moneyway," *Plant Cell, Tissue and Organ Culture*, vol. 37, no. 1, pp. 73–77, 1994.

[8] D. Delmail, P. Labrousse, and M. Botineau, "The most powerful multivariate normality test for plant genomics and dynamics data sets," *Ecological Informatics*, vol. 6, no. 2, pp. 125–126, 2011.

[9] M. Friedman, "The use of ranks to avoid the assumption of normality implicit in the analysis of variance," *Journal of the American Statistical Association*, vol. 32, no. 200, pp. 675–701, 1937.

[10] P. Labrousse, D. Delmail, M. C. Arnaud, and P. Thalouarn, "Mineral nutrient concentration influences sunflower infection by broomrape (*Orobanche cumana*)," *Botany*, vol. 88, no. 9, pp. 839–849, 2010.

[11] M. Scholz, F. Kaplan, C. L. Guy, J. Kopka, and J. Selbig, "Nonlinear PCA: a missing data approach," *Bioinformatics*, vol. 21, no. 20, pp. 3887–3895, 2005.

[12] E. G. Barrett-Lennard and M. Dracup, "A porous agar medium for improving the growth of plants under sterile conditions," *Plant and Soil*, vol. 108, no. 2, pp. 294–298, 1988.

[13] C. Newell, D. Growns, and J. McComb, "The influence of medium aeration on *in vitro* rooting of Australian plant microcuttings," *Plant Cell, Tissue and Organ Culture*, vol. 75, no. 2, pp. 131–142, 2003.

[14] L. P. R. Bidel, P. Renault, L. Pagès, and L. M. Rivière, "Mapping meristem respiration of *Prunus persica* (L.) Batsch seedlings: potential respiration of the meristems, O_2 diffusional constraints and combined effects on root growth," *Journal of Experimental Botany*, vol. 51, no. 345, pp. 755–768, 2000.

[15] M. E. Galway, J. W. Heckman, and J. W. Schiefelbein, "Growth and ultrastructure of Arabidopsis root hairs: the *rhd3* mutation alters vacuole enlargement and tip growth," *Planta*, vol. 201, no. 2, pp. 209–218, 1997.

[16] G. C. Douglas, C. B. Rutledge, A. D. Casey, and D. H. S. Richardson, "Micropropagation of floribunda, ground cover and miniature roses," *Plant Cell, Tissue and Organ Culture*, vol. 19, no. 1, pp. 55–64, 1989.

[17] M. S. Kim, N. B. Klopfenstein, and B. M. Cregg, "*In vitro* and *ex vitro* rooting of micropropagated shoots using three green ash (*Fraxinus pennsylvanica*) clones," *New Forests*, vol. 16, no. 1, pp. 43–57, 1998.

[18] A. V. Roberts, E. F. Smith, and J. Mottley, "The preparation of propagation of micropropagated plantlets for transfer to soil without acclimatization," in *Methods in Molecular Biology, Vol. 6, Plant Cell and Tissue Culture*, J. M. Pollard and J. M. Walker, Eds., pp. 227–236, 1990.

[19] A. V. Roberts, E. F. Smith, I. Horan, S. Walker, D. Matthews, and J. Mottley, "Stage III techniques for improving water relations and autotrophy in micropropagated plants," in *Physiology, Growth and Development of Plants in Culture*, P. J. Lumsden, J. R. Nicholas, and W. J. Davies, Eds., pp. 314–322, 1994.

[20] J. A. McComb and S. Newton, "Propagation of Kangaroo Paw using tissue culture," *The Journal of Horticultural Science and Biotechnology*, vol. 56, pp. 181–183, 1980.

[21] K. Gebhardt, "Development of a sterile cultivation system for rooting of shoot tip cultures (red raspberries) in duroplast foam," *Plant Science*, vol. 39, no. 2, pp. 141–148, 1985.

[22] T. D. Roche, R. D. Long, A. J. Sayegh, and M. J. Hennerty, "Commercial scale photo-autotrophic micropropagation applications in Irish agriculture, horticulture and forestry," *Acta Horticulturae*, vol. 440, pp. 515–520, 1996.

[23] E. Rugini and D. C. Verma, "Micropropagation of difficult to propagate almond (*Prunus amygalus* Batsch) cultivar," *Plant Science Letters*, vol. 28, pp. 273–281, 1982.

[24] C. Jay-Allemand, P. Capelli, and D. Cornu, "Root development of *in vitro* hybrid walnut microcuttings in a vermiculite-containing gelrite medium," *Scientia Horticulturae*, vol. 51, no. 3-4, pp. 335–342, 1992.

[25] K. Gebhardt and M. Friedrich, "Micropropagation of *Calluna vulgaris* cv. "H.E. Beale"," *Plant Cell, Tissue and Organ Culture*, vol. 9, no. 2, pp. 137–145, 1987.

[26] X. Lin, B. A. Bergann, and A. Stomp, "Effect of medium physical support, shoot length and genotype on in vitro rooting and plantlet morphology of Sweetgum," *Journal of Environmental Horticulture*, vol. 13, pp. 117–121, 1995.

[27] G. Gangopadhyay, S. Das, S. K. Mitra, R. Poddar, B. K. Modak, and K. K. Mukherjee, "Enhanced rate of multiplication and rooting through the use of coir in aseptic liquid culture media," *Plant Cell, Tissue and Organ Culture*, vol. 68, no. 3, pp. 301–310, 2002.

[28] R. Sánchez-Fernández, M. Fricker, L. B. Corben et al., "Cell proliferation and hair tip growth in the Arabidopsis root are under mechanistically different forms of redox control," *Proceedings of the National Academy of Sciences of the United States of America*, vol. 94, no. 6, pp. 2745–2750, 1997.

[29] C. Gapper and L. Dolan, "Control of plant development by reactive oxygen species," *Plant Physiology*, vol. 141, no. 2, pp. 341–345, 2006.

[30] A. Pitzschke and H. Hirt, "Mitogen-activated protein kinases and reactive oxygen species signaling in plants," *Plant Physiology*, vol. 141, no. 2, pp. 351–356, 2006.

[31] B. N. Hazarika, "Morpho-physiological disorders in *in vitro* culture of plants," *Scientia Horticulturae*, vol. 108, no. 2, pp. 105–120, 2006.

[32] S. B. Babbar and N. Jain, ""Isubgol" as an alternative gelling agent in plant tissue culture media," *Plant Cell Reports*, vol. 17, no. 4, pp. 318–322, 1998.

[33] R. K. Tyagi, A. Agrawal, C. Mahalakshmi, Z. Hussain, and H. Tyagi, "Low-cost media for *in vitro* conservation of turmeric (*Curcuma longa* L.) and genetic stability assessment using RAPD markers," *In vitro Cellular and Developmental Biology—Plant*, vol. 43, no. 1, pp. 51–58, 2007.

[34] A. Agrawal, R. Sanayaima, R. Tandon, and R. K. Tyagi, "Cost-effective *in vitro* conservation of banana using alternatives of gelling agent (isabgol) and carbon source (market sugar)," *Acta Physiologiae Plantarum*, vol. 32, no. 4, pp. 703–711, 2010.

[35] S. Saglam and C. Y. Cifici, "Effects of agar and isubgol on adventitious shoot regeneration of woad (*Isatis tinctoria*)," *International Journal of Agriculture and Biology*, vol. 12, no. 2, pp. 281–285, 2010.

[36] R. Mohan, C. R. Soccol, M. Quoirin, and A. Pandey, "Use of sugarcane bagasse as an alternative low-cost support material during the rooting stage of apple micropropagation," *In vitro Cellular and Developmental Biology—Plant*, vol. 40, no. 4, pp. 408–411, 2004.

[37] R. Mohan, E. A. Chui, L. A. Biasi, and C. R. Soccol, "Alternative *in vitro* propagation: use of sugarcane bagasse as a low cost support material during rooting stage of strawberry cv. Dover," *Brazilian Archives of Biology and Technology*, vol. 48, pp. 37–42, 2005.

[38] N. Jain and S. B. Babbar, "Gum katira—a cheap gelling agent for plant tissue culture media," *Plant Cell, Tissue and Organ Culture*, vol. 71, no. 3, pp. 223–229, 2002.

[39] S. B. Babbar, R. Jain, and N. Walia, "Guar gum as a gelling agent for plant tissue culture media," *In vitro Cellular and Developmental Biology—Plant*, vol. 41, no. 3, pp. 258–261, 2005.

[40] R. Jain and S. B. Babbar, "Xanthan gum: an economical substitute for agar in plant tissue culture media," *Plant Cell Reports*, vol. 25, no. 2, pp. 81–84, 2006.

[41] G. Gangopadhyay, T. Bandyopadhyay, S. B. Gangopadhyay, and K. K. Mukherjee, "Luffa sponge—a unique matrix for tissue culture of *Philodendron*," *Current Science*, vol. 86, no. 2, pp. 315–319, 2004.

[42] M. Lucchesini, A. Mensuali Sodi, S. Pacifici, and F. Tognoni, "Unconventional production of *Echinacea angustifolia* DC. Microplants," *Acta Horticulturae*, vol. 812, pp. 271–276, 2009.

[43] S. G. Dalvi, P. P. Gudhate, P. N. Tawar, and D. T. Prasad, "Low cost support matrix for potato micro-propagation," *Potato Journal*, vol. 38, no. 1, pp. 47–50, 2011.

[44] P. S. Naik and D. Sarkar, "Sago: an alternative cheap gelling agent for potato *in vitro* culture," *Biologia Plantarum*, vol. 44, no. 2, pp. 293–296, 2001.

[45] P. Kuria, P. Demo, A. B. Nyende, and E. M. Kahangi, "Cassava starch as an alternative cheap gelling agent for the *in vitro* micro-propagation of potato (*Solanum tuberosum* L.)," *African Journal of Biotechnology*, vol. 7, no. 3, pp. 301–307, 2008.

[46] A. Hoffmann, M. Pasqual, N. N.J. Chalfun, and S. S.N. Vieira, "Substrates for *in vitro* adventitious root induction and development on two apple rootstocks," *Pesquisa Agropecuaria Brasileira*, vol. 36, no. 11, pp. 1371–1379, 2001.

[47] C. Newell, D. J. Growns, and J. A. McComb, "A novel *in vitro* rooting method employing an aerobic medium," *Australian Journal of Botany*, vol. 53, no. 1, pp. 81–89, 2005.

[48] M. K. Goel, A. K. Kukreja, and S. P. S. Khanuja, "Cost-effective approaches for *in vitro* mass propagation of *Rauwolfia serpentina* Benth. Ex Kurz," *Asian Journal of Plant Sciences*, vol. 6, no. 6, pp. 957–961, 2007.

[49] H. Prknová, "The use of silica sand in micropropagation of woods," *Journal of Forest Science*, vol. 53, no. 2, pp. 88–92, 2007.

[50] E. Bunn, "Investigations of alternative *in vitro* rooting methods with rare and recalcitrant plants," *Acta Horticulturae*, vol. 829, pp. 279–282, 2010.

[51] N. Daud, R. M. Taha, N. N. M. Noor, and H. Alimon, "Potential of alternative gelling agents in media for the in vitro micropropagation of *Celosia* sp," *International Journal of Botany*, vol. 7, no. 2, pp. 183–188, 2011.

[52] A. Sharifi, N. Moshtaghi, and A. Bagheri, "Agar alternatives for micropropagation of African violet (*Saintpaulia ionantha*)," *African Journal of Biotechnology*, vol. 9, no. 54, pp. 9199–9203, 2010.

[53] G. Gangopadhyay, S. K. Roy, and K. K. Mukherjee, "Plant response to alternative matrices for *in vitro* root induction," *African Journal of Biotechnology*, vol. 8, no. 13, pp. 2923–2928, 2009.

20

Identification and Mechanism of *Echinochloa crus-galli* Resistance to Fenoxaprop-p-ethyl with respect to Physiological and Anatomical Differences

Amany Hamza,[1] Aly Derbalah,[1] and Mohamed El-Nady[2, 3]

[1] *Pesticides Department, Faculty of Agriculture, Kafrelshiekh University, Kafr el-Sheikh 33516, Egypt*
[2] *Department of Agricultural Botany, Faculty of Agriculture, Kafrelshiekh University, Kafr el-Sheikh 33516, Egypt*
[3] *Department of Biology, Faculty of Applied Science, Taibah University, P.O. Box 344, Al Madinah Al-Munawwarah, Saudi Arabia*

Correspondence should be addressed to Aly Derbalah, aliderbalah@yahoo.com

Academic Editors: S. Y. Morozov and R. Sarkar

Identification and mechanism of *Echinochloa crus-galli* (L.) resistance to fenoxaprop-p-ethyl via physiological and anatomical differences between susceptible and resistant were investigated. The physiological and anatomical differences that were take into account were growth reduction, chlorophyll content reduction, lamina thickness, and xylem vessel diameter in both susceptible and resistant biotypes of *E. crus-galli*. The results showed that the growth reduction fifty (GR_{50}) of resistant biotype was 12.07-times higher than that of the susceptible biotype of *E. crus-galli* treated with fenoxaprop-p-ethyl. The chlorophyll content was highly reduced in the susceptible biotype relative to the resistant one of *E. crus-galli* treated with fenoxaprop-p-ethyl. An anatomical test showed significant differences in the cytology of susceptible and resistant biotypes of *E. crus-galli* treated with fenoxaprop-p-ethyl with respect to lamina thickness and xylem vessel diameter. The resistance of *E. crus-galli* to fenoxaprop-p-ethyl may be due to the faster metabolism of fenoxaprop-p-ethyl below the physiologically active concentration or the insensitivity of its target enzyme (Acetyl-CoA carboxylase).

1. Introduction

E. crus-galli (L.) *Beauv* is a type of wild grass originating from tropical Asia that was formerly classified as a type of panicum grass. Considered as one of the world's worst weeds, it reduces crop yields and causes forage crops to fail by removing up to 80% of the available soil nitrogen. The high levels of nitrates it accumulates can poison livestock. It acts as a host for several mosaic virus diseases. Heavy infestations can interfere with mechanical harvesting. Individual plants can produce up to 40,000 seeds per year. Water, birds, insects, machinery, and animal feet disperse it, but contaminated seed is probably the most common dispersal method. More than 35% of grain yield in seeded rice was reduced by infestation with *E. crus-galli* [1].

Due to the great risk of this weed infestation in Egypt and worldwide, herbicide is becoming the most popular method of weed control in rice. However, while herbicide application

certainly controls the weeds, experience shows, however, that although herbicide use alleviates the problem of labor for weeding, incorrect use of herbicides may bring about other environmental problems such as selecting for resistance to herbicides. Weed resistance to herbicides concerns many sectors of the agricultural community: farmers, advisors, researchers, and the agrochemical industry in Egypt and worldwide. The fear exists that in an extreme case of resistance, farmers might lose a valuable chemical tool that had previously provided effective control of yield-reducing weeds. Resistance is often seen as a problem caused by a particular active ingredient. This is an oversimplification and a misconception. Resistance results from agronomic systems which have been developed to rely too heavily on herbicides as the sole method of weed control [2]. Without monitoring and rapid detection of the resistance evolution, interpretation of its mechanism and trying to find sustainable management strategies, the future usefulness of herbicides as a tool for weed control might be seriously jeopardized. Furthermore, the identification of resistance mechanism to herbicides is considered the key step toward developing appropriate solutions to overcome this phenomenon.

Resistance mechanism through evaluation of the activity of target site enzymes has been reported before [3–5]; however, characterizing the resistance mechanisms of weeds to herbicides via investigating the anatomical and physiological differences in susceptible and resistant biotypes considered a source of major concern and has not been studied before.

Therefore, this study attempted to identify the occurrence and mechanism of *E. crus-galli*-resistant biotype against fenoxaprop-p-ethyl via investigation of the physiological (chlorophyll content and growth reduction) and anatomical differences between the susceptible and resistant biotypes of *E. crus-galli* (barnyardgrass) treated with fenoxaprop-p-ethyl.

2. Materials and Methods

2.1. The Used Herbicide. Fenoxaprop-p-ethyl with the trade name of Whip-super EW 7.5% was obtained from Rice Weeds Research Department, Rice Research and Training Center, Sakah, and Kafr el-Sheikh, Egypt. This herbicide was applied at the filed rate of 49.5 gm a.i/hectare.

2.2. The Tested Weed. The susceptible biotype (SBT) of the *Echinochloa crus-galli* to fenoxaprop-p-ethyl (obtained from Rice Weeds Research Department, Rice Research and Training Centre, Sakah, Kafr El-Sheikh). The resistant biotype (RBT) of *Echinochloa crus-galli* used in this study was previously treated for several years with the tested herbicide by selection pressure and recorded resistance [6].

2.3. Whole Plant Bioassay. Dose-response experiments were conducted at the greenhouse of the Agricultural Botany Department, Faculty of Agriculture, Kafr El-Sheikh University, Egypt. The soil used in this experiment was fertilized with nitrogen at a rate of 360 kg/h of urea fertilizer (containing 46% nitrogen). Super phosphate fertilizer (phosphorus 15%) was added at a rate of 240 kg/ha before planting. Potassium was not added because the Egyptian soil is rich in this element. Seeds of susceptible and resistant biotypes of *Echinochloa crus-galli* were planted in 30×30 cm plastic pots filled with soil. Emerged seedlings were thinned to four uniform and equally distant-spaced plants per pot. These experiments were conducted at average daily temperatures ranging from 22 to 31°C and at a 16-h day length. Pots were immersed with water up to 4 cm above the soil surface. The tested herbicide, fenoxaprop-p-ethyl, was applied as a single application using a hand sprayer at the 4-leaf to 1-tiller stage of growth of the tested weed. The concentration levels used were 0.1, 0.5, 1, and 2 folds of fenoxaprop-p-ethyl recommended dose. After forty-eight hours of treatment, the plants were irrigated and water was raised up to 4 cm above the soil surface [4]. Experiments were done in a completely randomized design with six replicates. Fresh

weight of treated and untreated plant was determined after 14 days of fenoxaprop-p-ethyl application. Data were pooled and fitted to a log-logistic regression model [7, 8] as shown in (1)

$$ Y = c + \left\{ \frac{(d - c)}{\left[1 + (x/g)^b \right]} \right\}, \qquad (1) $$

where Y is the fresh weight of germinated seedling aboveground expressed as percentage of the untreated control, c and d are the coefficients corresponding to the lower and upper asymptotes, b is the slope of the line, g (GR_{50}) is the herbicide rate at the point of inflection halfway between the upper (d) and lower (c) asymptotes, and x (independent variable) is the herbicide dose.

Regression analysis was conducted using the Sigma Plot statistical software version 10.0 [4]. The herbicide rate used to reduce plant growth by 50% relative to the untreated control (GR_{50}) was calculated for resistant and susceptible biotypes of *E. crus-galli*. R/S ratios were calculated as the GR_{50} of the resistant (R) biotype divided by the GR_{50} of the susceptible (S) biotype.

2.4. Chlorophyll Measurements. Plant leaves differ from that used in fresh weight determination were used to determine the chlorophyll content of resistant and sensitive biotypes of *E. crus-galli* after 14 days of treatment with fenoxaprop-p-ethyl at the level applied in the real field conditions. Chlorophyll content of untreated controls was measured after 14 days also. Moreover, chlorophyll content of resistant treated and untreated biotypes were remeasured after 21 days of treatment with fenoxaprop-p-ethyl (treated leaves were regrown again and chlorophyll content increased). No chlorophyll data was taken for the suscible biotype after 21 days due to the plant is completely dead. Chlorophyll A, B, and total were determined in *E. crus-galli* lamina using the method described by Moran and Porath [9]. Data were subjected to statistical analysis of variance according to the method described by K. A. Gomez and A. A. Gomez [10].

2.5. Anatomical Test. The leaf specimens which included the midrib collected from plants differ from that used in fresh weight and chlorophyll determination were taken after 14 days of treatment from the second leaf of the resistant and susceptible biotypes of *E. crus-galli* treated with fenoxaprop-p-ethyl at the recommended dose level (1 fold). Moreover, leaf specimens of resistant treated and untreated biotypes were measured again after 21 days of treatment with fenoxaprop-p-ethyl (treated leaves were recovered again and chlorophyll content increased). No leaf specimens were taken for the susceptible biotype after 21 days due to the plant is completely dead. Specimens were fixed in a formalin, ethyl alcohol, and acetic acid mixture (1 : 18 : 1 v/v). Then specimens were washed and dehydrated in an alcohol series. The dehydrated specimens were infiltrated and embedded in paraffin wax (52–54°C m.p.). The embedded specimens were sectioned using a rotary microtome (Leica RM 2125) to a thickness of 8–10 μm. Sections were mounted on slides and

TABLE 1: Effect of fenoxaprop-p-ethyl on the susceptible and resistant biotypes of *E. crus-galli* expressed as the rates of the herbicide required for 50% reduction of the aboveground biomass (GR_{50}) and estimated resistance ratio.

Weed biotype	GR_{50} gm a.i/ha	b	c	d	R^2	R/S value	P value
Susceptible	3	1.47	0.72	97	0.99	—	<0.05
Resistant	36.21	1.79	1.35	100	0.99	12.07	<0.05

c: the mean response (fresh weight as percent of control) at very high herbicide rate.
d: the mean response (fresh weight as percent of control) at zero herbicide rate.
b: slope of the line.
GR_{50}: herbicide rate to reduce plant growth by 50% relative to untreated control.
R^2: the coefficient of determination.
R/S ratio: the GR_{50} of the resistant biotype divided by the GR_{50} of the susceptible biotype.
P value: the probability of the obtained results.

TABLE 2: Chlorophyll contents in susceptible and resistant biotypes of *E. crus-galli* after 14 days of treatment with fenoxaprop-p-ethyl compared with untreated ones.

Treatments	Chlorophyll pigments (mg/L)		
	A	B	Total
Susceptible (control)	3.001^a	1.785^a	5.211^a
Susceptible + F	1.872^c	0.552^d	2.425^d
Resistance (control)	2.494^b	1.461^b	3.956^b
Resistance + F	1.458^d	1.175^c	2.633^c

*F = fenoxaprop-p-ethyl.
[a,b,c,d] indicate the significance and non-significance between means using Duncan multiple range test.

deparaffinized. Staining was accomplished with safranine and azur II [11], cleared in xylol, and mounted in Canada balsam [12]. Ten readings from 3 slides from different leaves of the same plant were examined with electric microscope (Lieca DM LS) with digital camera (Lieca DC 300) and then photographed. The anatomical manifestation was calculated using Lieca IM 1000 image manager software. Lieca software was calibrated using 1 cm stage micrometer scaled at $100\,\mu m$ increment (Leitz Wetzler, Germany 604364) at a 4 and 10x magnifications.

2.6. Statistical Analysis. Data from the experiments were statistically analyzed using one-way repeated measurement analysis of variance according to the method described by K. A. Gomez and A. A. Gomez [10]. Duncan's multiple range test was used to separate means using SAS software (version 6.12, SAS Institute Inc., and Cary, USA).

3. Results

3.1. E. crus-galli Resistance to Fenoxaprop-p-ethyl by Means of Fresh Weight Reduction of Treated Plants. A dose-response experiment was conducted on whole plants of *E. crus-galli* treated with fenoxaprop-p-ethyl to detect its resistance against this herbicide. The response of the tested susceptible and resistant biotypes against this herbicide was determined as reduction in the fresh weight of the treated plants relative to the control after 14 days of fenoxaprop-p-ethyl

application. The results showed that the rates of fenoxaprop-p-ethyl required for 50% growth reduction were 3 and 36.2 gm a.i./ha for the susceptible and resistant biotypes of *E. crus-galli*, respectively (Table 1). Table 1 revealed that the GR_{50} of *E. crus-galli*-resistant biotype was 12.07-times higher than that required to obtain the same effect on the susceptible biotype.

3.2. Effect of Tested Herbicide on Chlorophyll Content of Susceptible and Resistant Biotypes of E. crus-galli. The chlorophyll content of *E. crus-galli* was measured after 14 days of herbicide application to evaluate the physiological conditions of the tested weed. Table 2 showed that the chlorophyll content after fenoxaprop-p-ethyl application was decreased either in the resistant or susceptible biotypes. The rate of reduction in chlorophyll content was higher in susceptible biotype than the resistant one of *E. crus-galli*. The chlorophyll content was higher in the untreated susceptible biotype of *E. crus-galli* relative to the treated one. The chlorophyll content was slightly higher in the untreated resistant biotype of *E. crus-galli* relative to the treated one. A very important action took place. Chlorophyll content of the resistant biotype treated with fenoxaprop-p-ethyl increased again relative to the untreated resistant plants after 21 days of treatment. This action was due to the re-growth of *E. crus-galli* leaves as shown in Table 3.

3.3. Anatomical Differences between Susceptible and Resistant Biotypes of E. crus-galli against Fenoxaprop-p-ethyl. The anatomical differences between the susceptible and resistant biotypes of *E. crus-galli* treated with fenoxaprop-p-ethyl with respect to lamina thickness and xylem vessel diameter are presented in Table 4 and Figure 1. The results showed that susceptible biotype (SBT) treated with fenoxaprop-p-ethyl had less laminal thickness and tissues intensively stained with azur II compared with the untreated plants. The normal internal leaf structure of treated SBT is more difficult to be identified, which may be due to cell death compared with untreated plants. Laminal thickness and xylem vessel diameter of treated biotypes were reduced compared with the untreated plants, but the lowest value was caused by treated SBT. In contrast, leaf tissues in treated resistant biotype (RBT) seem to be normal and easily identified. Intensively stained cells with azur II, though, were noticed in some local lesions (areas), which may be due to cell death.

TABLE 3: Chlorophyll contents in resistant untreated and resistant treated biotypes of *E. crus-galli* after 21 days of treatment.

Treatments	Chlorophyll pigments (mg/L)		
	A	B	Total
Resistance (control)	3.126[b]	1.927[b]	5.011[b]
Resistance + F	3.320[a]	2.083[a]	5.403[a]

*F = fenoxaprop-p-ethyl.
[a,b]indicate the significance and non-significance between means using Duncan multiple range test.

TABLE 4: Some anatomical parameters in the two sensitive and resistant biotypes of *E. crus-galli*, that is, laminal thickness and vessel diameters after 14 days of treatment with fenoxaprop-p-ethyl compared with untreated ones.

Treatments	Laminal thickness (μm)	Xylem vessels diameter (μm)
Susceptible (control)	130[c]	35[a]
Susceptible + F	34[a]	17[c]
Resistance (control)	137[c]	36[a]
Resistance + F	51[b]	25[b]

*F = fenoxaprop-p-ethyl.
[a,b,c]indicate the significance and non-significance between means using Duncan multiple range test.

TABLE 5: Some anatomical parameters in recovered treated resistant biotypes of *E. crus-galli*, that is, laminal thickness and vessel diameters after 21 days of treatment with fenoxaprop-p-ethyl compared with untreated resistant one.

Treatments	Laminal thickness (μm)	Vessels diameter (μm)
Resistance (control)	138[a]	34[b]
Resistance + F	155[b]	34[a]

*F = fenoxaprop-p-ethyl.
[a,b]indicate the significance and non-significance between means using Duncan multiple range test.

Concerning of recovery in RBT of *E. crus-galli* treated with fenoxaprop-p-ethyl, relative to the untreated one, data in Table 5 and Figure 2 indicated that lamina thickness was increased up to untreated RBT. Furthermore, no differences in xylem vessel diameter were found between treated and untreated RBT.

4. Discussion

The resistance of *E. crus-galli* to fenoxaprop-p-ethyl (ACCase inhibitor) was identified in this study and confirmed the occurrence of *E. crus-galli* resistance to fenoxaprop-p-ethyl in Egypt. This finding had been reported previously outside Egypt [3, 13–17]. The results of this study also implied that the physiological and anatomical dereferences as well as growth reduction help to identify the occurrence of resistant weed.

Chlorophyll content has been known as a typical parameter for evaluating the physiological conditions in common sense. The reduction in chlorophyll content of *E. crus-galli* resistant and susceptible biotypes after foliar application

of fenoxaprop-p-ethyl in this study is in agreement with the findings of [18, 19] who reported that the application of fenoxaprop-p-ethyl as ACCase-inhibitor leads to injury symptoms in the form of chlorosis (reduction in the chlorophyll content). The reduction in chlorophyll content after foliar application of fenoxaprop-p-ethyl is likely to be due to its incorporation into the cell membrane function through physiological processes, such as depolarization of membrane potential [20] which is not clarified yet. Subsequently, this makes the translocation of fenoxaprop-p-ethyl become more difficult and thus, relatively large amount of the herbicide retained in the treated leaf tissue and reduce the chlorophyll content [21]. Moreover, this reduction may be due to the enhanced activity of chlorophyll degrading enzyme chlorophyllase and/or disruption of the fine structure of chloroplast and instability of chloroplast or pigment-protein complex, which leads to oxidation of chlorophyll and decreased its concentration.

The results also showed that the reduction in chlorophyll content in resistant biotype treated with fenoxaprop-p-ethyl was lower than that of susceptible biotype of *E. crus-galli*. Furthermore, the results indicated that the chlorophyll content of the treated resistant biotype of *E. crus-galli* again increased more than the untreated one after 21 days of fenoxaprop-p-ethyl application. The possible mechanism of lower reduction or reincrease of chlorophyll content in the resistant biotype of *E. crus-galli* relative to the susceptible one, may be due to the relatively faster metabolism of fenoxaprop-p-ethyl through glycosylation [22] which occurs relatively high in the resistant biotype. That is, relatively small amount of the compound was retained in the treated leaf tissue and, thus, photosynthesis in the treated leaf was able to occur continuously [21]. The accumulation of carbohydrates that has been found in the leaves treated with ACCase inhibitors [23, 24] support, this point of view. Therefore, even if the plant is treated with a rate close to the lethal dose, the treated plants are still alive and likely to regrow when the phytotoxic compound is degraded below the physiologically active concentration [21].

In this study, there were anatomical differences between resistant and susceptible biotypes of *E. crus-galli* treated with fenoxaprop-p-ethyl with respect to leaf lamina thickness and xylem vessel diameter. Moreover, lamina thickness of treated RBT was increased up to untreated RBT. Moreover, no differences in xylem vessel diameter were found between treated and untreated RBT of *E. crus-galli*. Reduction in leaf lamina thickness in sensitive biotype of *E. crus-galli* treated with fenoxaprop-p-ethyl was a reflection of the decrease in mesophyll cells. The decrease in mesophyll cells may be attributed to inhibition of cell division and/or cell enlargement which subsequently may be due to the disruption in plasma membrane that mainly consists of phospholipids. Therefore, any reduction in fatty acids biosynthesis due to fenoxaprop-p-ethyl application that is known as fatty acids synthesis inhibitor [25–27] will affect the membrane formation and subsequently its functions such as cell division and/or cell enlargement. The inhibition in cell division and/or ell enlargement resulted in reduction in mesophyll cells and subsequently in leaf laminal thickness.

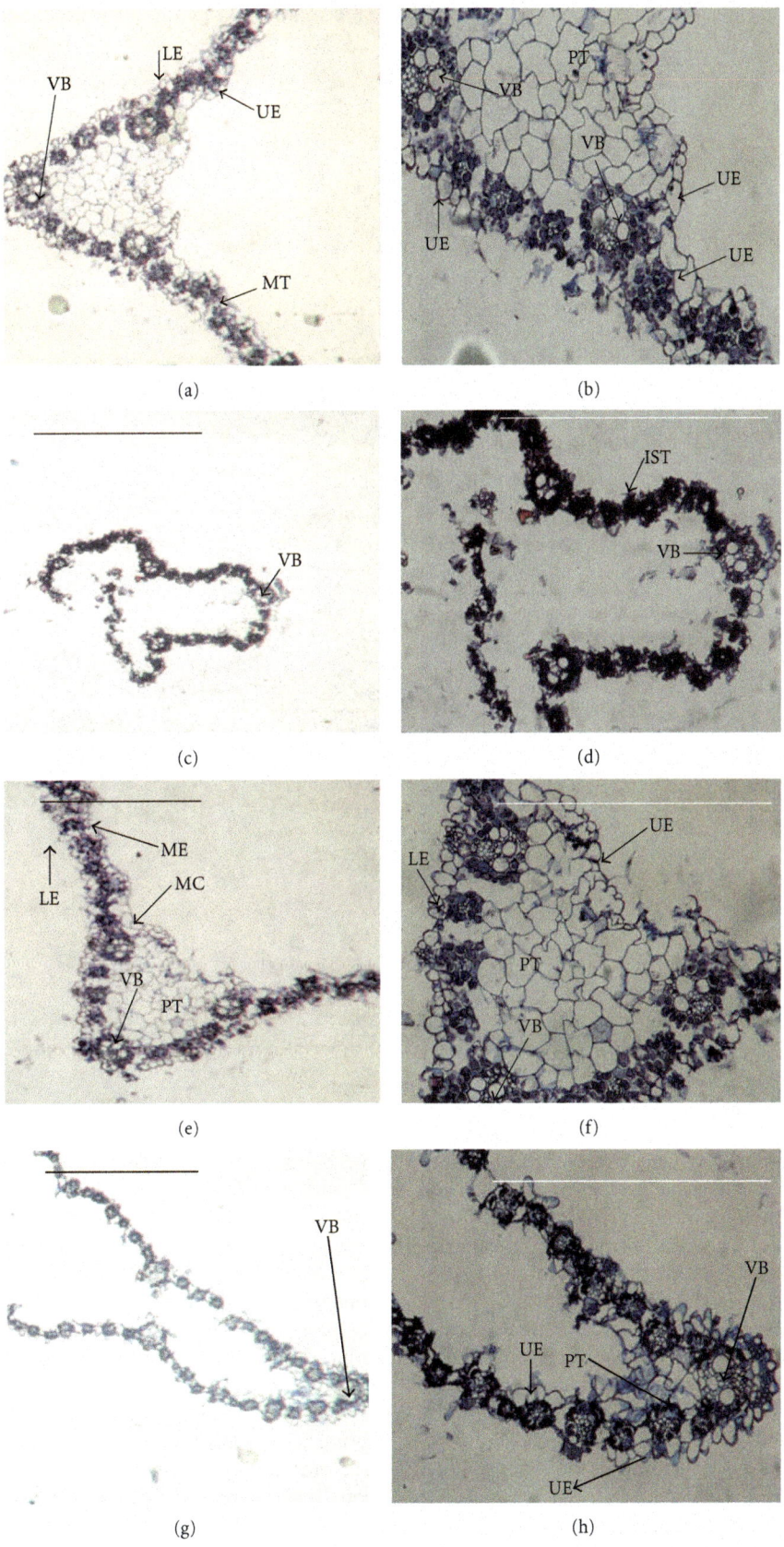

FIGURE 1: Cross-sections through the lamina of untreated susceptible biotype (a, b), treated susceptible biotype (c, d), untreated resistant biotype (e, f), and resistant biotype treated with fenoxaprop-p-ethyl (g, h) of *E. crus-galli* after 14 days. *Upper epidermis (UE), lower epidermis (LE), parenchyma tissue (PT), mesophyll tissue (MT), motor cells (MC), vascular bundle (VB), and intensive stained tissue (IST); Bar = 500 μm.

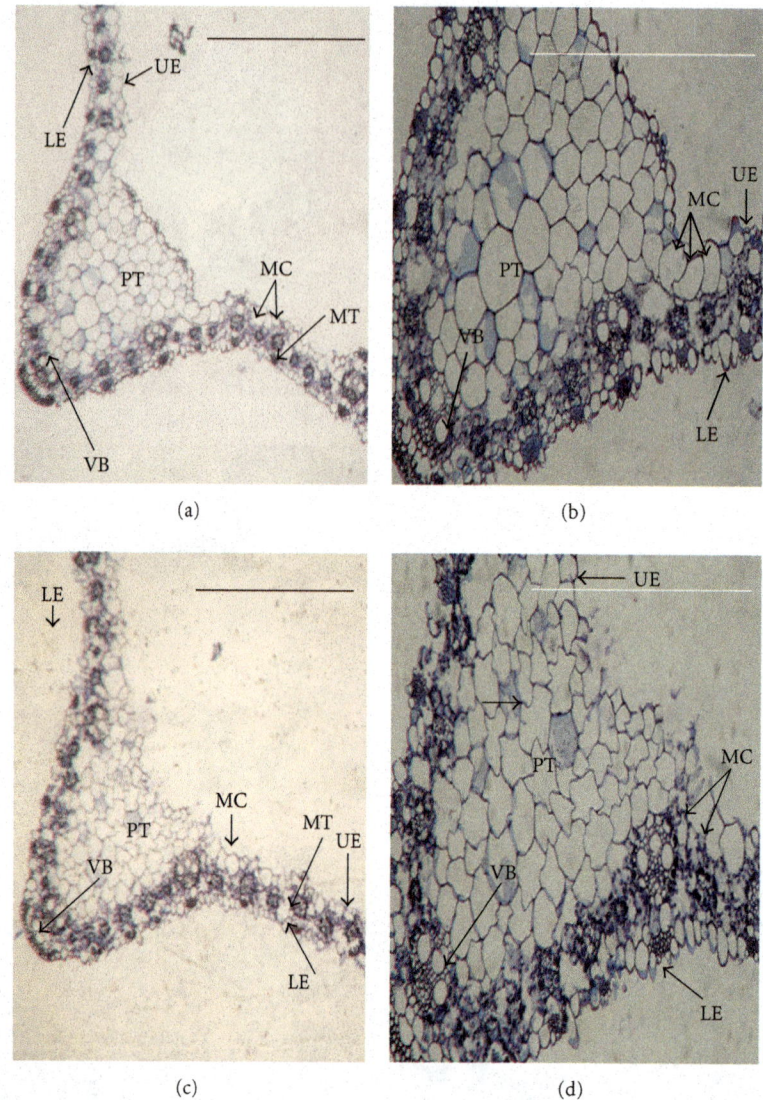

FIGURE 2: Cross sections through the lamina of untreated resistant (a, b) and treated resistant with fenoxaprop-p-ethyl biotypes (c, d) of *E. crus-galli* after 21 days.

From all previous data, fatty acids are critical components of cell membranes, therefore, reduction or inhibition of fatty acids biosynthesis will affect chlorophyll content and plasma membrane functions. Since that fenoxaprop-p-ethyl is known as fatty acids biosynthesis inhibitor [25–28], its foliar application lead to reduction in chlorophyll content and disruption in plasma membrane functions and subsequently the photosynthesis, number of mesophyll cells, and growth indicators such as growth reduction fifty. All of these parameters were recorded in this study for the treated sensitive biotype of *E. crus-galli* with fenoxaprop-p-ethyl. However, for the resistant biotype of *E. crus-galli* a slight reduction was recorded relative to the sensitive one and this may be due to the fact that the reduction in fatty acids biosynthesis was much lower than that of sensitive biotype.

Therefore, the resistance mechanism of *E. crus-galli* to the fenoxaprop-p-ethyl may be conferred by two proposed

mechanisms. Firstly, the mechanism may be due to an alteration in the gene(s) of target site enzyme (ACCase) protein likely the mechanism that confers resistance of *E. crus-galli* to the fenoxaprop-p-ethyl herbicide. The alteration or changes in the protein of ACCase enzyme (the target of fenoxaprop-p-ethyl) in the resistant biotype compared to the susceptible one, induced low affinity of fenoxaprop-p-ethyl herbicide to bind with the target enzyme and the enzyme became insensitive to the herbicide. Moreover, it is apparent that the altered sensitivity of ACCase of the resistant *Echinochloa* sp to fenoxaprop-p-ethyl may be due to a mutation of the target site enzyme which does not affect equally the bending of the other aryloxyphenoxypropionate [26]. Similarly, the relatively high growth reduction dose of the resistant *E. crus-galli* biotype in this study to the herbicide, chlorophyll content and anatomical differences of the two *E. crus-galli* biotypes provides additional support to this proposed mechanism of resistance.

The second mechanism of *E. crus-galli* resistant to fenoxaprop-p-ethyl may be due to the relatively faster metabolism of fenoxaprop-p-ethyl through glycosylation [22] occurs relatively high in the resistant biotype. That is which relatively small amount of the compound was retained in the treated leaf tissue and, thus, photosynthesis in the treated leaf was able to occur continuously [21]. There are some reports that revealed that an accumulation of carbohydrates has been found in the leaves treated with ACCase inhibitors [23, 24]. Therefore, even if the plant is treated with a rate close to the lethal dose, the treated plants are still alive and likely to re-grow when the phytotoxic compound is degraded below the physiologically active concentration which may be due to the faster metabolism of fenoxaprop-p-ethyl. This proposed mechanism was in agreement with chlorophyll data in this study. Both proposed mechanisms to fenoxaprop-p-ethyl and other ACCase inhibitors had been reported before for another weed [28]; however, against *E. crus-galli* based on physioanatomical differences, this is considered to be the first report.

5. Conclusions

There were significant differences between susceptible and resistant biotypes of *E. crus-galli* treated with fenoxaprop-p-ethyl with respect to chlorophyll content, growth reduction, and anatomical test. These differences concluded the ability to identify the occurrence of *E. crus-galli* resistant to fenoxaprop-p-ethyl and assumed that the resistance mechanism was explained either by target site insensitivity or by an enhanced rate of metabolism.

References

[1] M. L. Naples and P. J. A. Kessler, *Weeds of rain fed lowland rice fields of Laos & Cambodia*, M.S. thesis, University of Leiden, Cambodia, 2005.

[2] Weed Science Society of America, *Herbicide Handbook*, 9th edition, 2007.

[3] A. J. Fischer, C. M. Ateh, D. E. Bayer, and J. E. Hill, "Herbicide-resistant *Echinochloa oryzoides* and *E. phyllopogon* in California *Oryza sativa* fields," *Weed Science*, vol. 48, no. 2, pp. 225–230, 2000.

[4] M. D. Osuna, F. Vidotto, A. J. Fischer, D. E. Bayer, R. De Prado, and A. Ferrero, "Cross-resistance to bispyribac-sodium and bensulfuron-methyl in *Echinochloa phyllopogon* and *Cyperus difformis*," *Pesticide Biochemistry and Physiology*, vol. 73, no. 1, pp. 9–17, 2002.

[5] R. Busi, F. Vidotto, and A. Ferrero, "Resistance patterns to ALS-inhibitors in *Cyperus difformis* and *Schoenoplectus mucronatus* (abstract)," in *Abstract Book of the 4th International Weed Science Congress*, 2004.

[6] A. M. Hamza, *Evoloution and resistance mechanism of some rice weeds against some herbicides*, Doctoral thesis, Faculty of Agriculture, Kafr-El Sheikh University, Kafrelsheikh, Egypt, 2009.

[7] J. C. Streibig, M. Rudemo, and J. E. Jensen, "Dose response curves and statistical models," in *Herbicides Bioassays*, J. C. Streibig and P. Kudsk, Eds., pp. 30–55, CRC, Boca Raton, Fla, USA, 1993.

[8] S. S. Seefeldt, J. E. Jensen, and E. P. Feurst, "Log-logistic analysis of herbicide dose-response relationships," *Weed Technology*, vol. 9, no. 2, pp. 218–227, 1995.

[9] R. Moran and D. Porath, "Chlorophyll determination in intact tissues using N, N-Dimethyl formamide," *Plant Physiology*, vol. 69, pp. 1370–1381, 1980.

[10] K. A. Gomez and A. A. Gomez, *Statistical procedures for agricultural research*, John Wiley & Sons, New York, NY, USA, 2nd edition, 1984.

[11] M. Gutmann, "Improved staining procedures for photographic documentation of phenolic deposits in semithin sections of giant tissue," *Journal of Microscopy*, vol. 179, no. 3, pp. 277–281, 1995.

[12] S. E. Ruzin, *Plant Microtechniques and Microscopy*, Oxford University Press, New York, NY, USA, 1st edition, 1999.

[13] J. Gressel, "More non-target site herbicide cross-resistance in *Echinochloa* spp. in rice," *Residential Pest Management*, vol. 11, pp. 6–7, 2000.

[14] J. P. Ruiz-Santaella, Y. Bakkali, A. J. Fischer, and R. De Prado, "Is it possible to detect *Echinochloa* spp. tolerance to ACCase-inhibiting herbicides using a simple quick tolerance test?" *Communications in Agricultural and Applied Biological Sciences*, vol. 68, no. 4, pp. 331–334, 2003.

[15] J. P. Ruiz-Santaella, A. J. Fisher, and R. De Prado, "Alternative control of two biotypes of *Echinochloa phyllopogon* susceptible and resistant to fenoxaprop-ethyl," *Communications in Agricultural and Applied Biological Sciences*, vol. 68, no. 4, pp. 403–407, 2003.

[16] D. S. Kim, J. C. Caseley, P. Brain, C. R. Riches, and B. E. Valverde, "Rapid detection of propanil and fenoxaprop resistance in *Echinochloa colona*," *Weed Science*, vol. 48, no. 6, pp. 695–700, 2000.

[17] B. E. Valverde and K. Itoh, "World rice and herbicide resistance," in *Herbicide Resistance and World Grains*, S. B. Powles and D. L. Shaner, Eds., pp. 195–249, CRC Press, Boca Raton, Fla, USA, 2001.

[18] D. W. Lycan and S. E. Hart, "Cool-season turfgrass response to bispyribac-sodium," *HortScience*, vol. 40, no. 5, pp. 1552–1555, 2005.

[19] W. P. Anderson, *Weed Sciences: Principles and Application*, West Publishing, Minneapolis, Minn, USA, 3rd edition, 1996.

[20] J. P. Wright, "Use of membrane potential measurements to study mode of action of diclofop-methyl," *Weed Science*, vol. 42, pp. 285–292, 1994.

[21] J. S. Kim, J. I. Oh, T. J. Kim, Y. P. Jong, and Y. C. Kwang, "Physiological basis of differential phytotoxic activity between fenoxaprop-P-ethyl and cyhalofop-butyl-treated barnyardgrass," *Weed Biology and Management*, vol. 5, no. 2, pp. 39–45, 2005.

[22] H. Hosaka, H. Inaba, A. Satoh, and H. Ishikawa, "Morphological and histological effects of sethoxydim on corn (*Zea mays*) seedlings," *Weed Science*, vol. 32, pp. 711–721, 1984.

[23] H. Hosaka and M. Takagi, "Effects and absorption of sethoxydim in cell cycle progression of corn (*Zea mays*) and pea (*Pisum sativum*)," *Plant Physiology*, vol. 99, no. 4, pp. 1650–1656, 1992.

[24] A. Tal, M. L. Romano, G. R. Stephenson, A. L. Schwan, and J. C. Hall, "Glutathione conjuction: a detoxification pathway for fenoxaprop-p-ethyl in barley ,crabgrass , oat and wheat," *Pesticide Biochemistry and Physiology*, vol. 46, pp. 190–199, 1993.

[25] Y. I. Kuk, J. Wu, J. F. Derr, and K. K. Hatzios, "Mechanism of fenoxaprop resistance in an accession of smooth crabgrass

(*Digitaria ischaemum*)," *Pesticide Biochemistry and Physiology*, vol. 64, no. 2, pp. 112–123, 1999.

[26] Y. I. Kuk, J. Wu, J. F. Derr, and K. K. Hatzios, "Mechanism of fenoxaprop resistance in an accession of smooth crabgrass (*Digitaria ischaemum*)," *Pesticide Biochemistry and Physiology*, vol. 64, no. 2, pp. 112–123, 1999.

[27] D. S. Volenberg and D. E. Stoltenberg, "Giant foxtail (*Setaria faberi*) outcrossing and inheritance of resistance to acetyl-coenzyme A carboxylase inhibitors," *Weed Science*, vol. 50, no. 5, pp. 622–627, 2002.

[28] M. S. Yun, Y. Yogo, R. Miura, Y. Yamasue, and A. J. Fischer, "Cytochrome P-450 monooxygenase activity in herbicide-resistant and -susceptible late watergrass (*Echinochloa phyllopogon*)," *Pesticide Biochemistry and Physiology*, vol. 83, no. 2-3, pp. 107–114, 2005.

Physiological and Growth Responses of Six Turfgrass Species Relative to Salinity Tolerance

Md. Kamal Uddin,[1] Abdul Shukor Juraimi,[2] Mohd. Razi Ismail,[1] Md. Alamgir Hossain,[1] Radziah Othman,[3] and Anuar Abdul Rahim[3]

[1] Institute of Tropical Agriculture, Universiti Putra Malaysia, Selangor, 43400 Serdang, Malaysia
[2] Department of Crop Science, Universiti Putra Malaysia, Selangor, 43400 Serdang, Malaysia
[3] Department of Land Management, Universiti Putra Malaysia, 43400 Serdang, Malaysia

Correspondence should be addressed to Md. Kamal Uddin, mkuddin07@yahoo.com

Academic Editor: Pablo Abbate

The demand for salinity-tolerant turfgrasses is increasing due to augmented use of effluent or low-quality water (sea water) for turf irrigation and the growing turfgrass industry in coastal areas. Experimental plants, grown in plastic pots filled with a mixture of river sand and KOSAS[R] peat (9 : 1), were irrigated with sea water at different dilutions imparting salinity levels of 0, 8, 16, 24, 32, 40, or 48 dS m^{-1}. Salinity tolerance was evaluated on the basis of leaf firing, shoot and root growth reduction, proline content, and relative water content. *Paspalum vaginatum* was found to be most salt tolerant followed by *Zoysia japonica* and *Zoysia matrella*, while *Digitaria didactyla*, *Cynodon dactylon* "Tifdwarf," and *Cynodon dactylon* "Satiri" were moderately tolerant. The results indicate the importance of turfgrass varietal selection for saline environments.

1. Introduction

Salinity is a major abiotic environmental stress that is reported to be responsible for reducing plant growth across the globe. Sea water intrusion, in coastal states, has imposed salinity problems in turfgrass culture [1, 2]. Sodium chloride (NaCl) is the major compound contributing salinity in soils, and more salt-tolerant turfgrasses are required to cope this problem [3]. Therefore, development of salt-tolerant turfgrasses is becoming increasingly necessary in many parts of the world including Malaysia. Salt accumulation in soils, limitations on use of groundwater, and salt water intrusion into groundwater may restrict cultivation of glycophytic crops in these areas [4]. Salinity lowers water potential and restricts of water to plants [5]. Presence of excessive salt (NaCl) outside the cell can induce an osmotic stress, which may adversely affect the plant growth [6]. Hence, osmotic balance or osmoregulation is certainly a crucial factor for the survival of a plant under salt-stressed conditions. Generally, plants have developed different adaptive mechanisms to mitigate salinity under the saline environments [7–9]. Among these, salt exclusion is considered to be the most important adaptive feature of nonhalophytic plants, whilst most tolerant halophytes are salt accumulators [5]. Salt-accumulating halophytes are very crucial for osmotic adjustment. It could be achieved in the following ways: (i) by accumulating inorganic osmolyte (K$^+$) and/or (ii) accumulating organic osmolytes such as proline. Therefore, salt-tolerant halophytic plants have the capability to minimize the detrimental effects by morphological means and physiological or biochemical processes [10].

Some of the turfgrass species are halophytic in nature. So salt-tolerant turf varieties would allow landscape development in saline environments and would be ideal in such environments, where limited or no fresh water is available for irrigation and salt water is the only option for irrigation practices. In addition, the use of sea water is also a good strategy for weed control in seashore *paspalum* worldwide.

TABLE 1: Turfgrass species used in this study.

Scientific name	Common name	Salt tolerance
Paspalum vaginatum Sw.	Seashore paspalum	Salt tolerant
Zoysia japonica Steud.	Japanese lawn grass	Salt tolerant
Zoysia matrella (L.) Merrill	Manila grass	Salt tolerant
Cynodon dactylon x. *Cynodon transvaalensis*.	Hybridbermuda grass (Satiri)	Medium salt tolerant
Cynodon dactylon x. *Cynodon transvaalensis*.	Hybridbermuda grass (Tifdwarf)	Medium salt tolerant
Digitaria didactyla Willd.	Serangoon grass	Medium salt tolerant

The native bermudagrass (*Cynodon dactylon*) here is quite salt-tolerant and grows vigorously, other salt-tolerant turfgrass species may also grow in the saline environments. In our previous reports [11, 12], several turfgrass species were identified in the coastal areas of Malaysia. Interestingly, the development of turfgrass industry especially in the coastal areas of Malaysia is an emerging field. To the best of our knowledge, published literatures are very scanty on salt tolerance studies in turfgrass species, which have been or being conducted in Malaysia. Therefore, this study was framed to determine the relative salinity tolerance and growth response of six important turfgrass species to salinity.

2. Materials and Methods

Glasshouse experiments were conducted at Faculty of Agriculture, University Putra Malaysia. Plastic pots (14×15 cm) were filled up with sandy soil (a mixture of river sand and peat; $9:1$, v/v). The sandy soil had electrical conductivity (EC) 0.3 dS m^{-1}, organic carbon 0.69%, sand 97.93%, silt 1.89%, and clay 0% with pH 5.23. The glasshouse temperature, relative humidity, and light intensity in morning time were $32°C$, 80%, and 110 micromol m^{-2} s^{-1}, and after noon $36°C$, 70%, and 175 micromol m^{-2} s^{-1}, respectively. The temperature was measured using a laboratory thermometer, and light intensity was monitored using a heavy duty light meter (Extech model 407026). Based on earlier findings of [13, 14], the three most salt-tolerant and three medium salt tolerant turfgrass species (Table 1) were used in this study.

The native soil was washed off the sods, and the sods were then transplanted into the plastic pots and grown for 8 weeks under nonsaline irrigation to achieve full growth. Three plants were transplanted in each pot. All species were narrow leaf and were clipped weekly at a cutting height of 5 mm. After 8 weeks thereafter, salinity treatments were initiated. Salinity treatments of 0, 8, 16, 24, 32, 40, and 48 dS m^{-1} (sea water) were applied. The control grasses were irrigated with distilled water. Sea water was diluted by adding distilled water to achieve different treatments. To avoid salinity shock, salinity levels were increased gradually by 8 dSm^{-1} day^{-1} for each treatment until the final salinity levels were achieved. After that, irrigation water was applied daily upto four weeks. The amount of water applied was 200 mL per pot. Data on leaf firing, proline, chlorophyll, relative water content, shoot and root dry weight were recorded 4 weeks after application of salinity treatment.

2.1. Determination of Leaf Firing. Leaf firing was estimated as total percentage of chlorotic leaf area, with 0% corresponding to no leaf firing and 100% for total brown leaves [15].

2.2. Determination of Shoot and Root Dry Weight. At the end of experiment (four weeks after salt initiation), shoots above the soil surface were harvested and washed with tap water and then distilled water to remove all soil particles. After harvesting the shoots, roots were removed from the soil, washed with tap water, and rinsed with distilled water. The shoot and root samples were then oven-dried to a constant weight at 70°C for 3 days. The dry weight (g/plant) was recorded for each treatment.

2.3. Determination of Proline Content. Proline was estimated following method of [16]. Fresh leaf tissue (0.5 g) was homogenized in 10 mL of 3% sulfosalicylic acid, and the homogenate was filtered through Whatman no. 2 filter paper. Two milliliters of the filtrate were brought to reaction with 2 mL acid ninhydrin solution (1.25 g ninhydrin in 30 mL glacial acetic acid), 20 mL orthophosphoric acid (6 M), and 2 mL of glacial acetic acid for 1 h at 100°C. The reaction was terminated in an ice bath. The reaction mixture was extracted with 4 mL toluene, mixed vigorously by passing a continuous stream of air for 1-2 min. The chromophore containing toluene was aspirated from the aqueous phase, warmed at room temperature, and the absorbance was recorded spectrophotometrically (Model UV-3101PC, UV-VIS NIR) at 520 nm. The proline concentration was determined from a standard curve and calculated on fresh weight basis as follows:

$$\mu\text{mol proline g}^{-1} \text{ fresh weight}$$
$$= \frac{\mu\text{g proline mL}^{-1} \times \text{mL of toluene}/115.5}{\text{g of sample}}. \tag{1}$$

2.4. Determination of Chlorophyll Content. Chlorophyll content was estimated following method of [17]. Fresh leaves, from each pot, were cut into small pieces using a scissors and 200 mg of cut leaves were transferred into a plastic vial containing 20 mL of 80% acetone. The vial was quickly corked airtight and kept in the dark for 72 h. Absorbance of the solution was recorded at 645 and 663 nm spectrophotometrically (Model UV-3101PC, UV-VIS NIR). Chlorophyll

TABLE 2: Main effect and interaction effect on different variables by salinity and species.

Variable	Salinity	Species	Salinity × species
Leaf firing	1665.78***	513.16***	75.83***
Shoot dry weight	95.82***	1317.65***	4.01***
Root dry weight	79.83***	287.54***	1.15 ns
Proline	2176.10***	585.87***	58.07***
Relative water content	78.07***	13.85***	1.45 ns
Chlorophyll-a	30.03***	152.19***	0.89 ns
Chlorophyll-b	67.91***	78.03***	4.20***
Total chlorophyll	65.86***	206.75***	2.13***

Numbers are F values significant at ***$P < 0.0001$, ns: not significant.

content was estimated and expressed as mg g^{-1} of sample using the following formulae:

Chlorophyll a content (mg/g fresh leaf)

$$= \frac{12.7(A_{663}) - 2.69(A_{645})}{1000} \times \frac{V}{W},$$

Chlorophyll b content (mg/g fresh leaf)

$$= \frac{22.9(A_{645}) - 4.86(A_{663})}{1000} \times \frac{V}{W}, \qquad (2)$$

Total chlorophyll content (mg/g fresh leaf)

$$= \frac{20.2(A_{645}) + 8.02(A_{663})}{1000} \times \frac{V}{W},$$

where A_{645} and A_{663} represent absorbance of solution at 645 and 663 nm, respectively, V: volume of the solution in mL, W: weight of fresh leaf sample in gram, 12.7, 2.69, 22.9, 4.86, 20.2, and 8.02 are absorption coefficients.

2.5. Determination of Relative Water Content. Relative water content (RWC) was determined as described by [18] on leaf tissues excised in the morning (around 9.00 am). Excised leaves from each pot (0.2 g) were measured for fresh weight (FW), and leaf samples were rehydrated in a water-filled petri dish for 4 h at room temperature. Turgor weight (TW) was measured by allowing full rehydration, removing all water from leaf surface, and weighing. Leaf dry weights were recorded after oven drying for one week at 60°C. The leaf relative water content was determined using the following formula:

$$\text{RWC} = \frac{\text{Fresh weight} - \text{Dry weight}}{\text{Fully turgid weight} - \text{Dry weight}} \times 100. \qquad (3)$$

2.6. Root Histology Using Scanning Electron Microscopy. Roots were sampled from two root zones (root tips at 0–50 mm from tip, and mature roots) and were cut into 5 mm portions with a sharp blade. The excised roots were placed in formalin acetic acid (FAA) and vacuumed for 1 h at 650 mm Hg. Specimens were postfixed in 1% osmium tetraoxide for 2 h, dehydrated for 30 min in each graded ethanol series at 30, 50, 70, 90, 95, and 100%, and dried in Baltec CPD 030 critical point dryer apparatus. The tissues were mounted on stubs, coated with gold using auto fine coater (JEOL JFC-1600, Japan) for 20 min, and viewed under a scanning electron microscope (JEOL JSM-5610LV, Japan), at high vacuum and acceleration voltage of 15 kV with a working distance of 23 mm.

2.7. Statistical Analysis. Data were analyzed statistically following randomized complete block design using ANOVA procedure in SAS statistical software (SAS). The treatment means were compared using protected least significant differences (LSD) at 5% level. Data of leaf firing was proportionate, so arcsine square root transformation was done.

3. Results

3.1. Leaf Firing. Interaction of salinity and species had a significant effect on leaf firing (Table 2). Leaf firing (%) increased with increasing salinity in all turfgrass species (Table 3). However, comparatively less salinity injury was recorded in P. vaginatum, Z. japonica, and Z. matrella compared to D. didactyla, C. dactylon "Tifdwarf," and C. dactylon "Satiri" at all salinity levels. There was no injury (0%) recorded in all species up to 16 dS m^{-1} salinity, except for D. didactyla and C. dactylon "Tifdwarf" which showed light injury symptoms of 5 and 8%, respectively. At 24 dS m^{-1}, the highest injury (25%) was recorded in D. didactyla, while the lowest injury of 5% was observed in P. vaginatum. At 32 dS m^{-1}, leaf firing drastically increased to 79 and 75% in D. didactyla and C. dactylon "Tifdwarf," respectively. At the highest salinity level of 48 dS m^{-1}, the least leaf firing was observed in P. vaginatum (15%) followed by Z. japonica (25%) and Z. matrella (39%) compared to 80–100% leaf firing in D. didactyla, C. dactylon "Tifdwarf," and C. dactylon "Satiri." Overall, the highest leaf firing was recorded in D. didactyla, while the lowest in P. vaginatum.

3.2. Shoot Dry Weight. Interaction effect of salinity and species was significant ($P < 0.05$) on shoot dry weight (Table 2). Shoot dry weights (SDWs) of turfgrass species decreased as the level of salinity increased (Figure 1). Results showed that P. vaginatum was the most salt-tolerant species

TABLE 3: Effect of salinity on leaf firing of six turfgrass species.

EC_w (dS m^{-1})	Turfgrass species (% leaf firing)						
	P. vaginatum	*Z. japonica*	*Z. matrella*	*D. didactyla*	*C. dactylon* "Tifdwarf"	*C. dactylon* "Satiri"	LSD (0.05)
0	0 e (0.28)	0 e (0.28)	0 e (0.28)	0 f (0.28)	0 f (0.28)	0 e (0.28)	0.00
8	0 e (0.28)	0 e (0.28)	0 e (0.28)	0 f (0.28)	0 f (0.28)	0 e (0.28)	0.00
16	0 e (0.28)	0 e (0.28)	0 e (0.28)	5 e (12.79)	8 e (16.37)	0 e (0.28)	2.45
24	5 d (12.89)	10 d (18.26)	15 d (22.65)	25 d (29.90)	18 d (25.01)	15 d (22.65)	4.31
32	8 c (16.37)	15 c (22.65)	20 c (26.49)	79 c (63.17)	45 c (42.14)	25 c (29.95)	2.57
40	12 b (20.20)	20 b (26.52)	26 b (30.64)	93 b (76.80)	85 b (67.39)	69 b (56.37)	4.32
48	15 a (22.65)	25 a (29.95)	39 a (38.64)	100 a (89.75)	94 a (77.81)	80 a (63.83)	4.52
LSD (0.05)	2.31	2.34	2.30	6.19	4.36	5.11	

Means within columns followed by the same letter are not significantly different at $P = 0.05$ (LSD test).
Values in the parentheses indicate transformed by Arcsine square root.

TABLE 4: Effect of salinity on leaf proline content of six turfgrass species.

EC_w (dS m^{-1})	Turfgrass species (proline contents in mg g^{-1}, fresh weight)						
	P. vaginatum	*Z. japonica*	*Z. matrella*	*D. didactyla*	*C. dactylon* "Tifdwarf"	*C. dactylon* "Satiri"	LSD (0.05)
0	3.33 f	3.60 d	3.67 f	3.55 e	5.60 f	6.35 e	0.96
8	4.60 ef (1.4)	4.07 d (1.1)	4.62 f (1.3)	6.42 e (1.8)	7.25 f (1.3)	10.60 e (1.7)	2.31
16	7.80 ed (2.3)	6.50 d (1.8)	6.02e (1.7)	12.40 d (3.5)	15.35 e (2.7)	29.90 d (4.7)	3.65
24	11.61 d (3.5)	13.10 c (3.6)	9.24 d (2.5)	15.05 d (4.2)	26.35 d (4.7)	52.50 c (8.3)	1.77
32	26.90 c (8.1)	16.25 c (4.5)	11.30 c (3.1)	34.55 c (9.7)	37.57 c (6.7)	66.52 b (10.5)	3.53
40	51.20 b (15.4)	45.82 b (12.7)	25.57 b (7.0)	43.27 b (12.2)	65.15 b (11.2)	71.35 a (11.2)	4.09
48	77.90 a (23.4)	49.62 a (13.8)	43.52 a (12.0)	49.92 a (14.1)	62.57 a (11.6)	74.85 a (11.8)	5.18
LSD (0.05)	4.45	3.26	1.26	3.49	1.93	4.43	

Means within columns followed by the same letter are not significantly different at $P = 0.05$ (LSD test).
Values in the parentheses indicate x-fold increase relative to the control.

being statistically significant with others. At the highest salinity level (48 dS m^{-1}), SDW reduction in *P. vaginatum* was only 23% relative to control treatment. *Zoysia japonica* followed a similar trend as *P. vaginatum* for salinities upto 24 dS m^{-1}. At 48 dS m^{-1}, significantly higher SDW reductions were observed in *D. didactyla* (51%), *C. dactylon* "Tifdwarf" (53%), and *C. dactylon* "Satiri" (44%).

3.3. Root Dry Weight. The results showed that root dry weight (RDW) significantly ($P < 0.05$) decreased with increasing salinity (Figure 2). At 16 dS m^{-1}, a significant difference was noted among the species. However, *P. vaginatum*, *C. dactylon* "Tifdwarf," *Z. japonica*, and *Z. matrella* produced greater RDW than the others at 24 dS m^{-1} salinity. At the highest salinity (48 dS m^{-1}), RDW reduction was least in *P. vaginatum* (34%) followed by *Z. japonica* (46%); while highest in *C. dactylon* "Tifdwarf" (67%) followed by *C. dactylon* "Satiri" (54%), *Z. matrella* (53%), and *D. didactyla* (47%). However, there were nonsignificant effect on root dry matter yield when salinity and species were interacted (Table 2).

3.4. Leaf Proline Content. Proline accumulation in the leaves of all turfgrass species increased with increasing salinity (Table 4). There were two distinct trends in proline accumulation among the species analyzed. In all turfgrass species (except *C. dactylon* "Satiri"), proline accumulation increased gradually up to 24 dS m^{-1} but increased abruptly at 32 and 48 dS m^{-1}. At 48 dS m^{-1}, a significantly higher (23.4-folds over the control) accumulation of proline was observed in *P. vaginatum* compared to in *C. dactylon* "Tifdwarf" (11.6-folds). There was a difference between the grasses with respect to proline accumulation patterns at 32 and 48 dS m^{-1}. On the basis of proline accumulation ability, turfgrass species were ranked as *P. vaginatum* > *Z. matrella* > *D. didactyla* > *Z. japonica* > both of the *C. dactylon* entries. Interaction between salinity and species had also a significant ($P < 0.001$) effect on proline level (Table 7).

3.5. Leaf Relative Water Content (RWC). Interaction effect of salinity and species was not significant for relative water content (Table 5). Relative water content (RWC) of all turfgrass species was significantly ($P < 0.05$) influenced by salinity. As salinity increased, RWC decreased. However,

TABLE 5: Effect of salinity on leaf relative water content of six turfgrass species.

EC_w (dS m^{-1})	Turfgrass species (relative water contents in %, fresh weight)						
	P. vaginatum	Z. japonica	Z. matrella	D. didactyla	C. dactylon "Tifdwarf"	C. dactylon "Satiri"	LSD (0.05)
0	93.16 a	89.48 a	89.89 a	87.33 a	90.85 a	90.18 a	8.83
8	90.24 ab	87.57 a	90.97 a	86.28 a	90.14 a	90.39 a	6.29
16	90.23 ba	85.22 a	86.87 a	84.78 a	85.02 ba	86.19 a	6.64
24	87.84 ba	84.92 a	88.51 a	82.42 a	83.91 ba	78.59 b	9.92
32	86.04 b bc	78.39 b	84.09 a	68.06 b	78.70 b	76.42 b	9.97
40	79.77 dc	72.28 c	73.28 b	63.46 b	65.27 c	64.85 b	9.51
48	78.68 b	66.30 d	64.98 c	55.35 c	62.51 c	57.30 c	8.03
LSD (0.05)	6.49	5.94	8.05	5.85	8.76	6.97	

Means within columns followed by the same letter are not significantly different at $P = 0.05$ (LSD test).

TABLE 6: Effect of salinity on chlorophyll-a concentration of six turfgrass species.

EC_w (dS m^{-1})	Turfgrass species (chlorophyll-a contents in mg g^{-1}, fresh weight)						
	P. vaginatum	Z. japonica	Z. matrella	D. didactyla	C. dactylon "Tifdwarf"	C. dactylon "Satiri"	LSD (0.05)
0	0.49 a	0.40 a	0.36 a	0.30 a	0.49 a	0.57 a	0.070
8	0.47 a	0.39 ab	0.33 ab	0.29 a	0.48 a	0.57 a	0.059
16	0.46 ab	0.39 ab	0.31 abc	0.27 a	0.46 a	0.56 a	0.067
24	0.45 abc	0.38 ab	0.30 abc	0.26 a	0.45 ab	0.55 a	0.080
32	0.42 bc	0.35 bc	0.29 bc	0.20 b	0.40 bc	0.53 a	0.079
40	0.41 c	0.33 cd	0.26 c	0.19 b	0.35 cd	0.45 b	0.061
48	0.40 c	0.30 d	0.24 c	0.12 c	0.31 d	0.41 b	0.065
LSD (0.05)	0.051	0.042	0.066	0.063	0.052	0.077	

Means within columns followed by the same letter are not significantly different at $P = 0.05$ (LSD test).

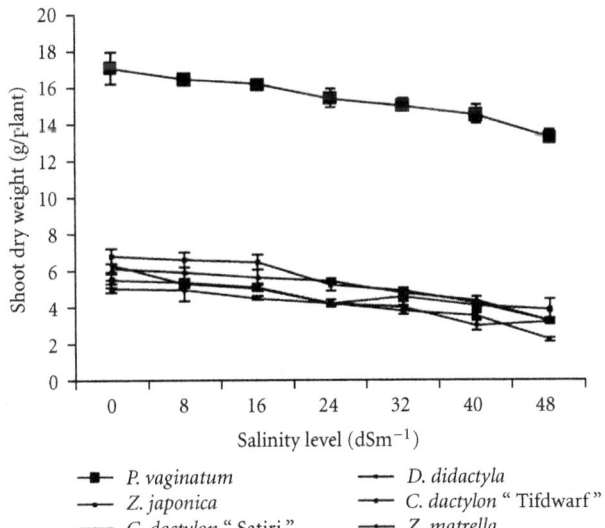

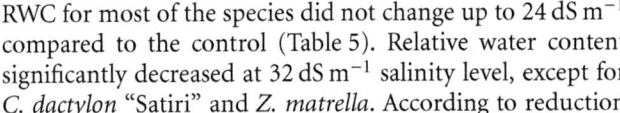

FIGURE 1: Shoot dry weight at different salinity levels of six turfgrass species.

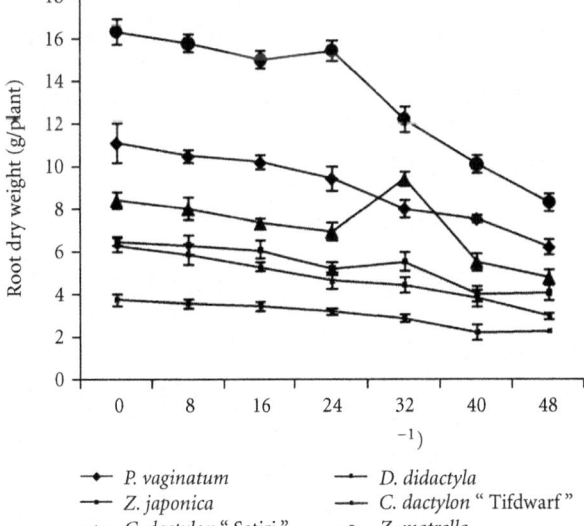

FIGURE 2: Root dry weight at different salinity levels of six turfgrass species.

RWC for most of the species did not change up to 24 dS m^{-1} compared to the control (Table 5). Relative water content significantly decreased at 32 dS m^{-1} salinity level, except for C. dactylon "Satiri" and Z. matrella. According to reduction

in RWC at 48 dS m^{-1} salinity level, species were ranked as D. didactyla (44.6%) > C. dactylon "Satiri" (42.7%) > C.

TABLE 7: Effect of salinity on chlorophyll-b concentration of six turfgrass species.

EC_w (dS m^{-1})	Turfgrass species (chlorophyll-b contents in mg g^{-1}, fresh weight)						
	P. vaginatum	Z. japonica	Z. matrella	D. didactyla	C. dactylon "Tifdwarf"	C. dactylon "Satiri"	LSD (0.05)
0	0.14 a	0.13 a	0.12 a	0.15 a	0.20 a	0.20 a	0.031
8	0.13 ab	0.13 a	0.11 b	0.13 ab	0.19 ab	0.19 ab	0.022
16	0.12 ab	0.10 b	0.10 b	0.13 ab	0.18 b	0.19 ab	0.017
24	0.12 ab	0.10 b	0.10 bc	0.12 ab	0.18 b	0.18 b	0.019
32	0.11 b	0.08 b	0.09 cd	0.10 bc	0.12 c	0.12 c	0.023
40	0.11 b	0.08 b	0.08 de	0.09 bc	0.09 d	0.11 cd	0.020
48	0.12 ab	0.08 b	0.06 e	0.08 c	0.11 cd	0.09 c	0.019
LSD (0.05)	0.024	0.026	0.014	0.033	0.018	0.018	

Means within columns followed by the same letter are not significantly different at $P = 0.05$ (LSD test).

TABLE 8: Effect of salinity on total chlorophyll concentration of six turfgrass species.

EC_w (dS m^{-1})	Turfgrass species (total chlorophyll contents in mg g^{-1}, fresh weight)						
	P. vaginatum	Z. japonica	Z. matrella	D. didactyla	C. dactylon "Tifdwarf"	C. dactylon "Satiri"	LSD (0.05)
0	0.62 a	0.53 a	0.48 a	0.45 a	0.68 a	0.77 a	0.069
8	0.60 a	0.52 a	0.44 ab	0.42 ab	0.67 a	0.76 a	0.065
16	0.59 ab	0.49 a	0.41 b	0.40 ab	0.66 a	0.74 a	0.068
24	0.57 bac	0.48 ab	0.40 bc	0.38 b	0.63 ab	0.73 a	0.075
32	0.53 bc	0.41 bc	0.37 bc	0.29 c	0.52 bc	0.65 b	0.083
40	0.52 c	0.43 dc	0.34 cd	0.29 c	0.45 c	0.54 c	0.064
48	0.52 c	0.38 d	0.31 d	0.20 d	0.42 c	0.52 c	0.068
LSD (0.05)	0.060	0.050	0.079	0.068	0.100	0.074	

Means within columns followed by the same letter are not significantly different at $P = 0.05$ (LSD test).

dactylon "Tifdwarf" (37.5%) > Z. matrella (35.0%) > Z. japonica (33.7%) > P. vaginatum (21.3%).

3.6. Leaf Chlorophyll Content. Interaction effect of salinity and species was not significant for chlorophyll-a content (Table 7). Increasing salinity up to 24 dS m^{-1} did not affect chlorophyll-a content (Table 6). There were also no differences between 40 and 48 dS m^{-1} treatments on chlorophyll-b content, except for D. didactyla. In P. vaginatum, the chlorophyll-b content (0.11 mg g^{-1} FW) at 32 and 40 dS m^{-1} salinity levels was significantly different from other salinity levels (average 0.126 mg g^{-1} FW) (Table 7). In Z. japonica, a significant reduction in chlorophyll-b content was observed at 16 dS m^{-1}, but there were no further reductions with increasing salinity.

Total chlorophyll content decreased under salt stress in different turfgrass species (Table 8). Interaction effect of salinity and species was significant ($P < 0.05$) for total chlorophyll (Table 7). Turf species with higher chlorophyll-a and chlorophyll-b contents, under control conditions, also had higher amounts of total chlorophyll. While C. dactylon "Satiri," C. dactylon "Tifdwarf," and D. didactyla had higher total chlorophyll under normal conditions, P. vaginatum,

and Z. japonica maintained comparatively higher amounts of total chlorophyll under salt stress with marginal reductions compared to other turf species.

3.7. Root Cell Histology. Differences in cell damage to root cortex of turfgrass species were observed. The damage resulted from cell collapse due to salt stress. Cortical cell of P. vaginatum, and Z. japonica did not show cell collapse in 24 and 48 dS m^{-1} salinity treatments (Figures 3 and 4). Zoysia matrella showed less cell collapse at 48 dS m^{-1} salinity treatment (Figure 5). Digitaria didactyla, C. dactylon "Tifdwarf," and C. dactylon "Satiri" showed severe cell collapse at the highest salinity level (48 dS m^{-1}) compared to the control (Figures 6, 7, and 8).

4. Discussion

The six turfgrass species in the present study exhibited a wide range in salinity tolerance in terms of dry matter production (Figures 1 and 2) and organic osmolyte accumulation (Table 4). In Malaysia, such type of research was not conducted ever before. Previously, we identified turfgrass species that were available in Malaysia and studied growth

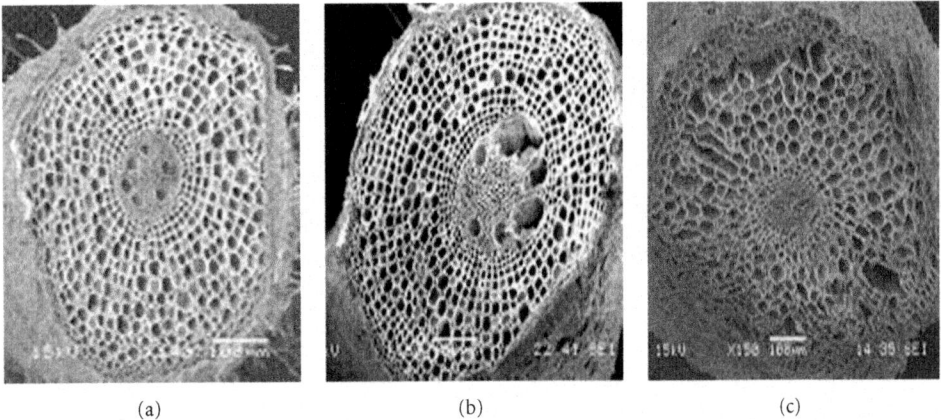

(a)   (b)   (c)

FIGURE 3: Scanning electron microscopy photographs showing root cortical tissue of *Paspalum vaginatum* under (a) 0, (b) 24, and (c) 48 dS m^{-1}.

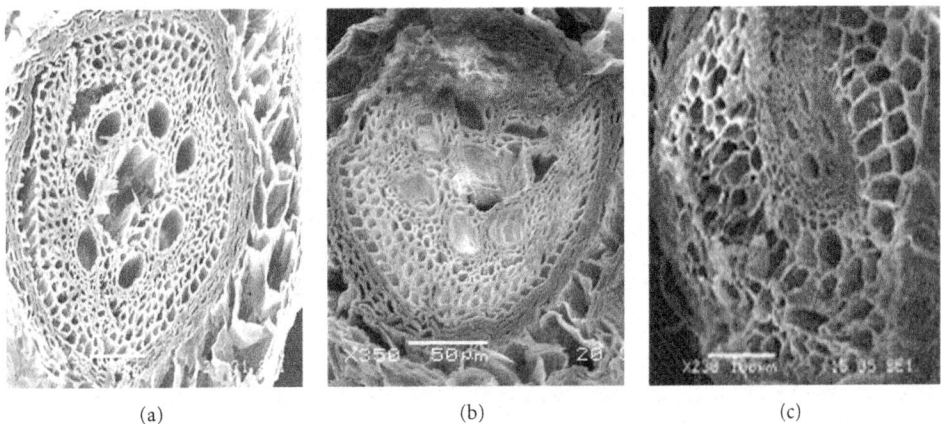

(a)   (b)   (c)

FIGURE 4: Scanning electron microscopy photographs showing root cortical tissue of *Zoysia japonica* under (a) 0, (b) 24, and (c) 48 dS m^{-1}.

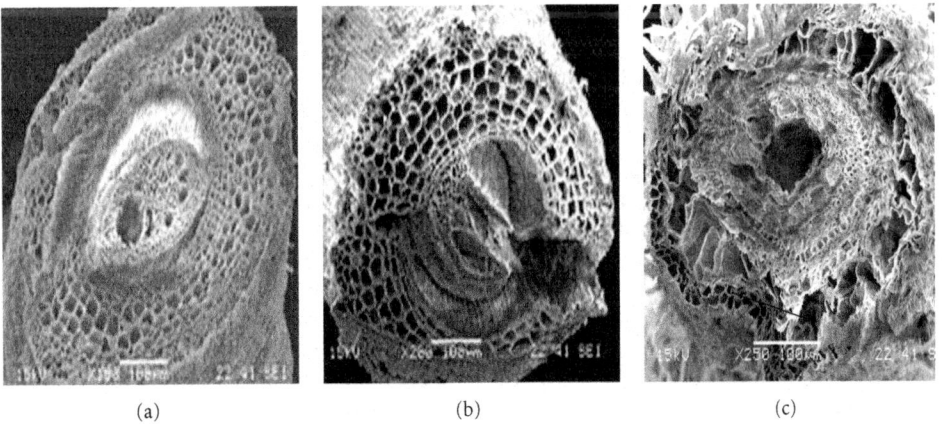

(a)   (b)   (c)

FIGURE 5: Scanning electron microscopy photographs showing root cortical tissue of *Zoysia matrella* under (a) 0, (b) 24, and (c) 48 dS m^{-1}.

performance under salinity-stressed conditions [13, 14]. Throughout the globe, seashore paspalum exhibits a wide range of salinity tolerance among ecotypes [19–22]. A wide intraspecific variation in salinity tolerance has been reported to be as great as the interspecific variations [23]. Several researchers have reported that halophytes, which are ion includers, often adapt to low water potential by accumulation of inorganic solutes to maintain turgor pressure and total water potential [24–26].

Salinity stressed plants certainly face osmotic challenges. This is in agreement with several previous reports [5, 19, 20, 27], which concur that osmotic adjustment is

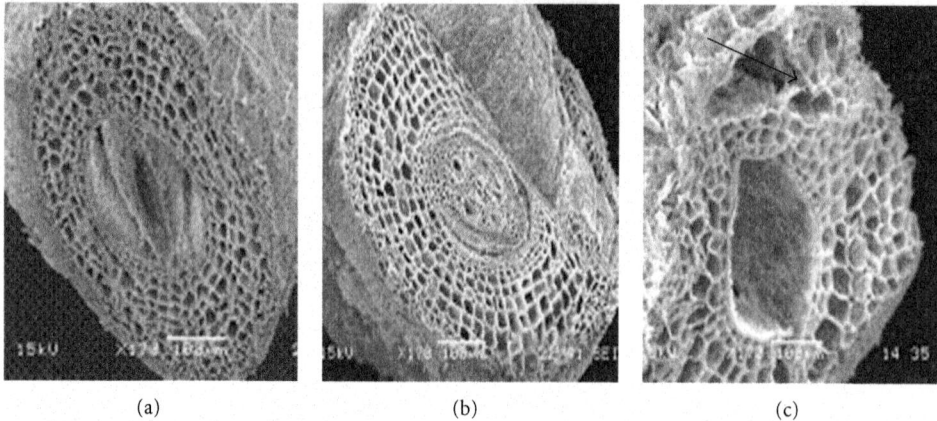

(a) (b) (c)

FIGURE 6: Scanning electron microscopy photographs showing root cortical tissue of *C. dactylon* "Tifdwarf" under (a) 0, (b) 24, and (c) $48 \, dS \, m^{-1}$. The arrow indicates cell damage (c) compared to control (a).

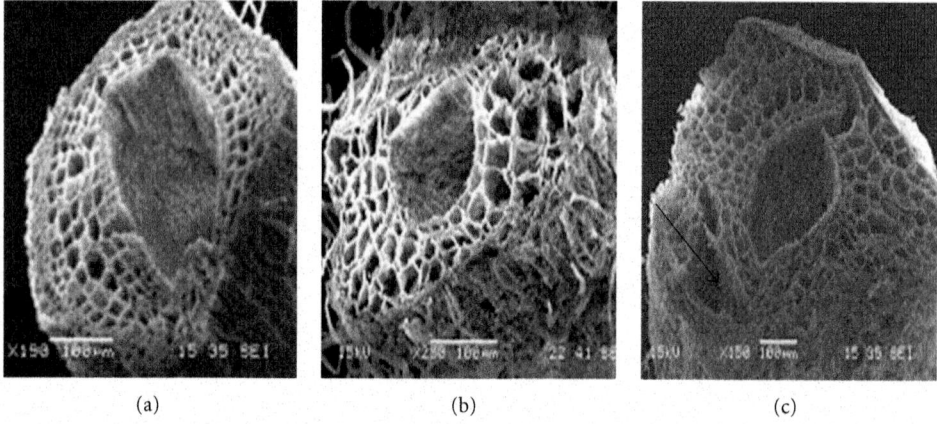

(a) (b) (c)

FIGURE 7: Scanning electron microscopy photographs showing root cortical tissue of *C. dactylon* "Satiri" under (a) 0, (b) 24, and (c) $48 \, dS \, m^{-1}$. The arrow indicates cell damage (c) compared to control (a).

the main mechanism for survival and growth of plants under salinity stress. The percentage relative water content (RWC) was determined as an indicator of osmotic status of turfgrass species studied (Table 5). Halophytes are often able to accumulate high charges of salts in their tissues for osmotic adjustment through the compartmentalization of ions in vacuoles and the production of compatible solutes, or osmotic, in the cytoplasm [27]. Some compatible solutes that show an increase in concentration under salinity stress may also play significant role in osmotic adjustment, and these include proline, glycine betaine, and sugars [28–31]. Glycine betaine and proline protect enzymes (proteins) from damages caused by salinity or dehydration stress [32, 33]. Interestingly, significant proline accumulation generally occurs only after exceeding a threshold of drought or salt stress [30]. In the current study, salinity triggered proline synthesis in response to salinity to turgor maintenance (Table 4). Osmotic adjustment through synthesis of organic compounds has been postulated to have a significant role in salt tolerance in *P. vaginatum* [34]. Our studies indicated that salinity damaged root structure as a result of cortical cell collapse in *C. dactylon* "Tifdwarf," *D. didactyla,* and *C.*

dactylon "Satiri." The structural damage in cortical tissue would interrupt radial water movement in the roots, thus limiting water uptake [35].

Chlorophyll degradation is the primary cause of photosynthetic degeneration/leaf firing and a main biochemical factor for the observed growth reduction [36]. The NaCl-induced decrease in chlorophyll level is widely reported in both glycophytes and halophytes [37–39]. In the present study, the chlorophyll damage was not recorded until $24 \, dS \, m^{-1}$ salinity level and thereafter chlorophyll damage increased with increasing salinity (Tables 6, 7, and 8). The chlorophyll degradation is associated with leaf firing (Table 3). Salinity-induced chlorophyll reduction may be related either to Mg deficiency and/or chlorophyll oxidation since reactive oxygen species (ROS) generation is common in salinity stressed conditions [40]. The chlorophyll-a content of all species decreased much more with increasing salinity (Table 6). However, [41] observed that salinity decreased chlorophyll-b content much more than chlorophyll-a. Chlorophyll content of *P. vaginatum* and *Z. japonica* seem to be insensitive to salinity up to $48 \, dS \, m^{-1}$. This is consistent with the earlier reports for other monocots including rice,

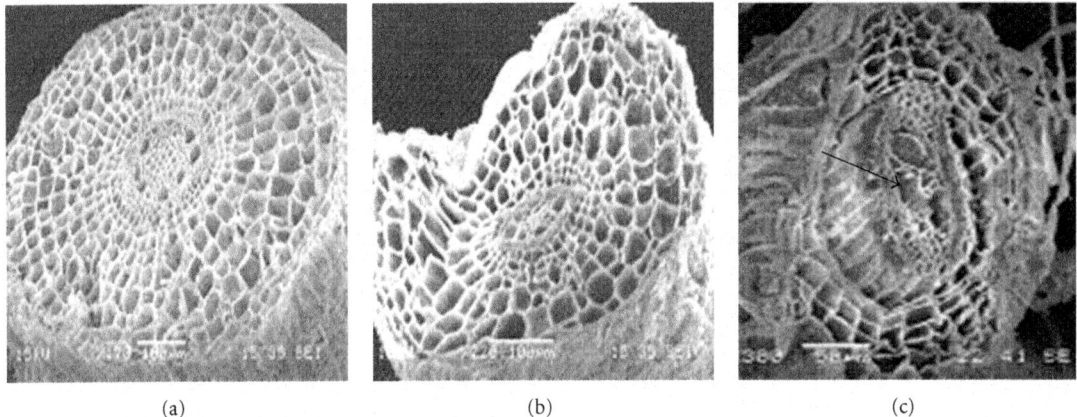

(a)   (b)   (c)

FIGURE 8: Scanning electron microscopy photographs showing root cortical tissue of *Digitaria didactyla* under (a) 0, (b) 24, and (c) 48 dS m^{-1}. The arrow indicates cell damage (c) compared to control (a).

wheat and maize chlorophyll-a by [42–44], chlorophyll-b and total chlorophyll contents decreased with increasing salinity [45], and salt-sensitive rice cultivars had lower chlorophyll content than salt-tolerant rice cultivars [45]. Similar observations were made by [46, 47].

5. Conclusion

The development of turfgrass industry in the coastal areas of Malaysia is challenging due to scarcity of fresh water for irrigation and salt tolerant weed species infestation. Sea water irrigation is a new technology widely used to suppress weed and maintaining the turfgrass growth simultaneously. Appropriate, realistic physiological criteria are essential to define the salinity tolerance and growth responses of turfgrass species. In the present study, salinity tolerance was evaluated on the basis of leaf firing, shoot and root growth reduction, proline content, and relative water content. We observed that *P. vaginatum* was highly salt tolerant at 48 dS m^{-1} followed by *Z. japonica* and *Z. matrella*, while *C. dactylon* "Tifdwarf" was least salt tolerant followed by *D. didactyla* and *C. dactylon* "Satiri." The conclusions are based on responses of six turfgrass species to salinity. Many of the principles can be employed to discuss issues related to development of better direct selection criteria for other turfgrass species.

Acknowledgments

Authors would like to acknowledge funding of this work by Malaysian Government Research Grant (Science Fund 05-01-04 SF0302) and University Putra Malaysia for Graduate Research Fellowship.

References

[1] L. B. McCarty and A. E. Dudeck, "Salinity effects on bentgrass germination," *HortScience*, vol. 28, no. 1, pp. 15–17, 1993.

[2] C. L. Murdoch, "Water the limiting factor for golf course development in Hawaii," *U.S.G.A. Green Section Record*, vol. 25, pp. 11–13, 1987.

[3] A. Harivandi, J. D. Bulter, and L. Wu, "Salinity and turfgrass culture," in *Turfgrass*, D. V. Waddington, R. N. Carrow, and R. C. Shearman, Eds., Agronomy Monograph, pp. 208–230, ASA. CSSA and SSSA, Madison, Wis, USA, 1992.

[4] A. C. Hixson, W. T. Crow, R. McSorley, and L. E. Trenholm, "Saline irrigation affects *Belonolaimus longicaudatus* and *Hoplolaimus galeatus* on seashore paspalum," *Journal of Nematology*, vol. 37, no. 1, pp. 37–44, 2005.

[5] R. Munns and M. Tester, "Mechanisms of salinity tolerance," *Annual Review of Plant Biology*, vol. 59, pp. 651–681, 2008.

[6] K. B. Marcum, "Use of saline and non-potable water in the turfgrass industry: constraints and developments," *Agricultural Water Management*, vol. 80, no. 1–3, pp. 132–146, 2006.

[7] D. Rhodes, A. Nadolska-Orczyk et al., "Salinity, osmolytes and compatible solutes," in *Salinity: Environment-Plants-Molecules*, A. Lauchli and U. Luttge, Eds., Kluwer Academic, Boston, Mass, USA, 2002.

[8] O. Borsani, V. Valpuesta, and M. A. Botella, "Developing salt tolerant plants in a new century: a molecular biology approach," *Plant Cell, Tissue and Organ Culture*, vol. 73, no. 2, pp. 101 115, 2003.

[9] R. K. Sairam, A. Tyagi, and V. Chinnusamy, "Salinity tolerance: cellular mechanisms and gene regulation," in *Plant–Environment Interactions*, B. Huang, Ed., CRC Press, Boca Raton, Fla, USA, 3rd edition, 2006.

[10] B. Jacoby, "Mechanism involved in salt tolerance of plants," in *Handbook of Plant and Crop Stress*, M. Pessarakli, Ed., pp. 97–124, Marcel Dekker, Inc., New York, NY, USA, 1999.

[11] M. K. Kamal-Uddin, A. S. Juraimi, M. Begum, M. R. Ismail, A. A. Rahim, and R. Othman, "Floristic composition of weed community in turf grass area of West Peninsular Malaysia," *International Journal of Agriculture and Biology*, vol. 11, no. 1, pp. 13–20, 2009.

[12] M. K. Uddin, A. S. Juraimi, M. R. Ismail, and J. T. Brosnan, "Characterizing weed populations in different turfgrass sites throughout the Klang Valley of Western Peninsular Malaysia," *Weed Technology*, vol. 24, no. 2, pp. 173–181, 2010.

[13] M. K. Uddin, A. S. Juraimi, M. R. Ismail, R. Othman, and A. A. Rahim, "Growth response of eight tropical turfgrass species to salinity," *African Journal of Biotechnology*, vol. 8, no. 21, pp. 5799–5806, 2009.

[14] M. K. Uddin, A. S. Juraimi, M. R. Ismail, R. Othman, and A. A. Rahim, "Relative salinity tolerance of warm season turfgrass

species," *Journal of Environmental Biology*, vol. 32, no. 3, pp. 309–312, 2011.

[15] S. F. Alshammary, Y. L. Qian, and S. J. Wallner, "Growth response of four turfgrass species to salinity," *Agricultural Water Management*, vol. 66, no. 2, pp. 97–111, 2004.

[16] L. S. Bates, R. P. Waldren, and I. D. Teare, "Rapid determination of free proline for water-stress studies," *Plant and Soil*, vol. 39, no. 1, pp. 205–207, 1973.

[17] F. H. Witham, D. F. Blaydes, and R. M. Devlin, *Exercises in Plant Physiology*, PWS, Boston, Mass, USA, 2nd edition, 1986.

[18] P. E. Whetherly, "Studies in the water relation of cotton plants. The field measurement of water deficit in leaves," *New Phytology*, vol. 49, pp. 81–87, 1950.

[19] G. Lee, R. N. Carrow, and R. R. Duncan, "Photosynthetic responses to salinity stress of halophytic seashore paspalum ecotypes," *Plant Science*, vol. 166, no. 6, pp. 1417–1425, 2004.

[20] G. Lee, R. N. Carrow, and R. R. Duncan, "Salinity tolerance of seashore paspalum ecotypes: shoot growth responses and criteria," *HortScience*, vol. 39, no. 5, pp. 1138–1142, 2004.

[21] G. Lee, R. N. Carrow, and R. R. Duncan, "Criteria for assessing salinity tolerance of the halophytic turfgrass seashore paspalum," *Crop Science*, vol. 45, no. 1, pp. 251–258, 2005.

[22] G. Lee, R. N. Carrow, and R. R. Duncan, "Growth and water relation responses to salinity stress in halophytic seashore paspalum ecotypes," *Scientia Horticulturae*, vol. 104, no. 2, pp. 221–236, 2005.

[23] M. W. Hester, I. A. Mendelssohn, and K. L. McKee, "Species and population variation to salinity stress in *Panicum hemitomon*, *Spartina patens*, and *Spartina alterniflora*: morphological and physiological constraints," *Environmental and Experimental Botany*, vol. 46, no. 3, pp. 277–297, 2001.

[24] E. P. Glenn, "Relationship between cations accumulation and water content of salt-tolerant grasses and a sedg," *Plant Cell Environment*, vol. 10, pp. 205–212, 1987.

[25] T. J. Flowers, S. A. Flowers, M. A. Hajibagheri, and A. R. Yeo, "Salt tolerance in the halophytic wild rice, *Porteresia coantata*," *New Phytology*, vol. 114, pp. 675–684, 1990.

[26] E. P. Glenn, M. C. Watson, J. W. O'Leary, and R. D. Axelson, "Comparison of salt tolerance and osmotic adjustment of low-sodium and high-sodium subspecies of the C$_4$ halophytes, *Atriplex canescens*," *Plant Cell Environment*, vol. 15, no. 6, pp. 711–718, 1992.

[27] J. Gorham, R. G. W. Jones, and E. McDonnell, "Some mechanisms of salt tolerance in crop plants," *Plant and Soil*, vol. 89, no. 1–3, pp. 15–40, 1985.

[28] R. G. Storey and R. W. Jones, "Response of *Atriplex spongiosa* and *Suaeda monoica* to salinity," *Plant Physiology*, vol. 63, no. 1, pp. 156–162, 1979.

[29] M. Briens and F. Larher, "Osmoregulation in halophytic higher plants: a comparative study of soluble carbohydrates, polyols, betaines and free proline," *Plant Cell Environment*, vol. 5, no. 4, pp. 287–292, 1982.

[30] A. J. Cavalieri and A. H. C. Huang, "Evaluation of proline accumulation in the adaptation of diverse species of marsh halophytes to the saline environment," *American Journal of Botany*, vol. 66, no. 3, pp. 307–312, 1979.

[31] A. J. Cavalieri and A. H. C. Huang, "Accumulation of proline and glycinebetaine in *Spartina alterniflora* Loisel. in response to NaCl and nitrogen in the marsh," *Oecologia*, vol. 49, no. 2, pp. 224–228, 1981.

[32] L. G. Paleg, G. R. Stewart, and J. W. Bradbeer, "Proline and glycine betaine influence protein salvation," *Plant Physiology*, vol. 75, no. 4, pp. 974–978, 1984.

[33] N. Smirnoff and Q. J. Cumbes, "Hydroxyl radical scavenging activity of compatible solutes," *Phytochemistry*, vol. 28, no. 4, pp. 1057–1060, 1989.

[34] K. B. Marcum and C. L. Murdoch, "Salinity tolerance mechanisms of six C$_4$ turfgrasses," *Journal of the American Society for Horticultural Science*, vol. 119, no. 4, pp. 779–784, 1994.

[35] M. G. Huck, B. L. Klepper, and H. M. Taylor, "Diurnal variations in root diameter," *Plant Physiology*, vol. 45, no. 4, pp. 529–530, 1970.

[36] J. D. Everard, R. Gucci, S. C. Kann, J. A. Flore, and W. H. Loescher, "Gas exchange and carbon partitioning in the leaves of celery (*Apium graveolens* L.) at various levels of root zone salinity," *Plant Physiology*, vol. 106, no. 1, pp. 281–292, 1994.

[37] Z. Abdullah, M. A. Khan, and T. J. Flowers, "Causes of sterility in seed set of rice under salinity stress," *Journal of Agronomy and Crop Science*, vol. 187, no. 1, pp. 25–32, 2001.

[38] C. Kaya, A. L. Tuna, M. Ashraf, and H. Altunlu, "Improved salt tolerance of melon (*Cucumis melo* L.) by the addition of proline and potassium nitrate," *Environmental and Experimental Botany*, vol. 60, no. 3, pp. 397–403, 2007.

[39] L. Shabala, T. A. Cuin, I. A. Newman, and S. Shabala, "Salinity-induced ion flux patterns from the excised roots of *Arabidopsis sos mutants*," *Planta*, vol. 222, no. 6, pp. 1041–1050, 2005.

[40] F. Moradi and A. M. Ismail, "Responses of photosynthesis, chlorophyll fluorescence and ROS-scavenging systems to salt stress during seedling and reproductive stages in rice," *Annals of Botany*, vol. 99, no. 6, pp. 1161–1173, 2007.

[41] M. S. Islam, *Morpho-physiology of blackgram and mungbean as influenced by salinity*, M.S. thesis, Department of Agronomy, BSMRAU, Salna, Gazipur, 2001.

[42] S. Lutts, J. M. Kinet, and J. Bouharmont, "NaCl-induced senescence in leaves of rice (*Oryza sativa* L.) cultivars differing in salinity resistance," *Annals of Botany*, vol. 78, no. 3, pp. 389–398, 1996.

[43] S. Krishna, B. T. Raj, E. C. Mawson, T. Yeung, and A. Thorpe, "Utilization of induction and quenching kinetics of chlorophyll a fluorescence for in vivo salinity screening studies in wheat (*Triticum aestivum* vars. Kharchia-65 and Fielder)," *Canadian Journal of Botany*, vol. 71, no. 1, pp. 87–92, 1993.

[44] S. N. Shabala, S. I. Shabala, A. I. Martynenko, O. Babourina, and I. A. Newman, "Salinity effect on bioelectric activity, growth, Na$^+$ accumulation and chlorophyll fluorescence of maize leaves: a comparative survey and prospects for screening," *Australian Journal of Plant Physiology*, vol. 25, no. 5, pp. 609–616, 1998.

[45] R. Pushpam and S. R. S. Rangasamy, "In vivo response of rice cultivars to salt stress," *Journal of Ecobiology*, vol. 14, pp. 177–182, 2002.

[46] M. M. Mohan, S. L. Narayanan, and S. M. Ibrahim, "Chlorophyll stability index (CSI): its impact on salt tolerance in rice," *International Rice Research Notes*, vol. 25, no. 2, pp. 38–39, 2000.

[47] M. P. Mandal and R. A. Singh, "Impact of salt stress on chlorophyll content in rice genotypes," *Journal Research Birsa Agriculture University India*, vol. 13, pp. 61–63, 2001.

22

Pollen, Tapetum, and Orbicule Development in *Colletia paradoxa* and *Discaria americana* (Rhamnaceae)

M. Gotelli, B. Galati, and D. Medan

Grupo de Biología Reproductiva en Plantas Superiores, Cátedra de Botánica Agrícola, Facultad de Agronomía, Universidad de Buenos Aires, Avenida San Martín 4453, 1417 Buenos Aires, Argentina

Correspondence should be addressed to M. Gotelli, gotelli@agro.uba.ar

Academic Editor: Harsh P. Bais

Tapetum, orbicule, and pollen grain ontogeny in *Colletia paradoxa* and *Discaria americana* were studied with transmission electron microscopy (TEM). The ultrastructural changes observed during the different stages of development in the tapetal cells and related to orbicule and pollen grain formation are described. The proorbicules have the appearance of lipid globule, and their formation is related to the endoplasmic reticulum of rough type (ERr). This is the first report on the presence of orbicules in the family Rhamnaceae. Pollen grains are shed at the bicellular stage.

1. Introduction

Rhamnaceae is a family of about 55 genera and 900 species, cosmopolitan in distribution, especially in warm temperate regions [1, 2]. The tribe Colletieae is a monophyletic group that comprises six genera that differ in flower and fruit traits [3–7]. Distribution of the tribe is associated with the Andes in South America, and usually found 30° South [7]. The traditional diagnostic characters of the tribe are decussate leaves, abundance of spines, and presence of serial meristems in the leaf axils [8].

Literature on pollen morphology of the Rhamnaceae is relatively abundant, and descriptions are often given [2, 9–11]. However, ultrastructural studies on the development of pollen grains in Rhamnaceae are rare and restricted to descriptions of the pollen grain wall. Microsporogenesis, microgametogenesis, and the sporophytic structures related, such as the tapetum, are not usually considered [12].

The aim of this paper is to describe the ultrastructure of pollen grains and microsporangium development of *Colletia paradoxa* and *Discaria americana* in order to broaden the current embryological knowledge of the Rhamnaceae family.

2. Materials and Methods

Samples of *Colletia paradoxa* and *Discaria americana* were collected from individuals cultivated in the Lucien Hauman Botanical Garden of the Facultad de Agronomía, Universidad de Buenos Aires. Reference specimens were deposited in the Herbarium Gaspar Xuarez (BAA).

For transmission electron microscopy, anthers at different developmental stages were prefixed overnight in 2.5% glutaraldehyde in phosphate buffer (pH 7.2) and then postfixed in OsO$_4$ at 2°C in the same buffer for 3 h. Following dehydration in ethanol series, the material was embedded in Spurr's resin. Ultrathin sections (750 to 900 nm) were made on a Sorvall ultramicrotome and then stained with uranyl acetate and lead citrate [13]. The sections were observed and photographed with a JEOL-JEM 1200 EX II TEM at 85.0 kV.

Resistant membranes with orbicules and pollen walls were isolated by acetolysis of whole anthers. The acetolysis was carried out following Erdtman's method [14]. Acetolysis-resistant structures were washed with water and mounted in glycerin-gelatin.

3. Results

3.1. Stage 1: Microspore Mother Cells (MMCs). In both species, the anther is bi-sporangiate and its wall consists of epidermis (ep), endothecium (en), two or three middle layers (ml), and a secretory type tapetum (t). Tapetal cells are binucleate. They present a cytoplasm with a few small vacuoles, mitochondria, and many dictyosomes and proplastids

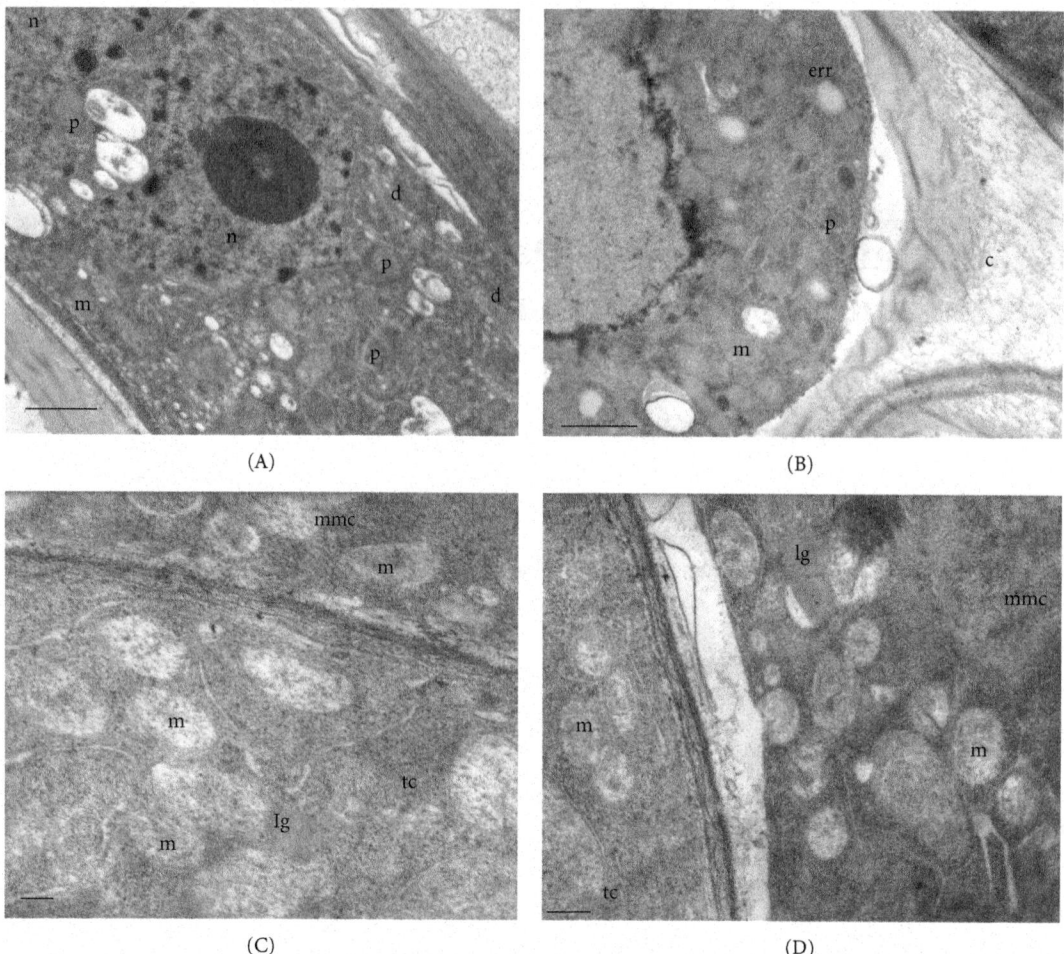

FIGURE 1: Stage 1: microspore mother cells (MMCs). (A)-(B) *Colletia paradoxa*. (C)-(D) *Discaria americana*. (A) Tapetal cell with two nuclei (n), plastids (p), dictyosomes (d), and mitochondria (m). (B) Details of the microspore mother cell: plastids (p), mitochondria (m), endoplasmic reticulum of rough type (err), and callose (c). (C) Microspore mother cell (mmc) and tapetal cell (tc) with mitochondria (m) and lipid globules (lg) (D) Microspore mother cell with lipid globules (lg). and mitochondria (m). Scale bar; (A)-(B) 1.3 μ; (C) 200 nm (D) 500 nm.

(Figure 1(A)) in *Colletia paradoxa*, and mitochondria and lipid globules in *Discaria americana* (Figure 1(B)).

The microspore mother cells are uninucleate and with a dense cytoplasm. In *Colletia paradoxa*, it is filled with abundant mitochondria, plastids, and endoplasmatic reticulum of rough type (ERR). A thick callosic wall is formed between the plasmalemma and the primary wall (Figure 1(C)). In *Discaria americana*, lipid globules and mitochondria are observed (Figure 1(D)).

3.2. Stage 2: Microspore Tetrads. The ultrastructure of the cytoplasm of tapetal cells of *Colletia paradoxa* is similar to that presented in the previous stage. Many dictyosomes, mitochondria, and plastids are present along with some endoplasmic reticulum of rough type (Figure 2(A)). In the inner tangential faces of these cells, proorbicules are observed as globular depositions of moderate electron density between the plasmalemma and the cell wall (Figure 2(A)). *Discaria americana* presents the cytoplasm of tapetal cells very dense with many mitochondria, and the cell wall is no longer distinguished in this stage (Figure 2(B)).

Microspore mother cells undergo simultaneous meiosis, forming tetrads with a tetrahedral arrangement. They remain surrounded by a thick callosic wall. At this stage, the primexine fibrillar matrix is starting to develop between the callosic wall and the plasmalemma (Figure 2(D)). The microspore cytoplasm shows many mitochondria and a few dictyosomes in *Colletia paradoxa* (Figure 2(D)), while in *Discaria americana* mitochondria, dictyosomes, endoplasmic reticulum of rough type, and lipid globules are observed (Figure 2(E)).

As the microspore tetrad matures, more mitochondria are found in the cytoplasm of tapetal cells of *Colletia paradoxa*. The wall of these cells appears laxer and degraded in some places. Orbicules are released and some of them are observed on the outer tangential cell wall of the tapetal cells (Figure 2(C)). An electron-dense deposition is observed on the surface of the proorbicules (Figure 2(C)).

3.3. Stage 3: Free Microspores. At this stage, the cell wall of tapetal cells in *Colletia paradoxa* is also dissolved and remains of it are observed between the orbicules, which

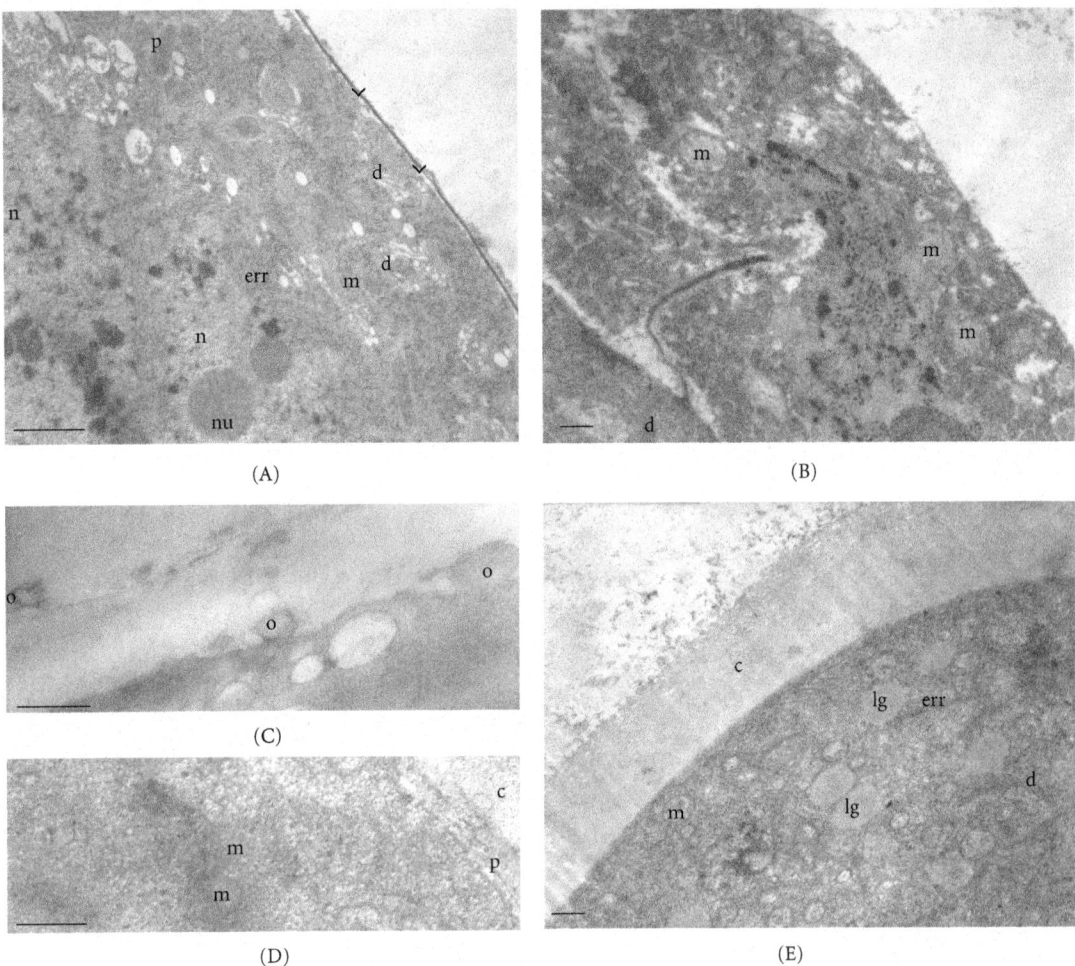

FIGURE 2: Stage 2: microspore tetrads. (A)–(C) *Colletia paradoxa*. (D)-(E) *Discaria americana*. (A) Tapetal cell at young microspore tetrad stage. Proorbicules (arrow head) are observed. Cytoplasm with endoplasmic reticulum of rough type (err), plastids (p), mitochondria (m), dictyosomes (d), and two nuclei (n), and nucleoli (nu). (B) Details of the tapetal cytoplasm and wall at an advanced microspore tetrad stage. Orbicules (o) are observed in the inner and the outer tapetal wall, which is partly degraded. Dictyosomes (d) and mitochondria (m) are also present. (C) Microspore with many mitochondria (m) and dictyosome (d). Callosic wall and primexine are observed. (D) Tapetal cell with many mitochondria (m) and dictyosomes (d). Cell wall degraded. (E) Microspore with mitochondria (m), lipid globules (lg), dictyosomes (d), endoplasmic reticulum of rough type (err), and callose (c). Scale bar: (A) 1.3 μ; (B) 300 nm; (C) 400 nm; (D)-(E) 500 nm.

have been released (Figure 3(B)). Many mitochondria and endoplasmatic reticulum of rough type are present in the cytoplasm of tapetal cells. There are vesicles near and in close contact to the plasmalemma (Figures 3(A)-3(B)). The ultrastructure of the cytoplasm of tapetal cells of *Discaria americana* is similar to that presented in the previous stage.

After the dissolution of the callose wall, the sporopollenin wall begins to form. The intine, endexine, foot layer, a tectum, and a granular infratectum are clearly distinguished in the two species (Figures 3(C)-3(D)). Microspores have a conspicuous nucleus, and their cytoplasm is limited to a parietal position due to the presence of a large vacuole. The structure of their cytoplasm is similar to the previous stage with many mitochondria and some endoplasmic reticulum of rough type (Figure 3(D)).

3.4. Stage 4: Young Pollen Grain. Tapetal cells are almost completely degraded in *Discaria americana* but in *Colletia*

paradoxa do not lose their individuality. Some mitochondria and numerous free ribosomes can still be observed on their cytoplasm. However, most membrane systems appear to be disintegrating. In this species, a tapetal membrane is formed in the inner tangential and radial faces of tapetal cells and the orbicules are observed on it. This membrane is partially resistant to acetolysis because fragments of it are observed after this treatment. Orbicules present a central core transparent to electrons, and a wall with the same electron density that is the pollen exine. They have a spherical to subspherical shape with a few irregularities and superficial invaginations in both species (Figures 4(A)-4(B)).

The generative cell formed by a mitotic division of the microspore occupies a parietal position. A conspicuous nucleus occupies almost all the cytoplasm. Some mitochondria can be observed. It is delimited by a thin wall transparent to electrons and shows a few connections with the vegetative cell, which has a very dense cytoplasm filled with amyloplasts

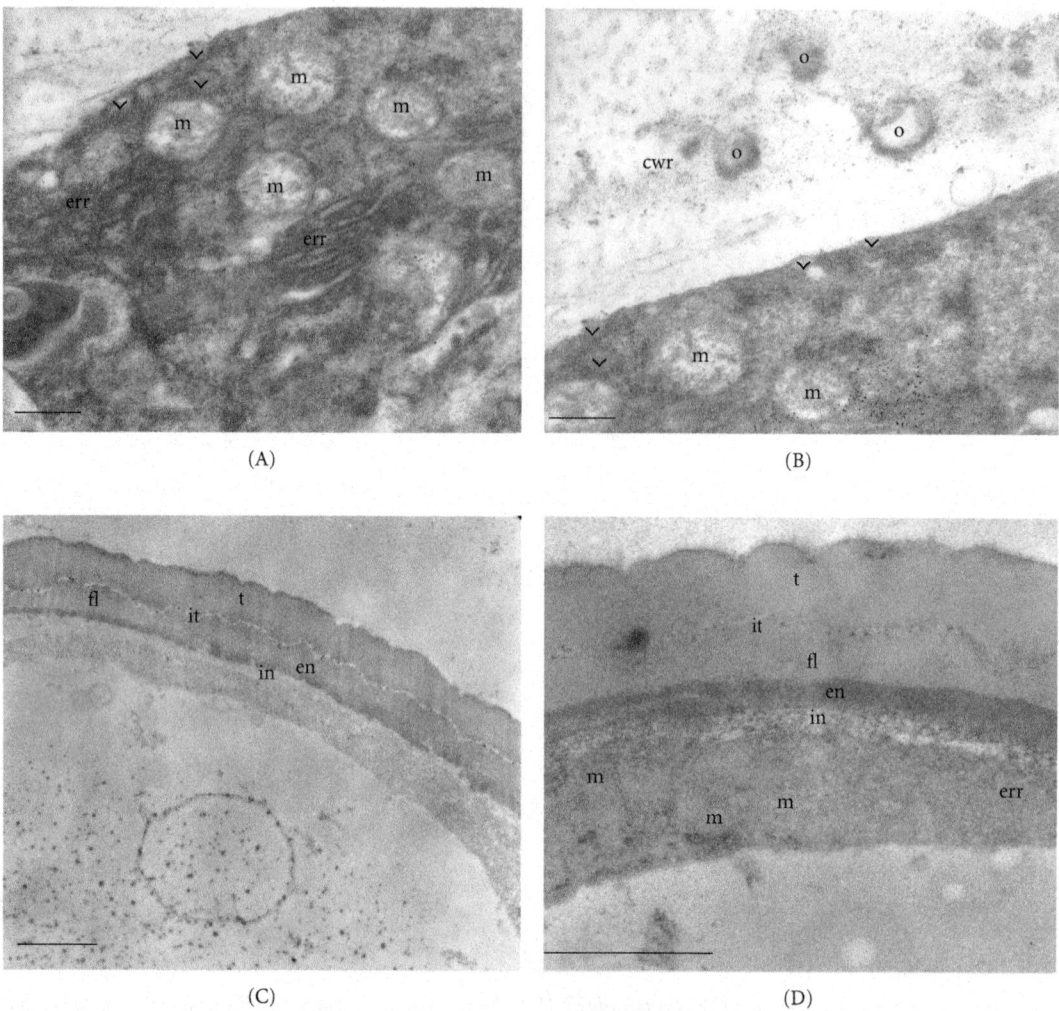

FIGURE 3: Stage 3: free microspores. *Colletia paradoxa.* (A) Tapetal cell with proorbicules (arrow head), many mitochondria (m), and rough endoplamic reticulum (err). (B) Details of tapetal cell, wall degraded, and orbicules are observed between its remains (cwr: cell wall remains). Mitochondria (m) and proorbicules (arrow head) are also observed. (C) Young microspore: intine (in), endexine (en), foot layer (fl), a tectum (t), and a granular infratectum (it) are clearly distinguished. (D) Details of the young microspore cytoplasm with mitochondria (m) and endoplasmic reticulum of tough type (err). Scale bar: (A) 280 nm; (B) 300 nm; (C)-(D) 1.3 μ.

and endoplasmic reticulum of rough type (Figures 5(A)-5(B)).

3.5. Stage 5: Mature Pollen Grain. At this stage, tapetal cells cytoplasm appears almost completely degraded in both species. The tapetal membrane can still be observed with some orbicules attached to it in *Colletia paradoxa.* Orbicules are also attached to the remains of the original tapetal cell wall, which presents a fibrillar aspect (Figure 4(C)). Remains of these cells in *Discaria americana* appear as electron-dense rests between the orbicules. The average size of orbicules in both species is 0.4 μm (Figure 4(D)).

The vegetative cell of the mature pollen grain presents many mitochondria, dictyosomes, starch grains, lipid globules, and endoplasmic reticulum of rough type (Figure 5(C)). The generative cell occupies a central position and has a reduced cytoplasm and cell wall practically absent only represented by the two adjacent membranes (Figure 5(D)).

The pollen grain wall is formed by three layers with different electron density: ectexine with a perforated tectum, a granular infratectum, and a foot layer thinner than the tectum; an electron-dense endexine; a thin intine of low electron density (Figures 5(A)–5(C)). Pollen grains of both species are suboblate, 3-colporate, tectum fossulate- perforate (Figures 6(A)–6(D)).

4. Discussion

Anther wall development in *Colletia paradoxa* and *Discaria americana* is of the basic type [15]. This feature was described as a general trait of the Rhamnaceae [16]. Medan and Aagesen [3] claim that in *Colletia paradoxa* there are often two or three pollen sacs in each anther, as here observed. The anther wall of both species comprises the epidermis, fibrous endothecium, two or three middle layers, and a secretory tapetum with binucleate cells as described for other members

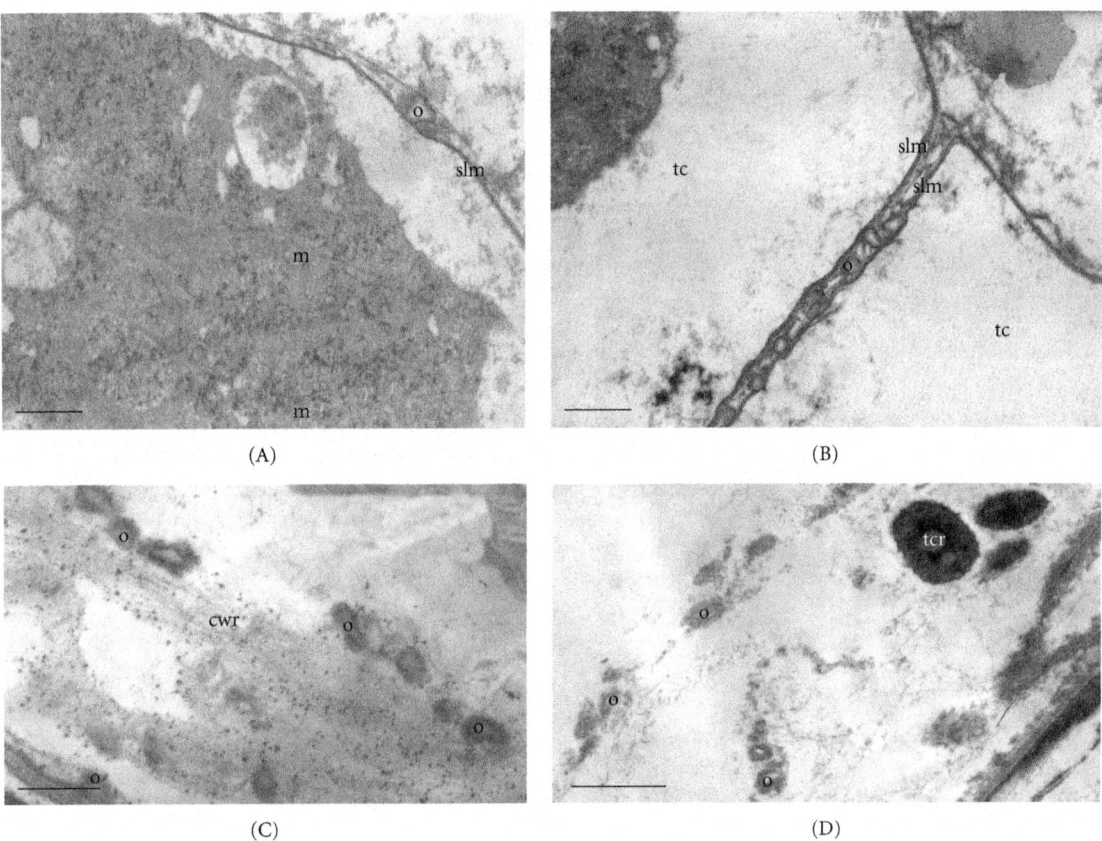

FIGURE 4: (A)-(B) Stage 4: young pollen grain. (C)-(D) Stage 5: mature pollen grain. (A)–(C) *Colletia* paradoxa. (D) *Discaria americana.* (A) Tapetal cell cytoplasm in degradation with some mitochondria (m) still recognizable and orbicules (o) attached to a sporopollenin-like. membrane (slm). (B) Orbicules (o) attached to the radial face of the sporopollenin-like membrane (slm) formed between two tapetal cells (tc) (C) Remains of the tapetal wall (cwr) with orbicules (o) and details of the sporopollenin-like membrane (slm) with orbicules (o) attached. (D) Orbicules (o) and tapetal cell remains (tcr). Scale bar: (A)-(B) 850 nm; (C) 500 nm; (D) $1\,\mu$.

of the family [17]. The third middle layer is originated by the division of any of the first two middle layers. Cytokinesis is simultaneous, and the arrangement of microspores in the tetrad is tetrahedral for both species. Tetrahedral and isobilateral tetrads were reported to appear in the family [17].

Orbicules are corpuscles of sporopollenin lining in the inner tangential and sometimes in the radial tapetal cell walls that appear during pollen grain development [18, 19]. Ubisch [20] gave a detailed description of them. Kosmath [21] called the Ubisch bodies and Erdtman et al. [22] orbicules. Its size is very variable, ranging from $0.14\,\mu$m to $20\,\mu$m [19]. The average size in the species studied is $0.40\,\mu$m. Huysmans et al. [18] studied the distribution of orbicules in Angiosperms, and there is no report of their presence in Rhamnales. This is the first report of orbicules in this taxonomic group.

Orbicules of *Colletia paradoxa* and *Discaria americana* have a central core transparent to electrons and a sporopollenin wall with a spheric to subspheric shape, with a few irregularities and superficial invaginations. This type of orbicule was previously described in the classification given by Galati [23]. The central core of orbicules in these two species appears as a proorbicule at the microspore tetrad stage. According to Clément and Audran [24] the orbicule

core represents remnants of the proorbicule and plays an important role in the orbicular wall ontogenesis. A fibrillar substance with the same electron density as sporopollenin accumulates on the proorbicules after they are released. Therefore, the accretion of sporopollenin on them appears extracellular. This is in concordance with Echlin [25], Christensen et al. [26], Clément and Audran [24], Galati and Rosenfeldt [27], Galati and Strittmatter [28], Amela García et al. [29], and Rosenfeldt and Galati [30].

In *Colletia paradoxa*, proorbicules are observed in the tapetal cells at an early microspore tetrad stage. Later on this stage, orbicules are released and pass through the degraded zones of the tapetal cell wall and some of them are observed at the outer tangential wall of tapetal cells. At the young microspore stage, tapetal cell wall starts to disintegrate and orbicules are observed between its remains. Therefore, orbicules are fully developed and become in contact with the locular fluid even before the complete loss of the tapetal cell wall. In *Jacaranda mimosifolia* [28] and *Pinus sylvestris* [31, 32] tapetal cell walls present a lax structure and orbicules are observed protruding it.

An increase in the amount of ERR in the cytoplasm of tapetal cells of *Colletia paradoxa* was observed at the young microspore stage, when most orbicules are being released.

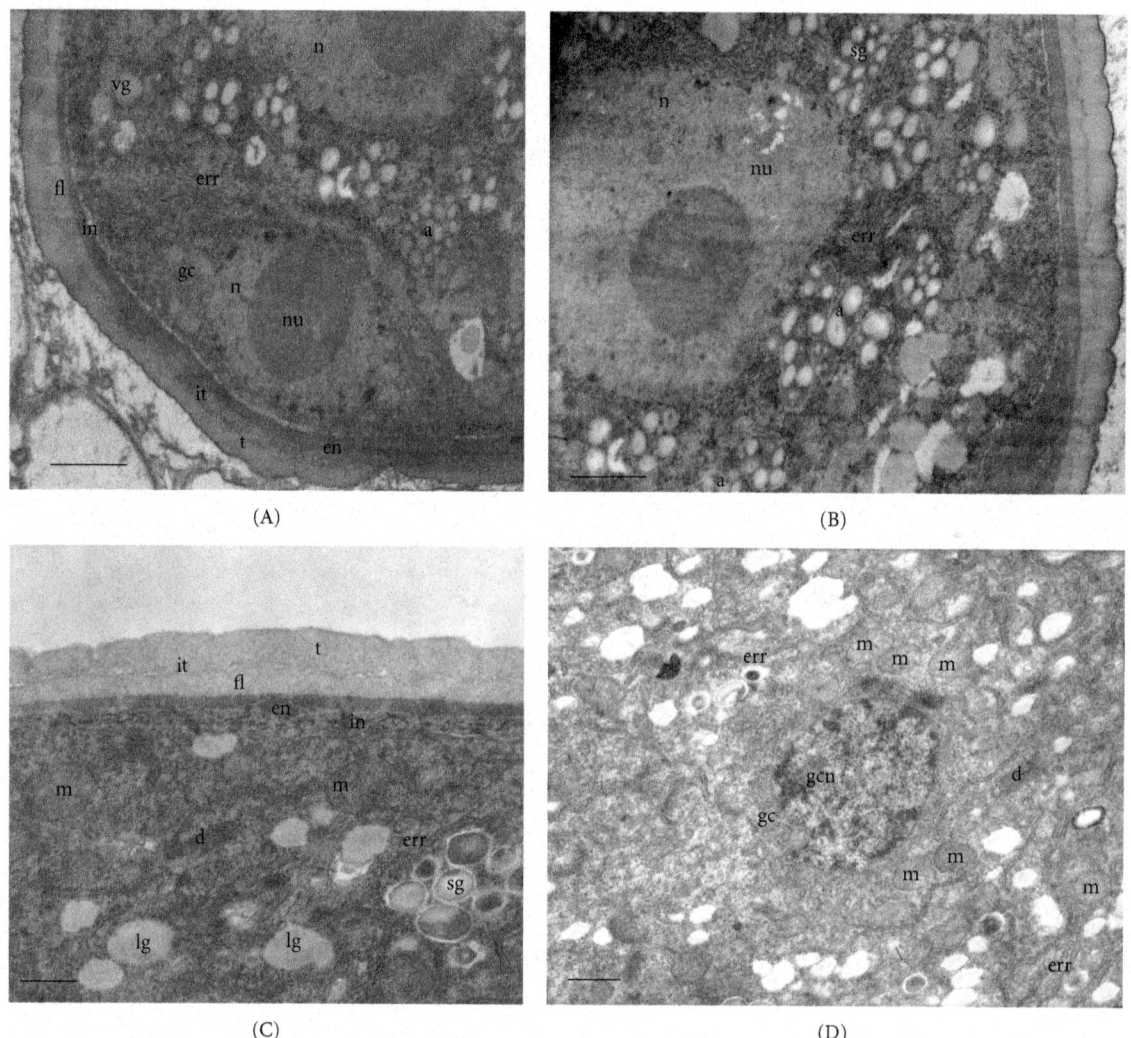

FIGURE 5: (A)-(B) Stage 4: young pollen grain of *Colletia paradoxa*. (C)-(D) Stage 5: mature pollen grain of *Discaria americana*. (A) Vegetative cell (VC), with endoplasmic reticulum of rough type (err), amyloplasts (a), and its nucleus (n) and nucleolus (nu); generative cell (GC) with its nucleus (n) and nucleolus (nu); tapetal cells (tc) and orbicules (o). (B) Details of the vegetative cell. (C) Generative cell with starch grains (sg), lipid globules, mitochondria (m), dictyosomes (d), endoplasmic reticulum of rough type (err). (D) Generative cell (gc) and its nucleus (gcn). Other references: t: tectum, it: infratectum, ft: foot layer, en: endexine, in: intine. Scale bar: (A)-(B) 1.3 μ; (C) 500 nm; (D) 1 μ.

ERR is usually found close to the inner tangential wall, and little vesicles similar to proorbicules are in association with it. This suggests that this organelle is related with the formation of these orbicules, as observed in *Ceiba insignis* [27], *Jacaranda mimosifolia* [28], *Prosopis juliflora* [33], *Catharanthus roseus* [34], *Anemarrhena asphodeloides* [35], *Lilium henryi* [36], *Lavandula dentata* [37], and *Allium cepa* [38].

At the young pollen grain stage, orbicules in *Discaria americana* appear lose in the anther locule, while in *Colletia paradoxa* they are attached to a tapetal membrane as observed in *Jacaranda mimosifolia* [28]. The presence of orbicules is generally associated with a tapetal membrane in species with secretor type tapetum and with a peritapetal membrane in species with intermediate or plasmodial type tapetum [19]. Tapetal membranes are usually described as tapetally secreted sporopollenin membrane that develops on the inner tangential face of the tapetum [39]. In *Colletia*

paradoxa, such structure is considered a membrane because it is partially resistant to acetolysis and also develops on the radial faces of tapetal cells. Therefore, orbicules are observed in the inner tangential and radial faces of this membrane.

In *Jacaranda mimosifolia*, the loss of tapetal cell walls occurs at the bicellular pollen grain stage, simultaneously with the formation of the tapetal membrane [28]. In *Colletia paradoxa*, tapetal cell walls start to degrade at the mature microspore tetrad stage. However, the sporopollenin-like tapetal membrane is formed at the young pollen grain stage. In *Colletia paradoxa*, tapetal cells degrade at the mature pollen grain stage, possibly due to the presence of such membrane, while in *Discaria Americana*, where no membrane was observed, tapetal cells degrade in the free microspore stage.

Many speculations were made trying to understand the role of orbicules [26, 34, 36, 40, 41]. However, none of them could be confirmed. Heslop-Harrison [42, 43] was the first

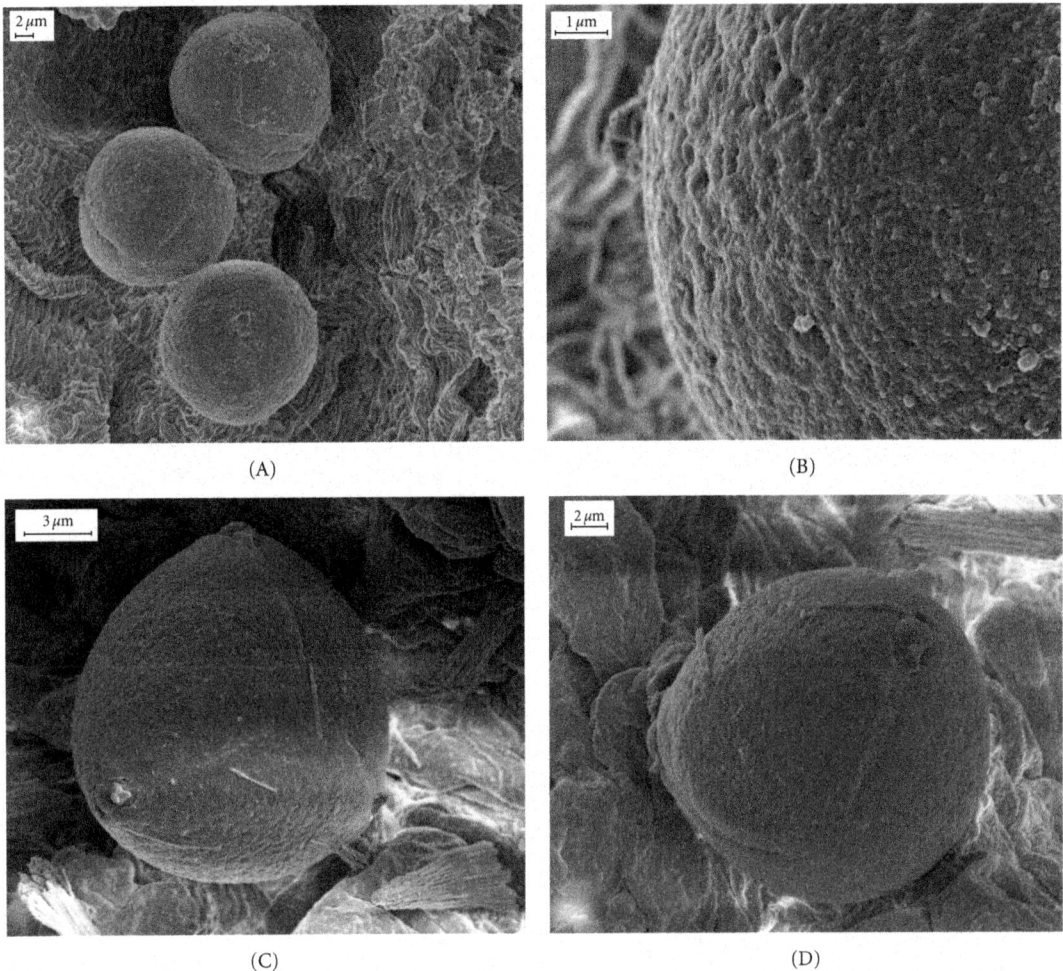

FIGURE 6: Stage 5: mature pollen grains. (A)-(B) *Colletia paradoxa*. (C)-(D) *Discaria americana*. (A) General view of pollen grains. (B) Details of a pollen grain. (C) Polar view of a pollen grain. (D) Equatorial view of a pollen grain.

to consider that the orbicules could be associated with pollen dispersal forming a not-wettable surface from which pollen can easily be freed. Later, Keijzer [44] supported this theory. Pacini and Franchi [45] wondered if pollen grains and orbicules would repulse each other, since they are formed by the same substance and, therefore, charged in the same way. Vaknin et al. [46] analyzed the role of electrostatic forces. Galati et al. [19] related orbicule morphology of representative species of diverse families of Angiosperms with different modes of pollen dispersal concluding that families, which may be taxonomically distant but share the mode of pollen dispersal, present similar orbicule morphology. All species of Colletieae studied to date show pollen transfer mediated by insects (Diptera, Hymenoptera, and Lepidoptera) [7]. This mode of pollen dispersal was described in species with orbicules spheric to subspheric, with a central core and an irregular surface, as here described. Therefore, these findings support the theory proposed by Galati et al. [19].

Pollen grain morphology is in accordance with the general morphological pattern of the family Rhamnaceae [2, 11, 17, 47]. In *Ziziphus lotus* L., the exine is characterized by a thick tectum, a thin infratectal layer that appears granular,

and a thick nexine [12]. Similar observations were made by Schirarend and Kohler [10] in *Colletia spinosissima*.

This is the first report of the ontogeny and ultrastructure of the pollen grain and related sporophytic structures of *Colletia paradoxa* and *Discaria americana*. Therefore, we consider it to be an important contribution to the general knowledge of the Rhamnaceae family.

References

[1] D. Medan and C. Schirarend, "Rhamnaceae," in *Flowering Plants. Dicotyledons: Celastrales, Oxalidales, Rosales, Cornales, Ericales (The Families and Genera of Vascular Plants)*, K. Kubitzki, Ed., vol. 6, Springer, Heidelberg, Germany, 2004.

[2] A. Perveen and M. Qaiser, "Pollen flora of Pakistan—XLIV. Rhamnaceae," *Pakistan Journal of Botany*, vol. 37, no. 2, pp. 195–202, 2005.

[3] D. Medan and L. Aagesen, "Comparative flower and fruit structure in the Colletieae (Rhamnaceae)," *Botanische Jahrbücher für Systematik, Pflanzengeschichte und Pflanzengeographie*, vol. 117, pp. 531–564, 1995.

[4] L. Aagesen, "Phylogeny of the tribe Colletieae, Rhamnaceae," *Botanical Journal of the Linnean Society*, vol. 131, no. 1, pp. 1–43, 1999.

[5] J. E. Richardson, M. F. Fay, Q. C. B. Cronk, and M. W. Chase, "A revision of the tribal classification of Rhamnaceae," *Kew Bulletin*, vol. 55, no. 2, pp. 311–340, 2000.

[6] J. E. Richardson, M. F. Fay, Q. C. B. Cronk, D. Bowman, and M. W. Chase, "A phylogenetic analysis of Rhamnaceae using *rbc*L and *trn*L-F plastid DNA sequences," *American Journal of Botany*, vol. 87, no. 9, pp. 1309–1324, 2000.

[7] D. Medan and A. M. Basilio, "Reproductive biology of *Colletia spinosissima* (Rhamnaceae) in Argentina," *Plant Systematics and Evolution*, vol. 229, no. 1-2, pp. 79–89, 2001.

[8] L. Aagesen, D. Medan, J. Kellermann, and H. H. Hilger, "Phylogeny of the tribe Colletieae (Rhamnaceae)—a sensitivity analysis of the plastid region *trn*L-*trn*F combined with morphology," *Plant Systematics and Evolution*, vol. 250, no. 3-4, pp. 197–214, 2005.

[9] E. Papagiannes, *Pollen studies of selected genera of Rhamnaceae*, M.S. thesis, University of Illinois at the Chicago, Chicago, Ill, USA, 1974.

[10] C. Schirarend and E. Kohler, "Rhamnaceae Juss," *World Pollen Spore Flora*, no. 17, pp. 1–53, 1993.

[11] W. Punt, A. Marks, and P. P. Hoen, "The Northwest European pollen flora 64. vitaceae," *Review of Palaeobotany and Palynology*, vol. 123, pp. 67–70, 2003.

[12] M. B. Nasri-Ayachi and M. A. Nabli, "Pollen wall ultrastructure and ontogeny in *Ziziphus lotus* L. (Rhamnaceae)," *Review of Palaeobotany and Palynology*, vol. 85, no. 1-2, pp. 85–98, 1995.

[13] T. P. O'Brien and M. E. McCully, *The Study of Plant Structure. Principles and Selected Methods*, Termarcarphi Pty Ltd., Melbourne, Australia, 1981.

[14] G. Erdtman, "The acetolysis method. A revised description," *Svensk Botanisk Tidskrift*, vol. 54, pp. 561–564, 1960.

[15] G. L. Davis, *Systematic Embryology of the Angiosperms*, John Wiley & Sons, New York, NY, USA, 1966.

[16] L. Watson and M. J. Dallwitz, *The Families of Flowering Plants: Descriptions, Illustrations, Identification and Information Retrieval*, 1992.

[17] B. M. Johri, K. B. Ambegaokar, and P. S. Srivastava, *Comparative Embryology of Angiosperms*, vol. 1-2, Springer, Berlin, Germany, 1992.

[18] S. Huysmans, G. El-Ghazaly, and E. Smets, "Orbicules in angiosperms: morphology, function, distribution, and relation with tapetum types," *Botanical Review*, vol. 64, no. 3, pp. 240–272, 1998.

[19] B. G. Galati, M. M. Gotelli, S. Rosenfeldt, J. P. Torretta, and G. Zarlavsky, "Orbicules in relation to the pollination modes," in *Pollen: Structure, Types and Effects*, B. J. Kaiser, Ed., Nova Science Publishers, Huntington, NY, USA, 2010.

[20] G. von Ubisch, "Kurze mitteilungen zur entwicklungsgeschiehte der antheren," *Planta*, vol. 3, pp. 490–495, 1927.

[21] L. Kosmath, "Studien über das Antherentapetum," *Osterreichische Botanische Zeitschrift*, vol. 76, no. 3, pp. 235–241, 1927.

[22] G. Erdtman, B. Berglund, and J. Praglowski, "An introduction to a Scandinavian pollen flora," *Grana*, vol. 2, pp. 3–92, 1961.

[23] B. G. Galati, "Ubisch bodies in Angiosperms," in *Advances in Plant Reproductive Biology*, A. K. Pandey and M. R. Dhakal, Eds., vol. 2, Narendra Publishing House, Delhi, India, 2003.

[24] C. Clément and J. C. Audran, "Cytochemical and ultrastructural evolution of orbicules in *Lilium*," *Plant Systematics and Evolution*, vol. 7, pp. 63–74, 1993.

[25] P. Echlin, "The role of the tapetum during microsporogenesis of angiosperms," in *Pollen Development and Physiology*, J. Heslop-Harrison, Ed., Butterworths, London, UK, 1971.

[26] J. E. Christensen, H. T. Horner, and N. R. Lersten, "Pollen wall and tapetal orbicular wall development in *Sorghum bicolor* (Gramineae)," *American Journal of Botany*, vol. 59, no. 1, pp. 43–58, 1972.

[27] B. G. Galati and S. Rosenfeldt, "The pollen development in *Ceiba insignis* (Kunth) Gibbs & Semir ex *Chorisia speciosa* St. Hil. (Bombacaceae)," *Phytomorphology*, vol. 48, no. 2, pp. 121–129, 1998.

[28] B. G. Galati and S. I. Strittmatter, "Correlation between pollen development and Ubisch bodies ontogeny in *Jacaranda mimosifolia* (Bignoniaceae)," *Beitrage zur Biologie der Pflanzen*, vol. 71, pp. 1–12, 1999.

[29] M. T. Amela García, B. G. Galati, and A. M. Anton, "Microsporogenesis, microgametogenesis and pollen morphology of *Passiflora* spp. (Passifloraceae)," *Botanical Journal of the Linnean Society*, vol. 139, no. 4, pp. 383–394, 2002.

[30] S. Rosenfeldt and B. G. Galati, "Orbicules diversity in *Oxalis* species from the province of Buenos Aires (Argentina)," *Biocell*, vol. 32, no. 1, pp. 41–47, 2008.

[31] J. R. Rowley and B. Walles, "Origin and structure of Ubisch bodies in *Pinus sylvestris*," *Acta Societatis Botanicorum Poloniae*, vol. 56, pp. 215–227, 1987.

[32] J. R. Rowley and B. Walles, "Cell differentiation in microsporangia of *Pinus sylvestris*. V. Diakinesis to tetrad formation," *Nordic Journal of Botany*, vol. 13, pp. 67–82, 1993.

[33] M. R. Vijayaraghavan and B. Chaundhry, "Structure and development of orbicules in the tapetum of *Prosopis juliflora* (Leguminosae, Mimosoideae)," *Phytomorphology*, vol. 43, pp. 41–48, 1993.

[34] G. El-Ghazaly and S. Nilsson, "Development of tapetum and orbicules of *Catharanthus roseus* (Apocynaceae)," in *Pollen and spores Systematics Association Special*, S. Blackmore and S. H. Barnes, Eds., vol. 44, Clarendon Press, Oxford, UK, 1991.

[35] Z. K. Chen, F. H. Wang, and F. Zhou, "On the origin, development and ultrastructure of the orbicules and pollenkitt in the tapetum of *Anemarrhena asphodeloides* (Liliaceae)," *Grana*, vol. 27, pp. 273–282, 1988.

[36] R. Herich and A. Lux, "Lytic activity of Übisch bodies (orbicles)," *Cytologia*, vol. 50, pp. 563–569, 1985.

[37] M. Suarez-Cervera and J. A. Seoane-Camba, "Ontogénese des grains de pollen de *Lavandula dentata* L. et évolution des cellules tapétales," *Pollen et Spores*, vol. 28, pp. 5–28, 1986.

[38] M. C. Risueño, G. Giménez-Martín, J. F. López-Sáez, and M. I. R. García, "Origin and development of sporopollenin bodies," *Protoplasma*, vol. 67, no. 4, pp. 361–374, 1969.

[39] U. Banerjee, "Ultrastructure of the tapetal membranes in grasses," *Grana*, vol. 7, pp. 2–3, 1967.

[40] P. Maheshwari, *Introduction to the Embryology of Angiosperms*, McGraw-Hill, New York, NY, USA, 1950.

[41] U. C. Banerjee and E. S. Barghoorn, "The tapetal membranes in grasses and Ubisch body control of mature exine pattern," in *Pollen Development and Physiology*, J. Heslop-Harrison, Ed., Butterworths, London, UK, 1971.

[42] J. Heslop-Harrison, "Pollen wall development," *Science*, vol. 161, no. 838, pp. 230–237, 1968.

[43] J. Heslop-Harrison, "Tapetal origin of pollen-coat substances in *Lilium*," *New Phytologist*, vol. 67, pp. 779–786, 1968.

[44] C. J. Keijzer, "Pollen dispersal and the function of the orbicules," *Acta Botanica Neerlandica*, vol. 33, p. 244, 1984.

[45] E. Pacini and G. G. Franchi, "Role of the tapetum in pollen and spore dispersal," *Plant Systematics and Evolution*, vol. 7, pp. 1–11, 1993.

[46] Y. Vaknin, S. Gan-Mor, A. Bechar, B. Ronen, and D. Eisikowitch, "The role of electrostatic forces in pollination," in *Pollen and Pollination*, Dafni, Hesse, and Pacini, Eds., Springer, New York, NY, USA, 1999.

[47] C. Schirarend, "Pollen morphology of the genus *Paliurus* (Rhamnaceae)," *GRANA*, vol. 35, no. 6, pp. 347–356, 1996.

23

Ephedra alte (Joint Pine): An Invasive, Problematic Weedy Species in Forestry and Fruit Tree Orchards in Jordan

Jamal R. Qasem

Department of Plant Protection, Faculty of Agriculture, University of Jordan, P.O. Box 13282, Amman 11942, Jordan

Correspondence should be addressed to Jamal R. Qasem, jrqasem@ju.edu.jo

Academic Editor: Ata Akcil

A field survey was carried out to record plant species climbed by *Ephedra alte* in certain parts of Jordan during 2008–2010. Forty species of shrubs, ornamental, fruit, and forest trees belonging to 24 plant families suffered from the climbing habit of *E. alte*. Growth of host plants was adversely affected by *E. alte* growth that extended over their vegetation. In addition to its possible competition for water and nutrients, the extensive growth it forms over host species prevents photosynthesis, smothers growth and makes plants die underneath the extensive cover. However, *E. alte* did not climb all plant species, indicating a host preference range. Damaged fruit trees included *Amygdalus communis, Citrus aurantifolia, Ficus carica, Olea europaea, Opuntia ficus-indica,* and *Punica granatum*. Forestry species that were adversely affected included *Acacia cyanophylla, Ceratonia siliqua, Crataegus azarolus, Cupressus sempervirens, Pinus halepensis, Pistacia atlantica, Pistacia palaestina, Quercus coccifera, Quercus infectoria, Retama raetam, Rhamnus palaestina, Rhus tripartita,* and *Zizyphus spina-christi*. Woody ornamentals attacked were *Ailanthus altissima, Hedera helix, Jasminum fruticans, Jasminum grandiflorum, Nerium oleander,* and *Pyracantha coccinea*. Results indicated that *E. alte* is a strong competitive for light and can completely smother plants supporting its growth. *A. communis, F. carica, R. palaestina,* and *C. azarolus* were most frequently attacked.

1. Introduction

The Ephedraceae family consists of species with varied growth forms, habits and habitat requirements. Shrubby species that belong to this family have erect stems, decumbent plants up to 1 m high; scarious leaf sheath, at least of young shoots, 1-2 mm long, as long as the diameter of the subtended stem and longer than leaf rudiments. Climbing or prostrate plants are often with long lignified stems but when forming a shrub (after grazing), the scarious sheath is usually shorter than the diameter of subtended stem or leaves. Fleshy ripe fruiting bracts are red; the free part of the leaf is mostly less than 3 mm long and flowering branchlets are always arise from thicker stems with green photosynthetic bark [1, 2].

Ephedra is a distinct genus that consists of 50–65 species among which are shrubs, vines, but rarely small trees [3–5]. It is a dioecious plant, heavily branched, with very short scale

leaves. It is a nonsucculent glycophyte that grows in natural habitats and is widely distributed in temperate regions in different parts of the world but usually common in dry and open habitats and in the deserts. It has been reported at elevations ranging from the near sea level (species around the Mediterranean Sea) to almost 5000 m (*E. gerardiana* in the Himalayas). Under drought, heat, and frost conditions in highlands in Asia, species have shown greater wood xeromorphy than do the lowland species [6]. *Ephedra* is recorded in mobile and stable sand dunes and wadis with sandy ground and is heavily consumed by camels and other grazing animals.

Ephedra is long known for its medicinal value in the Mediterranean because of the ephedrine alkaloid and other chemicals in the stems of most members of this genus [7, 8]. Ephedrine has been long known to have contact allergenic properties and as being valuable in the treatment of asthma and many other complaints of the respiratory system [9]. Its naturally occurring isomer pseudoephedrine also appears as a contact sensitizer. However, recently *Ephedra*-derived products have been found hazardous and may cause cardiac dysfunction and even death when excessively used [9]. Other chemicals isolated from the aerial parts of *E. alte* were vicenin II, ephedralone, p-coumaric, protocatechuic acids, herbacetin [10], ephedradine C, and hordenine [11]. However, *Ephedra* spp. are widely varied in alkaloid content while these chemicals are generally absent in roots, berries, and seeds of these species.

E. alte (synonym *Ephedra aphylla* Forssk) is one of the common species in different Middle East countries [1]. It is normally a shrub not more than 1.4 m in height, but when it grows with taller vegetation, such as along irrigation ditches, it may grow into those plants as a scandent liana [2]. It is found hanging from cracks in limestone cliffs or near wadis in sand and is often found growing in juniper forest with *Pistacia*, *Opuntia*, *Daphne linearifolia*, *Artemesia*, and *Thymelaea hirsuta*. It flowers and fruits from March to June. This species extends across the eastern Mediterranean to the Arabian Peninsula [12] and is the only *Ephedra* species within its range from dry to very dry habitats that may be somewhat more extensive, from cliffs, along wadis, and with phreatophytes along irrigation ditches [2]. It is able to grow in dry habitats to which few angiosperms have adapted and has been mentioned as somewhat peculiar in its growth habit. *E. alte* has been also mentioned as being of a high toxicity against *Aedes aegypti* larvae and thus may become an important plant in controlling disease-causing mosquitoes [12].

In Jordan, four *Ephedra* species, *Ephedra alata* Decne, *Ephedra alte* C. A. Mey, *Ephedra foliate* Boiss., and *Ephedra transitoria* Riedl., have been reported [13] to spread in natural habitats and recorded in areas in and close to the Jordan valley (tropical and subtropical), the Mediterranean, and Sahara Arabian. *E. alte* has been mentioned to grow in arid stony desert where average annual rainfall is between 1 and 15 mm [14]. Contrary to other species, *E. alte* is mobile and found climbing many fruit and forestry species, which resulted in severe growth damage and death of these in different regions in the country. It may be regarded as an ecologically dangerous and a threatening species to the survival of many woody species. The literature on its negative ecological impact and behavior as an agricultural pest is lacking; therefore, the following was thought important:

(i) recording species occurrence,

(ii) recording inflicted species by *E. alte* growth,

(iii) quantifying the effect of *E. alte* on fruit species,

(iv) visualizing any possible negative effects other than competition and smothering.

2. Materials and Methods

2.1. Study Procedure. This study aimed to survey *E. alte* in certain parts (covered by forestry and/or fruit trees) of Jordan at which this species was more frequently observed, record its growth status and woody plant species attacked, and accommodate its climbing habit. The survey was carried out during the period from 2008 to 2010 at which *E. alte* was recorded in cultivated fields, fruit tree orchards, forests, and range lands. The survey covered most regions in the central and northern parts of Jordan which located between 36°00′E longitude and 31°00′N latitude and included the central and northern Jordan Valley, Dead Sea, As-Salt, Wadi-Shu'aib, Zay, Baqqai, Nau'r, Amman, Ma'daba, Zarqa, Jerash, Irbid, and Ajloun, at which more than 150 vegetated sites were surveyed year round. The distance between sites was varied depending on the presence or absence of woody vegetation cover. At each site *E. alte* was checked and climbed host species were recorded. The bulk cover of vegetative mass of *E. alte* intermingled or laid over vegetation of other plants was subjectively rated as light, moderate, or high after being visually estimated in the field. Incidence was recorded as rare (found only on few plants in one geographical location or biogeographical region), limited (found on few plants localized in certain locations of 1 or 2 biogeographical regions), common (found on more than 10 plants in 1 or 2 biogeographical regions), or very common (found on many plants at multiple locations within more than two biogeographical regions). *E. alte* and plant species inflicted were all photographed.

In late 2010, additional search was carried out at which six sites in different biogeographical regions were selected to represent the total surveyed locations (Table 1). In each site, *E. alte* was checked on the climbed host species. Species and number of plants climbed from each in every site were recorded. Frequency of the attacked number of plants of each species and between all sites was determined and percentage of their aerial parts covered by *E. alte* vegetation was visualized within and between sites. *E. alte* cover observed over climbed species was estimated using a cover abundance scale [15, 16]. Presence percentage was obtained from the number of *E. alte* plants in a specific site out of the six sites studied. Cover was estimated from estimates of *E. alte* vegetative mass projected on climbed tree as a percentage of total vegetative area of attacked species [17]. Climbed species frequency was used to detect changes in attacked number of plants in each species between different sites of different biogeographical regions. It is used to describe *E. alte* distribution on species forming plant community and often used in combination with density or cover estimates to measure trend or condition.

Notes on the vegetation type and dominating woody species in each site were also recorded.

2.2. Statistical Analysis. Data on the number of plants attacked by *E. alte* from each species and the percentage coverage of their aerial parts by the climber vegetation within and between the selected sites were subjected to the analysis of variance (ANOVA) and performed on species, sites, and species by sites using GLM procedure of SAS [18]. Means of percentage coverage of each species within each site were separated for significance using the *t*-test at $P \leq 0.05$ and

TABLE 1: Representative sites description and plant number attacked by *E. alte* late in 2010 survey.

No.	Site name	Biogeographical region	Total plants climbed by *E. alte* per site	Estimated area checked (ha)	Approximate latitude (m above sea level)	Common vegetation
(1)	Amman	Mediterranean	23	4	1000	*Olea europaea*
(2)	Amman	Mediterranean	168	6	1000	*Olea europaea*
(3)	Zarqa	Mediterranean	26	5	900	*Olea europaea*
(4)	Jerash	Mediterranean	49	4	1100	*Citrus* spp.
(5)	Wadi-Shu'aib upper and As-Salt	Subtropical-Mediterranean	46	3	−300–750	*Olea europaea, Citrus* spp. and almonds
(6)	South Shuna and central Jordan Valley	Tropical	49	5	−255	*Citrus* spp. and *Zizyphus spina-christi* and mixtures

frequency of the attacked number of plants and coverage percentage of species attacked were calculated using the Chi-square test using Freq procedure of SAS.

3. Results

E. alte was found climbing/covering 40 plant species of fruit (9 species), and forest (13 species) trees, ornamentals (6 species) and shrubs (12 species) that belong to 24 plant families (Table 2). Wild and cultivated species smothered by *E. alte* included deciduous and evergreen species. Among severely affected fruit trees were *A. communis, C. aurantifolia, O. europaea, O. ficus-indica, Prunus persica,* and *P. granatum.* Peculiarly, *E. alte* was also found climbing other climbers including *H. helix* (a woody ornamental), *Jasminum* spp., *Galium* sp., *Asparagus stipularis,* and *V. vinifera.* Although *E. alte* attacked different species, but its growth development was substantially varied on different targets (Table 2). It formed a massive growth on *A. communis, O. ficus-indica,* and *R. palaestina* (Figure 1) but relatively smaller growth on *C. siliqua, P. halepensis, P. granatum,* and *Q. coccifera. E. alte* however, was destructive to *A. communis, C. azarolus, R. palaestina* and *Z. spina-christi.* It was more frequently observed on *A. communis, O. ficus-indica, R. palaestina,* and *C. azarolus* but rare on *H. helix, Jasminum* spp., *P. coccinea, P. halepensis* and *V. vinifera* (Table 3).

E. alte forms adventitious roots that enable it to climb and attach to stem and branches of host plant. These roots were found penetrating the cracked bark of old *P. halepensis* trees (Figure 2(a)). However, connection between these and internal host tissue was not observed indicating a commensalisms relationship.

Considering representative sites surveyed late in 2010, certain species were heavily attacked by *E. alte* with the highest number of smothered plants. Among all species, *A. communis* was most frequently climbed with a total number of 87 plants in all sites (Table 3) followed by *R. palaestina* (67 plants) and *C. monogyna* (49 plants). However, species such as *A. stipularis, O. europaea, Q. coccifera. P. palaestina,* and *R. raetam* showed moderate numbers of climbed plants. Other species had less than 10 plants attacked. Frequency of *E. alte* occurrence was the highest on *A. communis, R. palaestina,* and *C. monogyna* in different sites (Table 3). However, *E. alte*

vegetative cover was highly varied between climbed species with *P. persica, A. stipularis, Galium* sp., *I. viscose,* and *C. monogyna* showing the highest coverage frequency by *E. alte* vegetation.

A. communis and *R. palaestina* were attacked in all sites (Table 4) followed by *O. europaea* and *C. spinosa* (in five sites). Differences in species number and vegetative cover frequency by *E. alte* were found between and within the searched sites (Table 4).

4. Discussion

Four *Ephedra* species have been reported to occur in Jordan [13, 19], and found growing in a wide range of habitats. *E. alte* appeared the most problematic since rapidly spread, invading and climbing both forest and fruit tree species in different locations. It grows at elevations from 255 m b.s.l to 1500 m a.s.l. and found in humid to dry and very dry regions to which few plant species have adapted. It was reported to extend across the eastern Mediterranean to the Arabian Peninsula.

Although *E. alte* grows separately in open lands sparsely or hardly covered by vegetation, it tends strongly to climb other plant species in its surroundings. It emerges beside other woody species (Figure 2(b)), climbs them, and rapidly forms a massive vegetative growth with long rope-like stems extending over aerial parts of other species. *E. alte* has been reported as being of unusual morphology among the North American and the European-Mediterranean species in having a strongly climbing liana habit [4] and a relatively unusual morphological feature of partially twining habit [6, 20]. Thus it resembles many weed species in that its stems twining on the stem of climbed plant and its branches are extended from on, or in between those of host plants, compete for light, prevent photosynthesis, and become difficult to control by none highly selective herbicides or even through hand removal. Therefore, *E. alte* may be regarded as an aggressive weed that must be controlled if to avoid its damage to other species. *Ephedra* has been reported as a unique genus among gymnosperms in its high frequency of polyploidy found in about 65% of species studied [21], including 22% of species in which both diploid and polyploidy counts have been obtained. This, however, is

TABLE 2: Common, scientific, and family names, growth form, vegetative mass, incidence, and biogeographical regions of plant species attacked by *E. alte* in Jordan for the period 2008–2010.

Common name	Scientific name	Family name	Growth status	Vegetative mass of *E. alte*	Incidence	Biogeographical region
Fruit trees						
Almond	*Amygdalus communis* L.	Rosaceae	C and W	High	Common	Subtropical and Mediterranean
Mexican lime	*Citrus aurantifolia* (Christm.) Swingle	Rutaceae	C	High	Limited	Mediterranean
Lemon	*Citrus limon* L.	Rutaceae	C	Moderate	Rare	Mediterranean and Subtropical
Fig	*Ficus carica* L.	Moraceae	C	High	Limited	Mediterranean and Subtropical
Grape	*Vitis vinifera* L.	Vitaceae	C	Moderate	Rare	Mediterranean
Indian fig	*Opuntia ficus-indica* (L.) Miller	Cactaceae	C	High	Very common	Mediterranean, tropical, and Subtropical
Nectarine	*Prunus persica* L.	Rosaceae	C	High	Rare	Mediterranean
Olive	*Olea europaea* L.	Oleaceae	C	High	Limited	Mediterranean
Pomegranate	*Punica granatum* L.	Punicaceae	C	High	Limited	Subtropical
Shrubs						
Bedstraw	*Galium* sp.	Rubiaceae	W	High	Rare	Mediterranean
Fern-leaved clematis	*Clematis cirrhosa* L.	Ranunculaceae	W	Moderate	Rare	Mediterranean
Giant cane	*Arundo donax* L.	Gramineae	W	Light	Rare	Subtropical
Grey asparagus	*Asparagus stipularis* Forssk.	Liliaceae	W	Moderate	Rare	Mediterranean
Caper	*Capparis spinosa* L.	Capparidaceae	W	High	Rare	Subtropical
Indian fleabane	*Pluchea indica* (L.) Less.	Compositae	W	Moderate	Rare	Tropical and Subtropical
Inula	*Inula viscosa* (L.) Aiton	Compositae	W	Light	Rare	Subtropical
Jerusalem spurge	*Euphorbia hierosolymitana* Boiss	Euphorbiaceae	W	Light	Rare	Mediterranean
Syrian mesquite	*Prosopis farcta* (Banks and Soland.) J. F. Macbr	Leguminosae	W	Moderate	Rare	Subtropical
Sumac	*Rhus coriaria* L.	Anacardiaceae	W	High	Rare	Subtropical
Sumac	*Rhus tripartita* L.	Anacardiaceae	W	Moderate	Limited	Subtropical
Thorny burnet	*Sarcopoterium spinosum* (L.) Spach	Rosaceae	W	High	Rare	Mediterranean
Ornamental shrubs and climbers						
Common ivy	*Hedera helix* L.	Araliaceae	C	High	Rare	Mediterranean
Heaven tree	*Ailanthus altissima* (Mill.) Swingle	Simaroubaceae	C	Moderate	Rare	Mediterranean
Bush jasmine	*Jasminum fruticans* L.	Oleaceae	C	Light	Rare	Mediterranean
Jasmine	*Jasminum grandiflorum* L.		C	High	Rare	Mediterranean
Oleander	*Nerium oleander* L.	Apocynaceae	W	Light	Rare	Subtropical
Firethorn	*Pyracantha coccinea* Roem	Rosaceae	C	Light	Rare	Mediterranean
Forest trees						
Aleppo oak	*Quercus infectoria* Olivier	Fagaceae	W	High	Limited	Mediterranean
Aleppo pine	*Pinus halepensis* Mill.	Pinaceae	C	Light	Rare	Mediterranean
Carob	*Ceratonia siliqua* L.	Leguminosae	C	Moderate	Rare	Subtropical
Christ thorn jujube	*Zizyphus spina-christi* L.	Rhamnaceae	W	High	Very common	Subtropical
Cypress	*Cupressus sempervirens* L. var. *horizontalis* (Miller) Gordon	Cupressaceae	C	High	Limited	Mediterranean
Cypress	*Cupressus sempervirens* L. var. *pyramidalis*	Cupressaceae	C	High	Limited	Mediterranean

TABLE 2: Continued.

Common name	Scientific name	Family name	Growth status	Vegetative mass of E. alte	Incidence	Biogeographical region
		Forest trees				
Golden wreath wattle	*Acacia cyanophylla* Lindley	Leguminosae	C	Moderate	Limited	Subtropical
Hawthorn	*Crataegus monogyna* Jacq.	Rosaceae	W	High	Very common	Mediterranean
Kermes oak	*Quercus coccifera* L.	Fagaceae	W	High	Very common	Mediterranean
Mosphilla	*Crataegus azarolus* L.	Rosaceae	W	High	Very common	Mediterranean
Palestine buckthorn	*Rhamnus palaestina* Boiss	Rhamnaceae	W	High	Very common	Subtropical
Terebinth	*Pistacia atlantica* Desf.	Anacardiaceae	W	High	Rare	Mediterranean
White weeping broom	*Retama raetam* (Forskal) Webb and Berth	Leguminosae	C	High	Common	Subtropical and Mediterranean
Wild pistachio	*Pistacia palaestina* Boiss	Anacardiaceae	C	Moderate	Common	Mediterranean

Rare: only on few plants in 1-2 sites of a biogeographical region.
Limited: on few plants localized in certain locations of 1 or 2 biogeographical regions.
Common: on certain plant species in > one biogeographical region.
Very common: on many plant species in different locations of different biogeographical regions.
C: cultivated, W: wild.

FIGURE 1: Joint-pine (a) killing *Amygdalus communis*, (b) climbing *Cupressus sempervirens*, (c) on *Opuntia ficus-indica*, (d) killing *Rhamnus palaestina*, and (e) climbing *Olea europaea*. Photos Scale: 23% × 23%.

a character shared with many noxious weed species. *E. alte*, however, is considered as a weed in Egypt [12].

It is well established that weeds compete for water, light, and nutrients [22]. In dense forests, light may become very limited in quantity and quality and far-reaching short stature species or low vegetation layers. It seems that *E. alte* with a unique vegetative growth (scale-like leaves, lignified stems, massive growth, etc.) tends to climb other species, forming a mattress-like vegetative cover, making it difficult for the attacked plants to perform normal photosynthesis and producing enough food to maintain growth and survive under poor fertility and low light supply. *E. alte;* however, it does not only prevent light from reaching effected species but its multiple stems emerge from the same point of

TABLE 3: Total number and frequency of E. alte plants per host plant species attacked and coverage percentage on host plant in six randomly selected representative sites in Jordan late in 2010.

Plant species	Frequency of E. alte density (plants per host species)	Frequency of E. alte density (%)	E. alte coverage (%)
Simaroubacea A. altissima	2	0.55	77.0 ± 15.5 cd*
Rosaceae A. communis	87	24.10	77.9 ± 3.0 cd
Gramineae A. donax	1	0.28	67.0 ± 21.5 bcd
Liliaceae A. stipularis	10	2.77	88.3 ± 7.0 d
Capparidaceae C. spinosa	8	2.22	45.5 ± 7.7 ab
Rutaceae C. limon	1	0.28	24.9 ± 21.9 ab
Cupressaceae C. sempervirens L. var. pyramidalis	7	1.94	55.6 ± 8.9 abc
Rosaceae C. monogyna	49	13.57	81.1 ± 3.7 cd
Rutaceae C. aurantifolia	5	1.39	35.0 ± 10.3 ab
Euphorbiaceae E. hierosolymitana	8	2.22	71.2 ± 7.9 bcd
Moraceae F. carica	5	1.39	69.9 ± 9.8 bcd
Rubiaceae Galium sp.	1	0.28	87.0 ± 21.5 cd
Compositae I. viscosa	1	0.28	84.1 ± 21.5 cd
Apocynaceae N. oleander	1	0.28	14.9 ± 21.9 a
Oleaceae O. europaea	19	5.26	46.8 ± 5.2 ab
Cactaceae O. ficus-indica	4	1.11	72.9 ± 11.3 cd
Pinaceae P. halepensis.	6	1.66	52.0 ± 8.9 abc
Compositae P. indica	2	0.55	49.9 ± 16.0 abc
Rosaceae S. spinosum	5	1.39	79.2 ± 9.8 cd
Leguminosae P. farcta	4	1.11	72.7 ± 11.0 cd
Rosaceae P. persica	1	0.28	98.0 ± 21.5 d
Anacardiaceae P. palaestina	10	2.77	56.5 ± 7.0 abc
Punicaceae P. granatum	3	0.83	68.7 ± 12.5 bcd
Fagaceae Q. coccifera	15	4.16	79.4 ± 6.4 cd
Leguminosae R. raetam	26	7.20	58.9 ± 6.7 abc
Rhamnaceae R. palaestina	67	18.56	78.2 ± 3.1 cd
Anacardiaceae R. coriaria	2	0.55	54.1 ± 15.5 abc

TABLE 3: Continued.

Plant species	Frequency of *E. alte* density (plants per host species)	Frequency of *E. alte* density (%)	*E. alte* coverage (%)
Anacardiaceae *R. tripartita.*	4	1.11	47.4 ± 12.0 ab
Rhamnaceae *Z. spina-christi*	7	1.94	73.5 ± 9.8 cd

* Means within column followed by the same letter were not significantly different according to *t*-test at $P \leq 0.05$.
Numbers of % coverage represent mean values ± SE.

(a)   (b)   (c)

FIGURE 2: Joint-pine (a) aerial roots inserted in *Pinus halepensis* bark. (b) Emerged exactly with *Cupressus sempervirens var. horizontalis.* (c) Fruiting stage. Photos Scale: 36% × 36%.

host stem emergence (Figure 2(b)). This growth habit seems unusual, not fully understood [6], and somehow similar to that of certain parasitic species (e.g., *Cuscuta* spp.), while its tendency to grow alone as well as to climb other species is the same as that of certain hemiparasitic genera including *Osyris* and *Thesium* [23]. These parasites also grow separately, do photosynthesis, but tend strongly to parasitize other plant species. In addition, the shape, appearance, and structure of *E. alte* fruits more or less resemble those of some parasitic genera (*Viscum*, *Loranthus*, and *Osyris*). Fruit is a berry, enclosing a single seed surrounded by a sticky juicy bulb that facilitates dispersal by birds (Figure 2(c)).

Emergence of *E. alte* stem from the near host stem may question its self-dependence for food, nutrients, and/or water. *E. alte* has scale-like leaves, its photosynthate area is mainly the green stems extended over vegetative parts of host species, but photosynthetic materials produced may not be high enough to support the bulky growth it forms in dense-thick plant populations. Shoots of some plants of *E. alte* were found reddish in color which may indicate deficient mineral element/s or low chlorophyll content and thus photosynthate materials produced. However, its climbing habit frequently occurred under both dense and sparse plant populations.

Although no connection was detected between aerial parts of *E. alte* and attacked species, adventitious roots developed from stem nodes of this species were found inserted in the cracked bark of *P. halepensis* trees and sometimes hard to pull out from host tissues (Figure 2(a)). In addition, extension of *E. alte* stems beside host stem may suggest certain sort of connection or interrelationship between their

root systems. This hypothesis, however, was not examined in the present work since *E. alte* roots were deeply extended in rocky soils. This may remain possible in form of natural root crafting which could be tested by injecting a suitable translocated herbicide into the stem of climbed species and following up any changes that may occur on growth of *E. alte*. It may be also examined by growing *E. alte* with a preferred and usually climbed species in a container for certain period and examining their root systems. The ambiguity in the rooting among the major groups of *Ephedra* is also evidenced [24] and is only likely to be solved by an examination of a number of divergent sequence regions to obtain an adequate number of informative characters. There is a limited degree of correlation between putative derived character states such as dry, winged ovulate cone bracts, single seed per cone, or unusual habit types, suggesting considerable homoplasy in the genus [6].

E. alte seems to have host preference. It climbed *C. aurantifolia* but was rarely found on *Citrus limon* and not recorded on *Citrus orange* while the surrounding *C. sempervirens* plants were attacked. In contrast, *E. alte* was not observed on *Casuarina equisetifolia* (Australian pine), *Tamarix pentandra* (tamarisk), or *Melia azedarach* (Chinaberry) trees. It attacked *Z. spina-christi* but not *Zizyphus jujuba* (common jujube). The relatively high number and diversity of targeted species may suggest certain type of association, high phenotypic plasticity, and/or physiological adaptability. Compatibility with chemicals that some of these targeted species may release into the surrounding environment is another possibility although was not tested

TABLE 4: Distribution of climbed plant species by *E. alte* between six representative sites selected randomly in Jordan, showing percentage of climber number and their coverage percentage on each species and between sites late in 2010.

Plant species	Site 1		Site 2		Site 3		Site 4		Site 5		Site 6	
	No. of *E. alte* (%)	Cover by *E. alte* (%)	No. of *E. alte* (%)	Cover by *E. alte* (%)	No. of *E. alte* (%)	Cover by *E. alte* (%)	No. of *E. alte* (%)	Cover by *E. alte* (%)	No. of *E. alte* (%)	Cover by *E. alte* (%)	No. of *E. alte* (%)	Cover by *E. alte* (%)
A. altissima	—	—	—	—	—	—	4.1	80.0 abc	—	—	—	—
A. communis	13.0	80.0 b*	35.7	76.6 b	42.3	83.3 b	4.1	64.0 abc	21.7	73.0 ab	2.0	95.0 c
A. donax	—	—	—	—	—	—	2.0	70.0 abc	—	—	—	—
A. stipularis	8.7	85.0 b	4.2	90.4 b	—	—	—	—	2.2	50.0 ab	—	—
C. spinosa	—	—	0.6	5.0 a	3.9	10.0 a	4.1	50.0 ab	2.2	95.0 b	6.1	66.7 abc
C. lemon	—	—	—	—	—	—	—	—	—	—	2.0	40.0 b
C. sempervirens	—	—	—	—	—	—	14.3	58.6 abc	—	—	—	—
C. monogyna	34.8	67.0 b	20.8	82.2 b	19.2	71.6 b	2.0	100.0 c	—	—	—	—
C. aurantifolia	—	—	—	—	—	—	10.2	38.0 a	—	—	—	—
E. hierosolymitana	—	—	4.8	71.3 b	—	—	—	—	—	—	—	—
F. carica	—	—	—	—	—	—	6.1	56.7 abc	4.4	90.0 b	—	—
Galium sp.	—	—	—	—	—	—	2.0	90.0 c	—	—	—	—
I. viscosa	—	—	—	—	—	—	—	—	2.2	80.0 ab	2.0	30.0 a
N. oleander	—	—	—	—	—	—	—	—	—	—	4.1	35.0 a
O. europaea	4.4	15.0 a	3.0	47.0 a	3.9	20.0 a	20.4	59.5 abc	—	—	—	—
O. ficus-indica	—	—	—	—	—	—	—	—	8.7	68.8 ab	—	—
P. halepensis	—	—	3.0	47.0 a	—	—	2.0	80.0 abc	—	—	—	—
P. indica	—	—	—	—	—	—	—	—	—	—	4.1	65.0 abc
S. spinosum	8.7	65.0 b	1.8	80.0 b	—	—	—	—	—	—	—	—
P. farcta	—	—	—	—	—	—	6.1	73.3 abc	2.2	98.0 b	2.0	95.0 c
P. persica	8.7	12.0 a	—	—	—	—	—	—	—	—	—	—
P. palaestina	4.4	95.0 b	—	—	—	—	8.2	72.5 abc	8.7	55.0 ab	—	—
P. granatum	—	—	—	—	—	—	—	—	4.4	45.0 a	—	—
Q. coccifera	—	—	—	—	—	—	—	—	28.3	76.8 ab	4.1	85.0 bc
R. raetam	—	—	—	—	—	—	4.1	40.0 a	—	—	49.0	75.8 bc
R. palaestina	17.4	62.5 b	26.2	80.3 b	30.8	81.0 b	10.2	69.2 abc	10.9	64.0 ab	2.0	95.0 c
R. coriaria	—	—	—	—	—	—	—	—	4.4	50.0 ab	—	—
R. tripartita	—	—	—	—	—	—	—	—	—	—	8.2	62.5 b
Z. spina-christi	—	—	—	—	—	—	—	—	—	—	14.3	88.6 c

* Means within column followed by the same letter were not significantly different according to *t*-test at $P \leq 0.05$.
Numbers of % coverage represent mean values ± SE.

in the present work. Released chemicals may attract or repel *Ephedra* from climbing certain species and thus subsequently enhance or inhibit its growth. Root exudates may also have a role in whether consist beneficial/harmful compounds that stimulate/inhibit emergence and growth of *E. alte* seedlings nearby and close to these species. Root secretions may contain certain growth promoters or mineral nutrients that stimulate growth of *E. alte* and explain its association with certain woody species but not others. Although *Ephedra* spp. were not reported as parasites but their morphology, growth habit, and maybe responses to certain conditions are more or less similar to those of certain parasitic species. Some parasitic species are also stimulated to germinate and to grow by chemicals released from roots of their host plants. *E. alte* was found completely destructive to many attacked species, and its ultimate effect on host plants is more or less similar to that of different parasitic genera (*Cuscuta*, *Loranthus*, and *Viscum*). Moreover, similarity between *E. alte* and these species may be also demonstrated through the long-distance seed dispersal that is probably mediated by migratory birds [25] although the potential for overwater dispersal is also evidenced by its wide spread distribution [3].

Variations in growth of *E. alte* on different plant species may reflect differences in ecological adaptation, chemical, morphological, or physiological compatibility between these and *E. alte*.

Studies on *Ephedra* spp. control are lacking since they are not considered as weeds. However, *Ephedra* control may be achieved by cutting the bulky stems just emerging from the above soil level; this possibility, however, was not employed in the present work. *E. alte* was also found attacked by different natural enemies in Jordan including mille bugs and a scale insect from croccideae (Hemopterae) (unpublished data) which led to desiccation of its green stems. Injection of *E. alte* by a suitable translocated herbicide may be another control option while foliage application of herbicides can be practiced on separate *Ephedra* plants but not after climbing other species; otherwise low selective rates may be used or a directed treatment to avoid host injury. This, however, could be a future research line with other aspects of *E. alte* associations, host preference within and among plant species, male and female aggressiveness, and climbing habit in relation to indigenous chemicals of inflicted species. Some work on the competitive relationships between *E. alte* and its climbed species over nutrients and water is worth conducting.

It is worth indicating that this work is first of its kind in the country and at world level that treated *E. alte* as an agricultural pest of a serious threat to other woody species of different growth habit or forms. The potential harmful effect of this species on others in its vicinity was even not thought about by researchers before the conduction of this work. The present study is the first to predict possible negative association/interrelationship (through root crafting or other types) of this species with a large number of economic or ecologically important woody species in certain form of dependence on other species. In addition, *E. alte* is for the first time considered as a strong smothering weedy species that competes for light and maybe nutrients and/or water with

host species and proved highly successful in dense as well as in sparse vegetation under different environments.

In conclusion, *E. alte* may be regarded as of a potential threat to different fruit and forestry species in Jordan. Its ecological harmful effects may be serious considering that forestry area represents less than 1% of the total country's area while almost 90% is a desert not receiving more than 50–70 mm average annual rainfall. The ecological threat this species exerts on other plant species becomes clearer when interacting with other ecological stresses, for example, poor soil fertility, frequent grazing, fire hazard, drought, and housing activities. All exert an ecological stress on the existence of certain forestry species or on the area devoted to fruit trees plantation. However, there is still much information required, and more studies are needed on *E. alte* prevalence in other parts of the country. Questions on climbing habit of this species in relation to ecological, physiological, and biochemical interactions with target species remained to be addressed, while its ecological threat and severity of this effect under different ecological conditions merit further research.

Acknowledgments

The author thanks the Deanship of Academic Research, University of Jordan for covering the expenses of publishing this work and Mr. M. Al-Abadi for joining in field trips.

References

[1] A. Danin and I. C. Hedge, "Contributions to the flora of Sinai—I. New and confused taxa," *Notes from the Royal Botanic Garden Edinburgh*, vol. 32, pp. 259–271, 1973.

[2] H. Freitag and M. Maier-Stolte, "The Ephedra-species of P. Forsskal: identity and typification," *Taxon*, vol. 38, no. 4, pp. 545–556, 1989.

[3] K. Kubitzki, "Gnetatae," in *The Families and Genera of Vascular Plants, Pteriodophytes and Gymnosperms*, K. U. Kramer and P. S. Green, Eds., vol. 1, pp. 378–391, Springer, Berlin, Germany, 1990.

[4] D. W. Stevenson, "Ephedraceae," in *Flora of North America*, vol. 2, pp. 428–434, Oxford University Press, New York, NY, USA, 1993.

[5] R. A. Price, "Systematics of the Gnetales: a review of morphological and molecular evidence," *International Journal of Plant Sciences*, vol. 157, no. 6, supplement, pp. S40–S49, 1996.

[6] J. Huang, D. E. Giannasi, and R. A. Price, "Phylogenetic relationships in *Ephedra* (Ephedraceae) inferred from chloroplast and nuclear DNA sequences," *Molecular Phylogenetics and Evolution*, vol. 35, no. 1, pp. 48–59, 2005.

[7] S. Caveney, D. A. Charlet, H. Freitag, M. Maier-Stolte, and A. N. Starratt, "New observations on the secondary chemistry of world *Ephedra* (Ephedraceae)," *American Journal of Botany*, vol. 88, no. 7, pp. 1199–1208, 2001.

[8] A. Y. Ibrahim, K. Mahmoud, and S. M. El-Hallouty, "Screening of antioxidant and cytotoxicity activities of some plant extracts from Egyptian flora," *Journal of Applied Sciences Research*, vol. 7, no. 7, pp. 1246–1257, 2011.

[9] R. Guharoy and J. A. Noviasky, "Time to ban ephedra—now," *American Journal of Health-System Pharmacy*, vol. 60, no. 15, pp. 1580–1582, 2003.

[10] S. A. M. Hussein, H. H. Barakat, M. A. M. Nawar, and G. Willuhn, "Flavonoids from *Ephedra aphylla*," *Phytochemistry*, vol. 45, no. 7, pp. 1529–1532, 1997.

[11] M. S. Abdel-Kader, F. F. Kassem, and R. M. Abdallah, "Two alkaloids from *Ephedra aphylla* growing in Egypt," *Natural Product Sciences*, vol. 9, no. 2, pp. 52–55, 2003.

[12] http://threatenedplants.myspecies.info/sites/threatenedplants .myspecies.info/files/Species%20Summary%20-%20Ephedra %20aphylla.pdf.

[13] G. E. Post and J. E. Dinsmore, *Flora of Syria, Palestine and Sinai*, AUB, Beirut, Lebanon, 2nd edition, 1932.

[14] F. B. Pyatt, G. Gilmore, J. P. Grattan, C. O. Hunt, and S. McLaren, "An imperial legacy? An exploration of the environmental impact of ancient metal mining and smelting in southern Jordan," *Journal of Archaeological Science*, vol. 27, no. 9, pp. 771–778, 2000.

[15] J. Braun-Blanquet, *Plant Sociology*, McGraw-Hill, New York, NY, USA, 1932, Edited by G. D. Fuller and H. S. Conard.

[16] J. Braun-Blanquet, *Pflanzensoziologie*, Springer, Berlin, Germany, 3rd edition, 1964.

[17] D. Mueller-Dombois and H. Ellenberg, *Aims and Methods of Vegetation Ecology*, John Wiley and Sons, New York, NY, USA, 1st edition, 1974.

[18] SAS, *SAS User's Guide: Statistics*, SAS Institute, Cary, NC, USA, 1996.

[19] L. J. Musselman, "Checklist of plants of the Hashemite Kingdom of Jordan," 1998, Old Dominion University, http://www .odu.edu/~lmusselm/plant/jordan/index.php.

[20] I. F. Mussayev, "On geography and phylogeny of some representatives of the genus *Ephedra* L.," *Botanicheskii Zhurnal SSSR*, vol. 63, pp. 523–543, 1978.

[21] J. Huang, *Molecular systematics and evolution of the genus Ephedra*, Ph.D. dissertation, University of Georgia, Athens, Ga, USA, 2000.

[22] R. L. Zimdahl, *Weed Crop Competition: A Review*, International Plant Protection Center. Oregon State University, Corvallis, Ore, USA, 1980.

[23] J. R. Qasem, "Recent advances in parasitic weed research, an overview," in *Weed Management Handbook*, H. P. Singh, D. R. Patish, and R. K. Kohli, Eds., pp. 627–728, The Haworth Press, Binghamton, NY, USA, 2006.

[24] D. E. Soltis and P. S. Soltis, "Choosing an approach and an appropriate gene for phylogenetic analysis," in *Molecular Systematics of Plants II: DNA Sequencing*, D. E. Soltis, P. S. Soltis, and J. J. Doyle, Eds., pp. 1–42, Kluwer Academic Publishers, Dordrecht, The Netherland, 1998.

[25] R. F. Thorne, "Phytogeography," in *Flora of North America*, vol. 1, pp. 132–153, Oxford University Press, New York, NY, USA, 1993.

Permissions

The contributors of this book come from diverse backgrounds, making this book a truly international effort. This book will bring forth new frontiers with its revolutionizing research information and detailed analysis of the nascent developments around the world.

We would like to thank all the contributing authors for lending their expertise to make the book truly unique. They have played a crucial role in the development of this book. Without their invaluable contributions this book wouldn't have been possible. They have made vital efforts to compile up to date information on the varied aspects of this subject to make this book a valuable addition to the collection of many professionals and students.

This book was conceptualized with the vision of imparting up-to-date information and advanced data in this field. To ensure the same, a matchless editorial board was set up. Every individual on the board went through rigorous rounds of assessment to prove their worth. After which they invested a large part of their time researching and compiling the most relevant data for our readers. Conferences and sessions were held from time to time between the editorial board and the contributing authors to present the data in the most comprehensible form. The editorial team has worked tirelessly to provide valuable and valid information to help people across the globe.

Every chapter published in this book has been scrutinized by our experts. Their significance has been extensively debated. The topics covered herein carry significant findings which will fuel the growth of the discipline. They may even be implemented as practical applications or may be referred to as a beginning point for another development. Chapters in this book were first published by Hindawi Publishing Corporation; hereby published with permission under the Creative Commons Attribution License or equivalent.

The editorial board has been involved in producing this book since its inception. They have spent rigorous hours researching and exploring the diverse topics which have resulted in the successful publishing of this book. They have passed on their knowledge of decades through this book. To expedite this challenging task, the publisher supported the team at every step. A small team of assistant editors was also appointed to further simplify the editing procedure and attain best results for the readers.

Our editorial team has been hand-picked from every corner of the world. Their multi-ethnicity adds dynamic inputs to the discussions which result in innovative outcomes. These outcomes are then further discussed with the researchers and contributors who give their valuable feedback and opinion regarding the same. The feedback is then collaborated with the researches and they are edited in a comprehensive manner to aid the understanding of the subject.

Apart from the editorial board, the designing team has also invested a significant amount of their time in understanding the subject and creating the most relevant covers. They scrutinized every image to scout for the most suitable representation of the subject and create an appropriate cover for the book.

The publishing team has been involved in this book since its early stages. They were actively engaged in every process, be it collecting the data, connecting with the contributors or procuring relevant information. The team has been an ardent support to the editorial, designing and production team. Their endless efforts to recruit the best for this project, has resulted in the accomplishment of this book. They are a veteran in the field of academics and their pool of knowledge is as vast as their experience in printing. Their expertise and guidance has proved useful at every step. Their uncompromising quality standards have made this book an exceptional effort. Their encouragement from time to time has been an inspiration for everyone.

The publisher and the editorial board hope that this book will prove to be a valuable piece of knowledge for researchers, students, practitioners and scholars across the globe.

List of Contributors

Wendy Mercedes Rauw
Departamento de Mejora Genetica Animal, Instituto Nacional de Investigacion y Tecnologia Agraria y Alimentaria, 28040 Madrid, Spain

Michael Bela Teglas
Department of Agriculture, Nutrition and Veterinary Sciences, University of Nevada, Mail Stop 202, Reno, NV 89557, USA

Sudeep Chandra
Department of Natural Resources and Environmental Science, University of Nevada, Mail Stop 186, Reno, NV 89512, USA

Matthew Lewis Forister
Department of Biology, University of Nevada, Reno, Mail Stop 314, Reno, NV 89557, USA

Veronica De Micco and Giovanna Aronne
Laboratorio di Botanica ed Ecologia Riproduttiva, Dipartimento di Arboricoltura, Botanica e Patologia Vegetale, Universit`a degli Studi di Napoli Federico II, Viale Universita 100, 80055 Portici, Italy

Muhammad Zubair, Sadia Hassan, Komal Rizwan, Nasir Rasool and Muhammad Riaz
Department of Chemistry, Government College University, Faisalabad 38000, Pakistan

M. Zia-Ul-Haq
Department of Pharmacognosy, University of Karachi, Karachi 75270, Pakistan

Vincenzo De Feo
Department of Pharmaceutical and Biomedical Sciences, University of Salerno, 84100 Salerno, Italy

J. Lema-Rumińska, K. Goncerzewicz and M. Gabriel
Laboratory of Biotechnology, Department of Ornamental Plants and Vegetable Crops, University of Technology and Life Sciences in Bydgoszcz, Bernardynska 6, 85-029 Bydgoszcz, Poland

Atul K. Gupta, J.M. Seneviratne and Anil Kumar
Department of Molecular Biology and Genetic Engineering, College of Basic Sciences and Humanities, GB Pant University of Agriculture and Technology, Pantnagar 263 145, India

G. K. Joshi
Department of Biotechnology, HNB Garhwal University, Srinagar, Uttarakhand 246174, India

Fan-Fan Zhang, Yan-LiWang, Zhi-Zhe Huang, Xiao-Chen Zhu, Feng-Jiao Zhang, Fa-Di Chen, Wei-Min Fang and Nian-Jun Teng
College of Horticulture, Nanjing Agricultural University, Nanjing 210095, China

Heiko Mibus and Margrethe Serek
Faculty of Natural Sciences, Institute for Ornamental and Woody Plant Science, University of Hannover, Herrenhauser Street 2, 30419 Hannover, Germany

Agata Jedrzejuk
Faculty of Natural Sciences, Institute for Ornamental and Woody Plant Science, University of Hannover, Herrenhauser Street 2, 30419 Hannover, Germany
Department of Ornamental Plants, Faculty of Horticulture and Landscape Architecture, Warsaw University of Life Sciences, Nowoursynowska 166, 02-787 Warsaw, Poland

Laila Naher, Umi Kalsom Yusuf and Faridah Abdullah
Department of Biology, Faculty of Science, Universiti Putra Malaysia, Selangor, 43400 Serdang, Malaysia

Soon Guan Tan
Department of Cell and Molecular Biology, Faculty of Biotechnology and Biomolecular Sciences, Universiti Putra Malaysia, Selangor, 43400 Serdang, Malaysia

Chai Ling Ho
Department of Cell and Molecular Biology, Faculty of Biotechnology and Biomolecular Sciences, Universiti Putra Malaysia, Selangor, 43400 Serdang, Malaysia
Institute of Tropical Agriculture, Universiti Putra Malaysia, Selangor, 43400 Serdang, Malaysia

Siti Hazar Ahmad
Department of Crop Science, Faculty of Agriculture, Universiti Putra Malaysia, Selangor, 43400 Serdang, Malaysia

Zoltan Ay, Robert Mihaly, Éva Kótai and Janos Pauk
Department of Biotechnology, Cereal Research Non-Profit Ltd. Co., Also kikoto sor 9, 6726 Szeged, Hungary

Matyas Cserhati
Biological Research Centre, Institute of Plant Biology, Hungarian Academy of Sciences, Temesvari korut 62, 6726 Szeged, Hungary

Yan-Li Wang, Zhi-Yong Guan, Fa-Di Chen, Wei-Min Fang and Nian-Jun Teng
College of Horticulture, Nanjing Agricultural University, Nanjing 210095, China

Puneet Kumar and Vijay Kumar Singhal
Department of Botany, Punjabi University, Patiala 147 002, Punjab, India

Marina Leterrier, Jose M. Palma and Francisco J. Corpas
Departamento de Bioquımica, Biologıa Celular y Molecular de Plantas, Estacion Experimental del Zaidın, CSIC, Apartado 419, 18080 Granada, Spain

Juan B. Barroso and Raquel Valderrama
Grupo de Senalizacion Molecular y Sistemas Antioxidantes en Plantas, Unidad Asociada al CSIC (EEZ), Departamento de Bioquımica y Biologıa Molecular, Universidad de Jaen, 23071 Jaen, Spain

Fatemeh Abdollahi and Vahid Niknam
Department of Plant Sciences, School of Biology and Center of Excellence in Phylogeny of Living Organisms, College of Sciences, University of Tehran, Tehran 14155-6455, Iran

Faezeh Ghanati
Department of Plant Science, Faculty of Biological Science, Tarbiat Modares University, Tehran 14115-154, Iran

Faribors Masroor and Seyyed Nasr Noorbakhsh
Department of Chemistry, Engineering Research Institute, Sooliran Street, 16 km Tehran-Karaj Old Road, Tehran 13455-754, Iran

Agata Jedrzejuk, Julia Rochala and Julita Rabiza-Swider
Department of Ornamental Plants, Faculty of Horticulture and Landscape Architecture, Warsaw University of Life Sciences, Nowoursynowska 166, 02-787 Warsaw, Poland

Jacek Zakrzewski
Department of Forest Botany, Faculty of Forestry, Warsaw University of Life Sciences, Nowoursynowska 166, 02-787 Warsaw, Poland

Rosa M. Perez-Clemente, Almudena Montoliu, Sara I. Zandalinas, Carlos de Ollas, and Aurelio Gomez-Cadenas
Department of Agricultural Sciences, Universitat Jaume I, Campus Riu Sec, 12071 Castello de la Plana, Spain

M. M. Rahman, A. A. Khan, I. H.Mian and A. M. Akanda
Department of Plant Pathology, Bangabadhu Sheikh Mujibur Rahman Agricultural University, Gazipur 1706, Bangladesh

S. B. Abd Hamid and M. E. Ali
Center for Research in Nanotechnology and Catalysis, University of Malaya, 50603 Kuala Lumpuor, Malaysia

K. Kollarova, I. Zelko, P. Capek and D. Liskova
Institute of Chemistry, Slovak Academy of Sciences, Dubravska cesta 9, 845 38 Bratislava, Slovakia

M. Henselova
Department of Plant Physiology, Faculty of Natural Sciences, Comenius University, Mlynska dolina B-2, 842 15 Bratislava, Slovakia

Ygor Lucena Cabral de Oliveira, Alexandre Gomes da Silva and Marcia Vanusa da Silva
Laboratorio de Produtos Naturais, Departamento de Bioquimica, Centro de Ciencias Biologicas, Universidade Federal de Pernambuco, Avenida Professor Moraes Rego, s/n Cidade Universitaria, 50670-420 Recife, PE, Brazil

Alexandre Jose Macedo
Faculdade de Farmacia e Centro de Biotecnologia, Universidade Federal do Rio Grande do Sul, Avenida Ipiranga, CEP 90610-000 Porto Alegre, RS, Brazil

Luis Claudio Nascimento da Silva and Maria Tereza dos Santos Correia
Laboratorio de Glicoproteinas, Departamento de Bioquimica, Centro de Ciencias Biologicas, Universidade Federal de Pernambuco, Avenida Professor Moraes Rego, s/n Cidade Universitaria, 50670-420 Recife, PE, Brazil

Janete Magali de Araujo
Laboratorio de Genetica de Microrganismos, Departamento de Antibioticos, Centro de Ciencias Biologicas, Universidade Federal de Pernambuco, Avenida Professor Moraes Rego, s/n Cidade Universitaria, 50670-420 Recife, PE, Brazil

Pascal Labrousse
Groupement de Recherche Eau, Sol, Environnement (GRESE EA4330), Laboratoire de Botanique et Cryptogamie, Faculte de Pharmacie, Universite de Limoges, 2 rue du Docteur Marcland, 87025 Limoges Cedex, France
Laboratoire de Chimie des Substances Naturelles (LCSN EA 1069), Faculte des Sciences et Techniques, Universite de Limoges, 123 Avenue Albert Thomas, 87060 Limoges Cedex, France

Raphael Decou, Michel Carlue, Sabine Lhernould and Pierre Krausz
Laboratoire de Chimie des Substances Naturelles (LCSN EA 1069), Faculte des Sciences et Techniques, Universite de Limoges, 123 Avenue Albert Thomas, 87060 Limoges Cedex, France

David Delmail
Laboratoire de Pharmacognosie et de Mycologie, UMR CNRS 6226 SCR, Universite de Rennes 1, Equipe PNSCM, 2 Avenue du Professeur L´eon Bernard, 35043 Rennes Cedex, France

Amany Hamza and Aly Derbalah
Pesticides Department, Faculty of Agriculture, Kafrelshiekh University, Kafr el-Sheikh 33516, Egypt

Mohamed El-Nady
Department of Agricultural Botany, Faculty of Agriculture, Kafrelshiekh University, Kafr el-Sheikh 33516, Egypt
Department of Biology, Faculty of Applied Science, Taibah University, P.O. Box 344, Al Madinah Al-Munawwarah, Saudi Arabia

Md. Kamal Uddin, Mohd. Razi Ismail and Md. Alamgir Hossain
Institute of Tropical Agriculture, Universiti Putra Malaysia, Selangor, 43400 Serdang, Malaysia

Abdul Shukor Juraimi
Department of Crop Science, Universiti Putra Malaysia, Selangor, 43400 Serdang, Malaysia

Radziah Othman and Anuar Abdul Rahim
Department of Land Management, Universiti Putra Malaysia, 43400 Serdang, Malaysia

M. Gotelli, B. Galati and D. Medan
Grupo de Biología Reproductiva en Plantas Superiores, Catedra de Botanica Agrícola, Facultad de Agronomía, Universidad de Buenos Aires, Avenida San Martín 4453, 1417 Buenos Aires, Argentina

Jamal R. Qasem
Department of Plant Protection, Faculty of Agriculture, University of Jordan, P.O. Box 13282, Amman 11942, Jordan

CPSIA information can be obtained
at www.ICGtesting.com
Printed in the USA
BVOW07*0632130617

486758BV00007B/2143/P